Texas SCIENCEPLUS®
Technology and Society

Project Directors

International: **Charles McFadden**
Professor of Science Education
The University of New Brunswick
Fredericton, N.B., Canada

National: **Robert E. Yager**
Professor of Science Education
The University of Iowa
Iowa City, Iowa

This new United States edition has been adapted from prior work by the Atlantic Science Curriculum Project, an international project linking teaching, curriculum development, and research in science education.

HOLT, RINEHART AND WINSTON
Austin • New York • Orlando • Chicago • Atlanta
San Francisco • Boston • Dallas • Toronto • London

ACKNOWLEDGMENTS

Project Advisors

Herbert Brunkhorst
Director, Institute for Science Education
California State University,
 San Bernardino
San Bernardino, California

David L. Cross
Science Consultant
Lansing School District
Lansing, Michigan

Jerry Hayes
Associate Director, Scientific Outreach
Teacher's Academy, Mathematics
 and Science
Chicago, Illinois

William C. Kyle, Jr.
Director, School Mathematics
 and Science Center
Purdue University
West Lafayette, Indiana

Mozell Lang
Science Education Specialist
Michigan Department of
 Education
Lansing, Michigan

Project Authors

Earl S. Morrison (Author in Chief)
Alan Moore (Associate Author in Chief)
Nan Armour
Allan Hammond

John Haysom
Elinor Nicoll
Muriel Smyth

Project Associates

We wish to thank the hundreds of science educators, teachers, and administrators from the scores of universities, high schools, and middle schools who have contributed to the success of *SciencePlus*.

1995 Texas Edition by Holt, Rinehart and Winston, Inc.

Copyright © 1993 by Holt, Rinehart and Winston, Inc. and University of New Brunswick. All rights reserved. No part of this publication may be reproduced or transmitted in any form or by any means, electronic or mechanical, including photocopy, recording, or any information storage and retrieval system, without permission in writing from the publisher.
Requests for permission to make copies of any part of the work should be mailed to: Permissions Department, Holt, Rinehart and Winston, Inc., 6277 Sea Harbor Drive, Orlando, Florida 32887-6777.
Some material in this work was previously published in *SciencePlus: Technology and Society*, Pupil's Edition, Level Red, copyright © 1993 by Holt, Rinehart and Winston, Inc. and University of New Brunswick.
Acknowledgments: See page S164, which is an extension of the copyright page.

SciencePlus® is a registered trademark of Harcourt Brace & Company, Inc. licensed to Holt, Rinehart and Winston, Inc. Cover photo of horned toad by Wyman Meinzer.

Printed in the United States of America

ISBN 0-03-098840-3

1 2 3 4 5 6 7 8 9   071   97 96 95 94

Contents

To the Student x
About Safety xii

Unit 1 Interactions 2

THE PLAYERS AND THEIR ROLES 5
  1 You and Your Environment 5
  2 The Living Players 10
  3 The Movement of Energy in a Community 16
  4 Who Eats What? 20
  5 The Nonliving Players 26

PUTTING IT ALL TOGETHER 36
  6 Natural Changes 36
  7 Humans: Boss or Co-Worker? 50

Making Connections 58

Science in Action:
    Spotlight on Veterinary Science 60

Science and Technology:
    Oceanography 62

Unit 2 Diversity of Living Things 64

SUCH VARIETY! 66
1. Living Things in All Shapes and Sizes 66
2. Why Is There Diversity? 74

BEGINNINGS AND ENDINGS 86
3. The Origins of Diversity 86
4. The Loss of Diversity 90

KEEPING TRACK 94
5. Making Sense of Diversity 94
6. Ordering Diversity 112
7. Life Cycles: Diverse Stages 116

Making Connections 124

Science in Action:
Spotlight on Medicine 126

Science and Technology:
Extinction in the Modern World 128

Unit 3 Solutions 130

A SOLUTION OR NOT? 132
1. Solutions and Non-Solutions 132
2. Parts of a Solution—Solutes and Solvents 140
3. Soluble, Not Very Soluble, or Insoluble? 148

SEPARATING SOLUTE AND SOLVENT 152
4. The Salt Industry 152
5. Desalination and Distillation 156
6. Another Solution 160
7. Sugar Solutions 164

CONCENTRATION OF SOLUTIONS 168
    8 Getting Colder.. 168
    9 How Much Solute?....................................... 170
   10 Saturation.. 176
   11 Solutions in the Environment........................ 179
   12 Growing Crystals... 183

Making Connections.. 186

Science in Action:
    Spotlight on Chemistry................................. 188

Science and the Arts:
    The Perfume Maker's Art.............................. 190

Unit 4 Force and Motion 192

UNDERSTANDING FORCES 194
    1 Forces at Work 194
    2 Different Kinds of Forces 200
    3 What Makes Things Fall?.............................. 204

MEASURING FORCES ... 210
- 4 The Tools to Use ... 210
- 5 Estimating the Size of a Force ... 216
- 6 How Strong Are You? ... 220

FRICTIONAL FORCE ... 225
- 7 Take a Look at Friction ... 225
- 8 Causes and Consequences of Friction ... 231

GOING THROUGH THE MOTIONS ... 236
- 9 Thought Experiments ... 236
- 10 Puffball: A Game of Force and Motion ... 242
- 11 What Is Inertia? ... 246
- 12 Forces in Pairs ... 251

Making Connections ... 256

Science in Action:
Spotlight on Electrical Engineering ... 258

Science and Technology:
Rocket Science ... 260

Unit 5 Structures and Design ... 262

THE SCIENCE OF STRUCTURE ... 264
- 1 Planning the Structure ... 264
- 2 Response of Structures to Force ... 272
- 3 Engineering Structures ... 282

THE ART OF DESIGN ... 293
- 4 Beauty in Structures ... 293
- 5 A Question of the Environment ... 304

Making Connections ... 310

Science in Action:
Spotlight on Architecture ... 312

Science and the Arts:
>An Architectural Marvel **314**

Unit 6 The Restless Earth 316

SHAKE, RATTLE, AND FLOW **318**
>1 Making Mountains **318**
>2 The Earth Breaks Apart **325**
>3 Volcanoes—Holes in the Earth **332**

THE ROLE OF ROCKS **341**
>4 Rocks in the Past **341**
>5 Hot Rocks **346**
>6 Rocks From Sediments **351**
>7 Changed Rocks **355**

FOSSILS—RECORDS OF THE PAST **360**
>8 Rocks Reveal a Story **360**
>9 Telling Time With Rocks **365**

Making Connections **370**

Science in Action:
>Spotlight on Geology **372**

Science and the Arts:
>Jewelry **374**

Unit 7 Toward the Stars 376

THE OBSERVERS 378
    1 What Is Astronomy? 378
    2 An Ancient Science 382

THE EARTH MOVES 392
    3 A Scientific Revolution 392
    4 Motions and Their Effects 398

EXPLORING THE SOLAR SYSTEM 406
    5 Visitors from Space 406
    6 The Space Probes 411
    7 Colonizing the Solar System 416

OUR UNIVERSE 422
    8 Messages from the Stars 422
    9 Earth's Place in the Universe 429
   10 A Likely Beginning 436

Making Connections 440

Science in Action:
   Spotlight on Astronomy 442

Science and Technology:
   The Most Powerful Objects in the Universe 444

Unit 8 Growing Plants 446

GARDEN INGREDIENTS 448
    1 A Garden in Space 448
    2 The Beginning of the Garden 450
    3 Breaking Ground 454

INNER ACTIONS OF PLANTS 463
    4 Water, Water, Everywhere! 463
    5 Giving Plants a Hand 469

6 Flowers and Pollination **473**
7 Plants from Plant Parts **479**

PLANTS IN YOUR ENVIRONMENT **484**
8 Medicine or Poison? **484**
9 Make a Green World! **486**

Making Connections **494**

Science in Action:
Spotlight on Agricultural Engineering **496**

Science and Technology:
Genetic Engineering **498**

CONCEPT MAPPING **500**
TEXAS EXPLORATIONS **501**
GLOSSARY **S145**
SPANISH GLOSSARY **S150**
INDEX **S156**

SourceBook **527**

Unit 1 .... **S1**	**Unit 5** **S77**
1.1 The Biosphere **S2**	5.1 Architecture in Other Cultures **S78**
1.2 Succession and the Biomes **S8**	
1.3 Humans and the Environment **S17**	**Unit 6** **S89**
	6.1 Rocks and the Rock Cycle **S90**
Unit 2 .... **S23**	6.2 Rock-Forming Minerals **S99**
2.1 The Diversity of Modern Life **S24**	6.3 Stories in Rocks **S105**
2.2 The Evolution of Diversity **S35**	
	Unit 7 .... **S111**
Unit 3 .... **S45**	7.1 Upward and Outward **S112**
3.1 Solutions, Suspensions,	7.2 Stars: A Universe of Suns **S118**
and Colloids **S46**	7.3 One Small Step **S126**
3.2 Acids, Bases, and Salts **S51**	
	Unit 8 .... **S133**
Unit 4 .... **S61**	8.1 Plant Structure **S134**
4.1 Motion and Force **S62**	8.2 Uses of Plants **S141**
4.2 Laws of Motion **S68**	
4.3 Gravitation **S74**	

To The Student

This book was written with you in mind! There are many things to try, to create, and to investigate — both in and out of class. There are stories to be read, articles to think about, puzzles to be solved, and even games to play.

GET INVOLVED!

The best way to learn is by doing. In the words of an old Chinese proverb:

> Tell me — I will forget.
>
> Show me — I may remember.
>
> Involve me — I will understand.

The authors of this book want **you** to **get involved** in science.

The activities in this book will allow you to make some basic and important scientific discoveries on your own. You will be acting much like the early investigators in science who, without expensive or complicated equipment, contributed so much to our knowledge.

What these early investigators had, and had in abundance, was **curiosity** and **imagination**. If you have these qualities, you are in good company! And if you develop sharp scientific skills, who knows, you might make your own contributions to science someday.

THE LEADING EDGE

Scientists are usually interested in understanding things that happen in **nature**. However, the discoveries that scientists make are often used by inventors and engineers. The end result is our most sophisticated **technology**, including such things as computers, laser discs, nuclear reactors, and instant global communication.

There is an interaction between science and technology. Science makes technology possible. On the other hand, the products of technology are used to make further scientific discoveries. In fact, much of the scientific work that is done today has become so technically complicated and expensive that no one person can do it entirely alone. But make no mistake, the creative ideas for even the most highly technical and expensive scientific work still come from **individuals**.

GO FOR IT!

Science is a process of discovery: a trek into the unknown. The skills you develop as you do the activities in this book — like observing, experimenting, and explaining observations and ideas — are the skills you will need to be a part of science in the future. There is a universe of scientific exploration and discovery awaiting those who take the challenge.

Keep a Journal

A Journal is an important tool in creative work. In this book, you will be asked to keep a Journal of your thoughts, observations, experiments, and conclusions. As you develop your Journal, you will see your own ideas taking shape over time. This is often the way scientists arrive at new discoveries. You too may log some discoveries as you develop your own Journal.

About Safety

Science investigations and experiments should be both enjoyable and safe. If you follow the safety guidelines listed here, as well as any others mentioned in the explorations, you should have no problems. You should **always** follow these guidelines, even when you think that there is little or no danger.

The major causes of laboratory accidents are carelessness, lack of attention, and inappropriate behavior. These all spring from a person's **attitude**. With a proper attitude and **consistent** safety habits, you should be able to feel quite comfortable and at home in a science laboratory.

Safety Guidelines

● Eye Safety

Wear goggles when handling acids or bases, using an open flame, or performing any other activity that could harm the eyes. If a substance gets into your eyes, wash them with plenty of water and notify your teacher at once. Never place a chemical substance near your unprotected eyes. Never use direct sunlight to illuminate a microscope.

● Safety Equipment

Know the location of all safety equipment, such as fire extinguishers and first aid kits.

● Neatness

Keep work areas free of all unnecessary books and papers. Tie back long, loose hair and button or roll up loose sleeves when working with chemicals or near a flame.

● Chemicals

Chemicals and other dangerous substances can be dangerous if they are handled carelessly. When handling certain chemicals, such as acids or bases, you should protect your eyes and clothes with safety glasses and an apron.

Heat

Whenever possible, use an electric hot plate instead of an open flame. If you must use an open flame to heat a glass container, shield the flame with a wire screen that has a ceramic center. When heating chemicals in a test tube, do not point the test tube toward anyone.

Electricity

Be cautious around electrical wiring. When using a microscope with a lamp, do not place its cord where it can cause someone to trip. Do not let cords hang over a table edge in a way that will cause equipment to fall if a cord is pulled. Do not use equipment with frayed cords.

Glassware

Examine all glassware before using. Glass containers for heating should be made of heat-resistant material. Never use cracked or chipped glassware. Use caution when inserting glass tubing into a rubber stopper. Hold the tubing and stopper with a towel to prevent puncture wounds.

Never taste chemicals, unless you are specifically instructed to do so.

Never pour water into a strong acid or base. The mixture produces heat. Sometimes the heat causes splattering. The correct procedure is to pour the acid or base slowly *into the water*. This way the mixture will stay cool.

If any solution is spilled, wash it off with plenty of water. If a strong acid or base is spilled, neutralize it first with an agent such as baking soda (for acids) or boric acid (for bases).

If you are instructed to note the odor of a substance, wave the fumes toward your nose with your hand rather than putting your nose close to the source.

Sharp/Pointed Objects

Use knives, razor blades, and other sharp instruments with care. Do not use double-edged razor blades in the laboratory.

Cleanup

Wash your hands immediately after handling hazardous materials. Before leaving the laboratory, clean up all work areas. Put away all equipment and supplies. Make sure water, gas, burners, and electric hot plates are turned off.

The instructions for the explorations in this book will include warning statements where necessary. In addition, you will find one or more of the following safety symbols when a procedure requires specific caution.

 Wear Safety Goggles

 Wear a Laboratory Apron

 Flame/ Heat

 Sharp/ Pointed Object

 Dangerous Chemical/ Poison

 Corrosive Substance

Unit 1 Interactions

THE PLAYERS AND THEIR ROLES
1. You and Your Environment, p. 5
2. The Living Players, p. 10
3. The Movement of Energy in a Community, p. 16
4. Who Eats What? p. 20
5. The Nonliving Players, p. 26

PUTTING IT ALL TOGETHER
6. Natural Changes, p. 36
7. Humans: Boss or Co-Worker? p. 50

For Your Journal

Write a paragraph summarizing your thoughts about each of the following:
1. How do environments change over time, and what causes them to change?
2. In what ways do people influence the environment?
3. How are all of the things in this picture related?
4. Is every relationship in this picture visible? Explain.

THE PLAYERS AND THEIR ROLES

1 You and Your Environment

Your **environment** is everything that surrounds you—trees, grass, soil, air, water, animals, insects, and other people too. How important do you think the environment is for your life? Could you live without it? Are there some parts you could live without? Are different parts of the environment related to other parts? Give these questions some thought.

To help your thinking, there is a game on page 7 that will help you find some answers.

The Players and Their Roles 5

EXPLORATION 1

Relationships in Nature

You Will Need

- one index card per student
- a ball of string or yarn

What to Do

Choose something from the environment that you'd like to be. You might be the wind, the sun, an ant, a daisy, the soil, or any one of many, many other things. Each person should be a different thing. Use a small card to label what you are.

The first person will hold on to the end of the string and pass the ball to another student in the circle with whom she or he can be "related." The first student will then explain to the whole group what this relationship is.

For example, the "daisy" student holds the end of the string. He or she passes the ball to the "water" student and says, "I need water to grow." The "water" student holds on to the ball of string, passes it to the "fish" student, and says, "I am your home." This continues, one move at a time, showing relationships in the circle.

You may end up holding portions of the string coming from and going to many different things. Now, what happens if one part of the environment is removed? Test your prediction. Have the person representing this part let go of the string.

What part(s) of the environment would you want not to let go of? Which parts seem to be the most important for maintaining the relationships in the circle? Look at the connections at air and water. Why are there so many?

The drawing above shows the connections a group of students made in playing the game. What does the circle look like? Does any living thing in the circle exist

alone? Would removing mosquitoes from the circle make any difference? How?

Why does the string connect the earthworm and the soil? The answer is quite simple: the earthworm lives in the soil and gets its food there. The earthworm and the soil *interact*. In the game, wherever one part of the environment is connected by the string to another part, there is an **interaction**. What interaction takes place between the earthworm and the bird? between the air and the soil?

Questions

1. What are some of the different ways in which the connected pairs interact? In other words, describe the various ways in which one member of a pair can need the other member.

2. Are *all* of the possible interactions shown in the diagram?

In this unit, you're going to learn about many different kinds of interactions.

Biotic and Abiotic

Ms. Wilkie's class played the interaction string game. Look at the list of the parts they chose. These parts are divided into two groups: **biotic** and **abiotic.** What do all of the biotic parts have in common? The abiotic parts?

Biotic	**Abiotic**
perch	air
spider	temperature
bacteria	water
insect	sand
deer	rocks
crow	light
maple tree	
girl	
dandelion	
Venus' flytrap	

Some connections made by Ms. Wilkie's class:

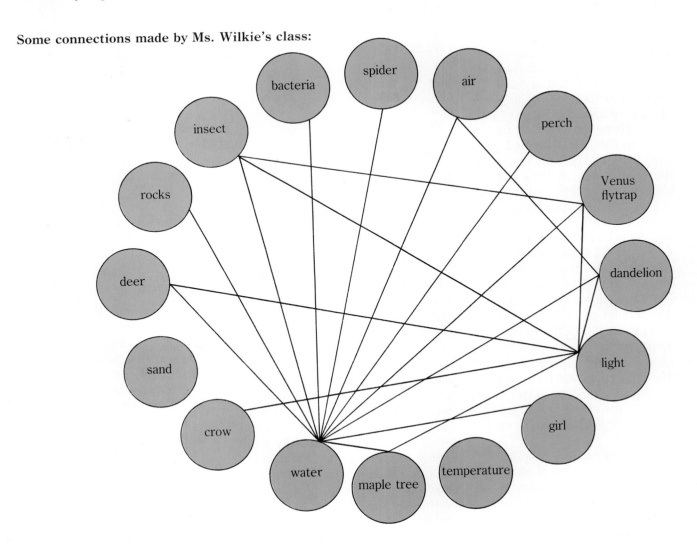

If you say that the biotic parts of the environment are living things, you are right. But here are a few more examples of biotic parts of the environment:

- A rotting log
- A dead animal
- A bone

Together with a classmate, devise a new statement that best describes *biotic* and *abiotic*.

Look around your schoolyard. Which parts of the schoolyard environment are biotic? Which parts are abiotic? Arrange the various parts in a circle and connect the ones that you think interact. Draw lines only for those interactions that you are sure of.

Now you know something more about this unit: it provides a look at various biotic and abiotic parts of your environment and at some of the interactions between them.

By the way, is the interaction string game of Ms. Wilkie's class complete? What parts of that environment (both biotic and abiotic) are not connected, but might be? Give a reason for each new connection you propose.

Some Detective Work

Take a close look at the large drawing on pages 4 and 5. How many examples of interaction can you find? Classify each interaction under one of the following headings:

- Animal–Animal
- Animal–Plant
- Plant–Plant
- Living–Nonliving
- Living–Once Living

What abiotic parts of the environment did you include in your lists?

Which photographs show an abiotic part of the environment? Which show a biotic part?

2 The Living Players

Communities of Living Things

You would probably think it strange to find orchids growing in the Arctic, or polar bears roaming the beaches of Florida. It would also seem a bit odd to find a mouse chasing a cat, or carrots attacking a rabbit!

Your experiences have shown you that every organism has its own way of living. It exists in the environment that best suits its needs. Also, every living thing relates to other organisms around it in specific ways. These relationships between an organism, its physical surroundings, and its neighbors make up an organism's **niche.** Its niche is more than just the place where an organism lives. The niche is the organism's entire way of life: it indicates the organism's *role* in the community where it lives. The *place* where an organism lives is called its **habitat.** For instance, your habitat might be San Antonio, Sacramento, Portland—any city or town. Your niche (right now) is student, son or daughter, nephew or niece, member of the school choir or basketball team, and so on. Being a student in school, for example, reveals a great deal about your way of life, your environment, and those with whom you interact.

Do you understand the difference between *habitat* and *niche?* In your Journal, complete this dictionary definition:

A ___?___ *is the way of life that an organism adopts to survive in a particular* ___?___.

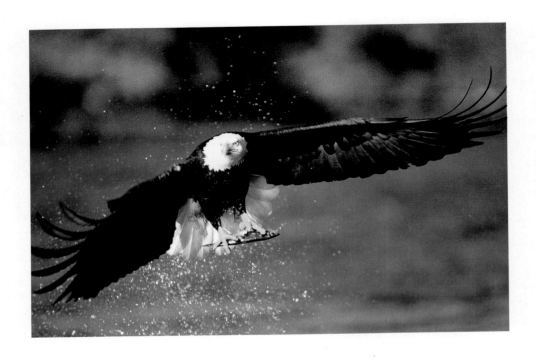

Copy this table into your Journal. Complete it by describing the niches of the organisms listed as thoroughly as you can.

Organism	Habitat	Description of Niche	
		How It Uses Other Things (Living and Nonliving)	**How It Is Used by Other Things in Its Environment**
robin	Open areas, lawns, bushes, trees, air, or ground	Uses bush or tree for a nesting site or for protection. Uses twigs and grass for nest. Eats worms and insects for food.	Eaten by larger birds, snakes, or cats. Eggs eaten by other birds. Parasites—mites and lice—live in its feathers.
turtle			
spruce tree			
beetle			

The Players and Their Roles

Types of Interactions in Communities

"One-Way" Benefits

The scene in the photo at the right is probably familiar to you. Robins like these might even live in your back yard, where trees provide a convenient place for building their nests. Let's analyze the relationship between a robin and a tree.

In this situation, two different kinds of organisms live closely together. One organism—the robin—benefits or gains from the relationship. The robin benefits because the tree provides a place for its nest. The other organism—the tree—is neither helped nor harmed by the nest. This kind of relationship is called **commensalism.**

Now let's look at another example of commensalism. Adult barnacles cannot move from place to place on their own. But they often attach themselves to the shell of a horseshoe crab and travel with the crab. The crab is unaffected by the presence of the barnacles getting the free ride. How do the barnacles benefit from traveling with the crab? Why is this an example of commensalism?

Think of still other examples of commensalism.

"Two-Way" Benefits

The bee and the flowering plant provide an example of another kind of relationship. Can you think how this relationship might differ from commensalism? Here are the facts:

- The bee gets nectar (a liquid food that it uses to make honey) from the flower.

- While the bee is feeding from the flower, some of the flower's pollen sticks to the bee's body.

- When the bee flies to another flower, it deposits some of the pollen on that flower. *Pollination* is part of the process by which a flowering plant produces seeds.

How is the bee helped in this relationship? How is the plant helped? A relationship between two organisms in which both benefit is called **mutualism.**

Next, consider the *lichens,* which are plant-like organisms that can be found almost everywhere. They may look like grayish-green scales on rocks and on the bark of trees. One kind is called "old man's beard" because that's what it looks like. It hangs from tree branches. Another kind of lichen can be found on the ground in the shape of little red-tipped, vertical sticks or stems. This type is called "matchstick lichen" or "British soldiers."

But a lichen is not a single organism. It is really two kinds of

12 Interactions

Various lichens

organisms living together for mutual benefit. You need a microscope to see both organisms.

One organism in lichen is a *fungus* (plural, fungi) with many thread-like branches. These branches provide both protection and moisture for green, single-celled organisms called *algae* (singular, alga). The algae make food by a process called *photosynthesis*. The fungi in lichen do not make their own food. Instead, they use the food made by the algae.

Name as many other examples of mutualism as you can.

Benefit and Harm

Compare commensalism and mutualism with a relationship called **parasitism**. In parasitism, one organism—the *parasite*—lives on and obtains its food from another organism—the *host*. However, in this relationship, the host is harmed instead of helped. Sometimes the host is even killed by the parasite.

Do research on the following pairs of living things that illustrate parasitism:

- People and tapeworms
- Trichina worms and pigs
- Rust and wheat
- Spruce trees and spruce budworms

Which organisms are the hosts? Which are the parasites?

alga   fungus

A diagram showing fungi and algae that make up lichen, as seen with a microscope.

The Players and Their Roles

Problems to Consider

1. Make three diagrams that will help you remember these three relationships: commensalism, mutualism, and parasitism.
2. Examine each of the pictures on this page and the next, and read the accompanying facts. Then identify the type of relationship being exhibited in each case.

a

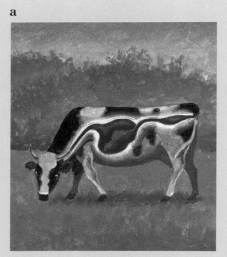

The stomach of a cow is home to a type of bacteria that digests the grass eaten by the cows.

b

A suckerfish attaches itself to a shark by a sucker on the top of its head. It gets a free ride and shares some of the scraps of food left over from what the shark has eaten.

c

Orchids grow on other plants for mechanical support only.

d

Small organisms called protozoa live inside the body of a termite and digest wood for the termite and for themselves.

e

Dodder, a plant with no leaves, winds around wheat and other plants for both support and food.

f

Ichneumon wasps pierce the eggs or larvae of other insects and lay their own eggs inside. Once hatched, the larvae of the ichneumon wasps then feed on the other insects' eggs or larvae.

14   Interactions

g

Lungworms thrive at the expense of the bighorn sheep that live in the Rocky Mountains.

h

Mistletoe grows on trees such as oak, obtaining both support and food while weakening the tree.

i

People carry burdock seeds (burs) on their clothing when walking through the brush. The seeds are eventually deposited somewhere else.

j

Certain bacteria find a comfortable shelter in the human digestive tract. The bacteria help remove water from indigestible material. They also help produce vitamin K, which is important in blood clotting.

k

Barnacles often take trips on whales and end up in a new part of the ocean.

l

Certain ants actually "herd" aphids. The ants move the aphids from place to place and protect them from predators. The ants collect the sweet liquid (honeydew) produced by the aphids.

3 The Movement of Energy in a Community

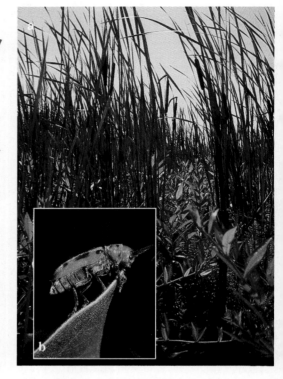

You live in a community. Your community is made up of your family, neighbors, friends, pets, and all the other living things in your neighborhood. Similarly, a forest contains many kinds of organisms. In a forest, some organisms cooperate, some compete, and some never even interact with each other. We can call this a forest, or woodland, community.

Every community of living things, whether in a forest or in a city, needs energy for its members to live. You know that an automobile gets its energy from burning a fuel, usually gasoline. Well, a community of living things also gets its energy from "burning" fuel, but it gets energy in a different way and in a different form: food.

Think about how energy "moves" through a forest community. Some animals get their energy (or food) by eating other animals. Some animals eat plants. But what do plants eat? What is their source of energy?

Energy → Plants → Animals → Animals

A few plants, such as the pitcher plant and the Venus' flytrap, eat animals (insects). But this is not the usual niche that plants occupy. So how do plants get their energy? You read earlier about **photosynthesis**—the process by which green plants change carbon dioxide (a gas they get from the air) and water (from the soil) into food. This food supplies energy not only for the plants, but also for the animals that eat the plants. The process of photosynthesis also *requires* some energy. Green plants use energy from sunlight to manufacture their food, as you can see from this energy diagram:

Sun's Energy → Plants → Animals → Animals

Types of Roles in a Community

Green plants play a particular and valuable role in a community. They make food energy, which they use themselves. But this food energy is also distributed throughout the community when plants are eaten by other organisms.

Because green plants produce food for themselves and for others, they are called **producers**. This is their niche.

The organisms that eat green plants and other organisms are called **consumers**. Why?

Which of the organisms on this page and the next are producers? Which are consumers?

16 Interactions

Both the deer and the fox are consumers. But because their diets are different, they are different types of consumers. The deer is called a **herbivore** (plant eater) because its main food is plants. The fox is called a **carnivore** (meat eater) because its main food is other animals. Which name fits humans? If you eat a turkey sandwich with lettuce and tomato, you are eating parts of both plants and animals. That would make you an **omnivore.** What do you think the term means? Make an energy diagram like the ones on page 16 to contain the following items: producer, sunlight, carnivore, herbivore.

Consumer Relations

Some consumers, such as the fox, kill other animals for food. The fox is a **predator,** and its food is its **prey.** What other examples of the predator-prey relationship can you think of?

Unlike the fox, some consumers do not simply eat and run. The tapeworm (a parasite) can live comfortably in a human (a host) for long periods of time, consuming food that the person eats. Certain parasites spend their entire lives in a single host. Other parasites spend part of their lives in one host and the rest of their lives in a second host. Still other organisms are parasites for part of their lives but are free-living the rest of the time.

Some animals wait until their food source is already dead instead of killing it. A vulture eating a dead bird is called a **scavenger.** An earthworm eating dead organic material in the soil is another example of a scavenger. Look back at the interaction diagram on page 6. What lines could you draw to show the earthworm's role as a scavenger? (Remember that all of the organisms in the diagram will eventually be dead.)

At right are some pictures of other consumers. Can you name them? Where might they get their food?

Not all living things are killed and eaten by consumers. Nor do scavengers eat every last bit of the dead organisms they feed on. What happens to the remains? Something must happen; otherwise the earth would be covered with the bodies of dead plants and animals. Can you imagine what this situation would be like?

Fortunately, some consumers clean up the earth by breaking down plant and animal bodies into substances that enrich the soil, which then supports the growth of new plants. Such consumers are called **decomposers.** Mushrooms growing on tree stumps and the black fuzzy mold on tomatoes are examples of decomposers that you can see easily.

The most common decomposers, however, are ones you can't see without a microscope. They are found in almost all environments on land and water. These are **bacteria.**

Now let's look at an example of where many decomposers are at work. Have you ever seen a compost pile—a mixture of leaves, grass, potato peelings, soil, and other biotic materials? If so, you may have noticed a matted layer of fine white threads on top of the pile. These threads are parts of fungi. Molds and mushrooms are also fungi. These organisms obtain their food from dead plants and animals. They also give off substances that break down the dead material into soluble substances that, in turn, can be absorbed by new plants.

Deeper down in a compost pile is a whole world of bacteria also feeding on the dead material and making it usable by plants.

Bacteria *(above)* and fungi *(below)* both help break down, or decompose, dead plants and animals. Both are called decomposers.

A Relationship Puzzle

Below is a list of interactions. Can you match each numbered event on the left with the type of interaction it represents?

1. A bird eats a caterpillar.
2. A mosquito sucks blood.
3. Mold grows on oranges.
4. An aphid lives on the leaves of trees.
5. A toad eats a fly.
6. A millipede eats dead plants.
7. A dog has fleas.
8. A human kills and eats a deer.
9. Bacteria live on a dead sparrow.
10. A sowbug eats rotting wood and leaves.

(a) predator-prey
(b) parasite-host
(c) scavenger and its food
(d) decomposer and its food

Tell Me About It!

So far in this unit, you have learned many new terms that are used to describe an organism's niche. Take another look at the picture on pages 4 and 5. Suppose you were asked to describe this scene to a friend who is blind. Use your new scientific vocabulary to describe the picture to your friend. Tell your friend what is happening or is about to happen, using such words as these:

- producer
- consumer
- herbivore
- carnivore
- parasite
- host
- predator
- prey
- scavenger
- decomposer
- mutualism
- commensalism
- energy

The Players and Their Roles

4 Who Eats What?

If you stand or sit still in a field or in the woods and watch closely, you can usually spot a number of animals. What are you most likely to find them doing? Did you say that they would be gathering or eating food? It's true—most animals spend a great deal of time looking for and consuming food, which is their source of energy.

Each type of animal prefers certain kinds of foods. And in most cases, each animal becomes food for other animals. Patterns of relationships among living things are easily found if you look for them.

Food Chains

Look at the pictures below and think about which organisms would complete this statement:

Plants are eaten by ___?___, which are eaten by ___?___, which are eaten by ___?___, which are eaten by ___?___.

Interactions

These relationships can be shown in a linked diagram or written out as a "chain." Such a diagram or statement describes a **food chain**. It shows how certain living things depend on one another for food energy. Each organism can be called a "link" in the chain. The food chain for the organisms shown on page 20 would have five links and would look like this:

Arrows are used to show the direction in which food energy moves. What would happen to the other organisms if the green plants were removed?

Green plants begin almost all food chains. Is it therefore true to say, "All members of food chains that begin with plants depend on the sun"?

Once again, study the food chain shown above and see if you can answer the following questions.

1. Which links in the food chain are producers? Which are consumers?
2. The grasshopper is called a *primary consumer*. Why?
3. Animals that eat animals are called *secondary consumers*. Why? Which organisms in the food chain above are the secondary consumers?
4. Which consumers are herbivores? Which consumers are carnivores?

At the end of the chain is the hawk. It apparently has no natural enemies in this situation. In this case, the hawk is called the *top carnivore*.

Construct a food chain showing what may happen or is happening in the diagram below. Identify the producer(s) and any primary or secondary consumers. Which organism is the top carnivore?

The Players and Their Roles

Complicating the Food Chain: Food Webs

Real situations are not as simple or straightforward as the food chains you have studied so far. Robins get energy from organisms other than just grasshoppers; snakes eat other animals besides mice. If you were to include all of the food for every single animal in an area, your food chain would be very complicated.

Let's examine what it might look like. At the right is more information about the animals you have been discussing. (Keep in mind that even this expanded food list is far from complete.)

Circles containing the names of the various organisms are randomly arranged on the page. Sketch this arrangement in your Journal. Draw arrows showing the movement of food energy from one organism to another. A few arrows have already been drawn. (Arrows can cross one another.)

Animal	Food
earthworm	leaves (dead)
grasshopper	grass, leaves
robin	grasshoppers, other insects, earthworms
snake	mice, insects, young rabbits
hawk	snakes, insects, mice, rabbits
rabbit	grass, leaves

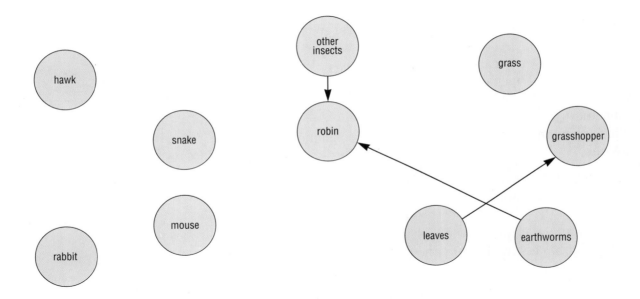

This kind of diagram is called a **food web**. A food web shows as many food relationships as possible between living things in an area. In what kind of area (environment) might you find the animals named above? Since there are many more animals than these in such an environment, you can see that a food web including all of them would be very large. It would show many links and a complex pattern of connections. To be complete, a food web must include all scavengers, parasites, and decomposers as well. In the following Exploration, you'll be analyzing a complex food web.

22 Interactions

EXPLORATION 2

Analysis of a Terrestrial Food Web

Organism	Food
hawk	squirrels, grasshoppers, mice
coyote	squirrels, grasshoppers, mice, deer
bobcat	squirrels, mice, deer
squirrel	seeds, tree buds
grasshopper	grass
mouse	seeds, grass
deer	tree buds and twigs, grass
fungi and bacteria	hawks, coyotes, bobcats, squirrels grasshoppers, mice, deer, seeds, buds, twigs, grass

In your Journal, draw a food web for the 13 organisms (or parts of organisms) listed below.

hawk   seeds
bobcat   twigs
coyote   grass
squirrel   buds
grasshopper fungi and
mouse   bacteria
deer

List the producers, the primary consumers (herbivores), and the secondary consumers (carnivores).

Did you notice all the arrows going to the fungi and bacteria? You may have wondered how all of these animals and plants could be food for fungi and bacteria. What role do fungi and bacteria play in this community?

Suppose that due to disease, the mice were removed from the community. What effect would this have? The chart below will help you in considering this question. Copy the chart into your Journal and complete it.

When mice are removed . . .	
These animals have lost a source of food:	These are the remaining food sources for each of the animals in the first square:
These organisms (or parts of organisms) are less likely to be eaten:	These primary consumers are left:

The Players and Their Roles   23

A Visit to a National Park

The Great Smoky Mountains National Park in Tennessee and North Carolina has a wide variety of animals. Here is a list of some of them and a few of the foods they eat:

Animal	Food
deer	wild apples, poplar twigs
porcupine	inner bark of spruce and pine trees
mouse	grain, seeds of wild apples
rabbit	grass, twigs
red squirrel	seeds from pine and spruce trees
weasel	mice, frogs, rabbits
fox	mice, frogs, rabbits
bear	wild apples, blueberries, mice
insect	grass, leaves, smaller insects
snake	frogs, salamanders, mice
frog	insects
hawk	snakes, weasels, mice, red squirrels
centipede	insects, worms
sowbug	dead materials from all organisms
earthworm	dead materials from all organisms
salamander	worms, insects

Construct a food web for these animals and their food sources, and then answer the following questions.

1. Name the organisms that fill each of these niches: producer, herbivore, carnivore, omnivore.
2. Which animals are in competition for the same food? Name the animals and their food.
3. Construct a food chain consisting of four links in which the final link is the top carnivore in the food web. Which animal is the top carnivore? Does it have any enemies? Explain.
4. What would happen to your food web if no more frog eggs hatched?
5. The environment containing your food web is under attack! There is an enormous increase in the population of the spruce budworm. In no more than six lines, write a report on what immediate effect this increase will have on the food web.

5 The Nonliving Players

Tolerance

If you were wearing a sweatshirt in your classroom, and the temperature rose above 30°C, how would you feel? What about if the temperature fell below 10°C?

There is a range of temperatures that you can tolerate comfortably. It is called your *range of tolerance for temperature*. Predict what your range of tolerance for temperature would be for the place where you are now.

Within the range of tolerance, there is a temperature that you like best—one that is most comfortable for you. That temperature is called the *optimum temperature*. What would you guess the optimum temperature is for doing science in a classroom?

You have a range of tolerance for other conditions too. What might some of these conditions be? You and all living things are affected by conditions in the environment. At the beginning of this unit, you identified the nonliving, or abiotic, parts of the environment, such as temperature, moisture, light, and wind. How do these factors affect various living things? For each of the abiotic factors, how does the range of tolerance vary for different organisms? You will investigate these questions in the next few pages.

Temperature Tolerance in Plants

How well do seeds grow at different temperatures? Try the following Exploration to find out.

EXPLORATION 3

Too Hot or Too Cold?

You Will Need

- 30 radish seeds
- 3 Petri dishes
- 3 pieces of cardboard
- paper towels
- a thermometer
- water
- scissors

What to Do

1. Cut four pieces of paper towel into the shape of a Petri dish and layer them in the bottom of the dish. Do this for the other Petri dishes.

2. Now pour water into each Petri dish until the layers of paper towels are soaked. Pour off any extra water.

3. Put 10 radish seeds on the paper towels in each dish. Spread out the seeds so that they do not touch each other.
4. Cover each Petri dish with a piece of cardboard.
5. Put one Petri dish in a refrigerator; place another in a warm location (such as near a radiator or in an incubator); place the third dish in the classroom in a safe place where it won't be disturbed.
6. Measure the temperature at each place.
7. Over several days, count the number of seeds that sprout each day in each Petri dish. Add water as needed. In your Journal, make a table like the one shown below and record your observations.

Day	In Refrigerator ____°C	In Warm Place ____°C	In Classroom ____°C	No. of Seeds Sprouted
0				
1				
2				

Judging from your results, at what temperature do radish seeds sprout best? How would you redesign this experiment to find the range of temperature tolerance shown by radish seeds?

The Players and Their Roles 27

Other Projects

To find out more about temperature tolerance in plants, select one of these research projects. Keep in mind that the first two need to be investigated at a specific time of year.

1. During the fall season, observe which plants survive the cold nights. Which plants survive frost the longest? Which plants do you think will survive the winter? Make a chart showing temperature and plant survival.

2. A spring follow-up to the fall activity: Which plants are the first to grow in the spring? Where are these plants located? What have you discovered about the temperature tolerance of different plants?

3. You are interested in growing a fruit tree in a planter or in your yard. Look in some seed catalogs to determine what kinds of fruit trees will grow successfully in your area.

4. Compare plants that are native to your area with plants that are native to an entirely different type of habitat. For example, if you live in a desert-like environment, you might want to compare native plants to the plants in a tropical rain forest. What makes plants well suited to where they live?

Temperature Tolerance in Animals

"Today's a real scorcher!"

What is the hottest temperature ever recorded for your area?

"Snowstorm warnings in effect. School's canceled!"

What is the lowest temperature you are likely to experience where you live?

What, then, is the range of temperatures your body has to endure? How do you prepare yourself for such temperature extremes? The most obvious answer is "by the clothes I wear." Can you think of some body responses that help you tolerate different temperatures? What are some examples?

How do other living things tolerate temperature extremes? To help answer this question, form small groups of three or four and choose at least two of the following questions or activities. You may need to look in books on the subject and ask informed people, such as zookeepers or veterinarians.

1. Discuss and list all the possible hardships experienced by living things (both animals and plants) in the winter. Check the ones that you think are the most stressful.

2. List ways in which living things cope with the difficulties you have identified.
3. In the winter, where would you expect to find the following animals: deer, bear, frog, robin, woodpecker, skunk, snake, and squirrel? How does each one survive the winter?
4. Of what value is snow in aiding temperature tolerance in plants and animals? Name some specific examples.
5. Noreen lives in Bermuda, where the temperature never goes below 15 °C. Write a letter telling her about the changes she might see in the animals and plants where you live as winter approaches.

Moisture Tolerance

Moisture is another vital abiotic factor in the lives of living things. Living things have both a range of tolerance for moisture and an optimum amount of moisture they need. Compare the range of moisture tolerance for each plant in the following pairs.

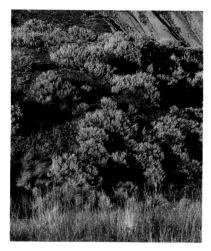

EXPLORATION 4

Experimental Design for Moisture Tolerance

Here is a challenge to your experimental abilities. Plan experiments to answer the following questions.

- Is there an optimum amount of moisture needed for radish seeds to sprout?
- Do all seeds have the same optimum moisture level for sprouting?

Show your experimental design to your teacher to make sure that you are on the right track and to check that your plan is safe. If there is time and the materials are available, perform your experiments and find the answers to the questions.

Who Needs Water Most?

An emergency meeting of the animals was called. Present were a mouse, a frog, some earthworms, a fish, and a snake. They met to discuss the long period of drought they had been experiencing that summer. Water levels were very low. It was a crisis for most, but particularly dangerous for some.

Compose a conversation among the animals that shows the relative importance of water to each. Include the amount of tolerance that each has to the water shortage.

Water Needs—Some Questions to Discuss

1. Humans can survive for several weeks without food, but not nearly so long without water. In fact, humans need almost 3 L of water every day! In a hot desert, you would need 11 L a day because of water lost through perspiration. Why is water essential to humans?

2. Even when you eat solid food and drink liquids other than water, you are taking in part of the 3 L of water that you need daily. Suppose you have just finished this delicious dinner:

Quantity	Food	Mass of Food	Percentage of Water in Food	Mass of Water
1 small glass	orange juice	90 g	87%	
1 slice	bread	30 g	35%	
1 serving	corn kernels	85 g	79%	
1 serving	roast beef	100 g	59%	
1	baked potato	100 g	75%	
1 serving	squash	110 g	96%	
2 slices	tomato	30 g	95%	
2 leaves	lettuce	50 g	94%	
1 slice	watermelon	800 g	93%	
1 glass	milk	250 g	87%	

Can you calculate how many liters of water you ate and drank in this meal? This will require you to perform a little arithmetic. First, you will need to calculate the mass of water for each food listed above. Remember that 1 mL of water has a mass of about 1 g.

3. Look at the graph below, and then answer the following questions. The amounts of water given are total starting quantities.

   (a) How long can a human live at 35 °C with 4 L of water? with no water?

   (b) How long can a human survive at 25 °C with 10 L of water?

   (c) What are the two factors on the graph that determine length of survival time?

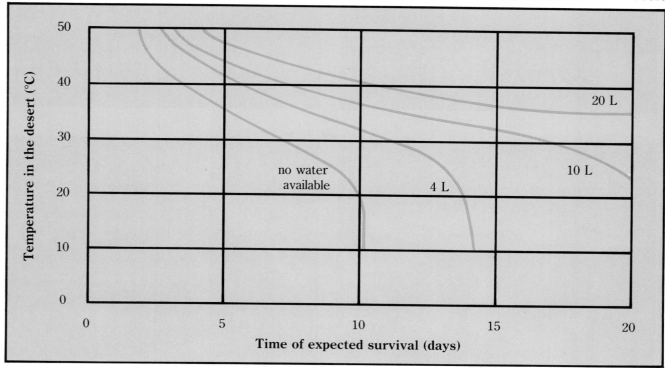

Problems to Ponder

1. Marine fish and sea gulls can drink salt water. Why can't you?

2. In the desert you find plants such as cacti, and animals such as the kangaroo rat, pocket mouse, and desert-dwelling lizards. How do they survive long periods without water?

3. Animals that live along shorelines are covered with water only half the time (when the tide is in). How do they survive those times when the tide is out?

4. Plants need water. What do they use it for?

5. Find out what the average monthly rainfall for your area was last year. Compare it with the rainfall in other parts of the country. On how many days did you have rain? Is there any difference in the amount and frequency of the rainfall in places near the ocean? inland? near mountains? in prairie regions?

Light Tolerance

Which of the organisms shown in (a) through (d) need the most light? the least light?

a

b

c

d

Which of the pictures below was taken at dawn, and which was taken at noon? Explain.

e

f

Light, especially sunlight, clearly plays a major role in the growth of plants. Can you explain the following situations?

- Geranium plants stored in the basement over the winter develop very long stems.
- The grass under a garbage can becomes yellowish.
- Most woodland flowers grow quickly and bloom very early each spring.
- Plants that grow in a forest under the branches of trees are small and scarce compared to those in open spaces.
- Moss tends to grow more abundantly on the north side of trees than on the other sides.

What conclusions can you draw about plants and light from your explanations for these interesting facts?

Some Light Problems to Investigate

The following questions can help you learn even more interesting facts about living things and light. Work in groups to find the answers.

1. Some animals are described as "nocturnal." What does this mean? What would you say about their range of light tolerance? Make a list of all the nocturnal animals you can think of. Then discover what the opposite of nocturnal is.
   Name some animals of this kind.

2. Find out how the amount of light affects each of the following:
   (a) Bats
   (b) Houseplants
   (c) Migrating animals and birds
   (d) Seed germination (sprouting)
   (e) Poinsettias
   (f) Human beings

3. Observe the changes that occur in the leaves, stems, and flowers of plants at different times of the day, evening, and night. Record your observations.

4. Are there animals that can get along without any light at all? Research information about animals that live in the ground and in the depths of the sea.

5. How does the amount of light affect the activity of different kinds of insects? If possible, observe the activities of some insects at different times of the day.

The Players and Their Roles

Abiotic Factors and the Seasons

Although you may not have been thinking about it, you've been reading about some of the effects that seasons have on abiotic factors. For example, the temperature decreases in winter and increases in summer. There is more light in summer than in winter. The seasons also affect the amount of moisture available.

To help you track seasonal changes in abiotic factors, create a chart like the one below and complete it in your Journal. Note that two abiotic factors that you haven't discussed have been added to the chart. As you fill out the chart, also think about the biotic factors that change as the seasons change.

Place (your choice):				
Abiotic Factors	Spring	Summer	Autumn	Winter
Temperature				
Moisture				
Light				
Wind				
Soil				

Challenge Your Thinking

1. **Going in Circles**

    (a) What abiotic or biotic part of the environment might "?" represent?

    (b) Tell the story of the dotted line.

    (c) Six lines originate at "tree." What might the interactions be?

    (d) "Grass" has two lines leaving it. How many more lines could you put in?

    (e) Suppose a line joined "Mike" and "tapeworm." How would you describe this interaction?

    (f) Locate the following:
    • A predator–prey interaction
    • A relationship involving a carnivore
    • A relationship involving a herbivore
    • An interaction involving a scavenger

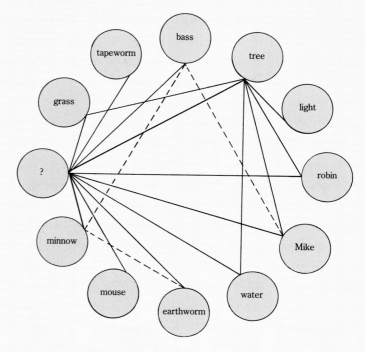

2. **Imagination Please!**

    Write a haiku describing any interactions that occur in nature. A haiku is a Japanese poem that has three lines of 5, 7, and 5 syllables, respectively. Ask a classmate to read your haiku to discover what type of interaction you've described.

    Here are some examples of haiku.

Hummingbirds hover
To drink the orchids' nectar
And to pollinate.

(Do you recognize mutualism?)

Crows like it, owls don't.
Plants cannot live without it,
Though bats will shun it.

(What is it? Hint: It's an abiotic factor.)

The Players and Their Roles

PUTTING IT ALL TOGETHER

6 Natural Changes

Succession—Environmental Change

Mt. St. Helens erupted in 1980. The blast destroyed hundreds of square kilometers of forest. Hillsides were stripped clean and then buried under several meters of volcanic ash. Almost nothing remained alive in the blast area.

But nature has a strong ability to preserve its original state and, if it is disturbed, to return to that state. Over the years, many changes have taken place in the area around the volcano. Plants and animals are gradually returning to the charred land.

Have you ever noticed changes in the plant and animal communities where you live? Suppose you made a visit to one of the following places. Describe your experience in either poetry or prose.

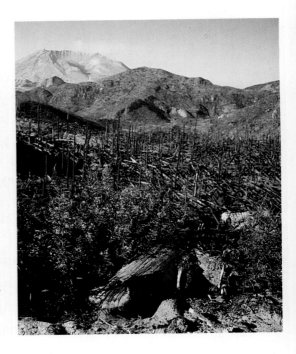

- an abandoned farm
- a pond that you had not seen for many years
- a place where loggers had cleared trees a few years before
- a garden left unattended for several weeks
- a field left unplowed for several years
- an abandoned lot

Changes in which one group of organisms replaces another may take days, weeks, years, or centuries, depending on the type of environment that is changing. All of the changes that occur in an area are collectively known as **succession**. The photos below illustrate succession in a pond community. During succession, one group of organisms follows, or *succeeds*, another. Succession is a slow, natural process. If possible, study an area where succession is taking place. Make regular trips to the area on your own, and record any changes you find in the plant and animal life there.

Can you observe succession occurring in these scenes?

EXPLORATION 5

Field Trip to a Changing Community

Where?
A rotting-log community in a nearby wooded area.

When?
Before the ground freezes. It may take quite a few hours to complete your visit.

Why?
You are going to discover what goes on in a fallen tree—who lives there, what work they do, and what they eat.

You Will Need
For each group:
- a field guide
- a paring knife
- your Journal and a pencil
- a thermometer
- a magnifying glass
- a trowel
- a white plastic tray

What to Do
1. Choose one or two well-rotted logs to examine.
2. Approach the log slowly and quietly. You might not be the only visitor! Look carefully for signs of animals that may be using the log.
3. Divide into four groups, with each group examining just one zone of the rotting-log community.
   - Zone A: Bark
   - Zone B: Wood beneath bark
   - Zone C: Bottom of log
   - Zone D: Soil beneath log
4. For each zone, describe and record the abiotic factors of temperature, light intensity, and dampness. For Zone D, investigate the top 6 cm of soil beneath the log.
5. For each zone, consider the tolerance ranges of the organisms for the abiotic factors listed in step 4.

Zone A
Bark

Describe the size, shape, and color of any fungi on the log.

Describe any green plants growing on the bark. The plants could be mosses, ferns, or tree seedlings.

Examine the bark closely for lichens. With a magnifying glass, observe the lichens, and make a drawing of what you see. Remember, this organism has two parts: a fungus (consumer) and an alga (producer). Since these are not visible without a microscope, you may wish to review the diagram of the lichen on page 13.

Look carefully for animal life or evidence of life (such as spider webs, animal droppings, and seed shells) on the bark. Are there holes in the bark to suggest animal life beneath?

Zone B
Wood Beneath Bark

Carefully remove some of the outer bark. Use a paring knife to break apart the wood underneath. With your magnifying glass, search it carefully for any small animals that make their home in the log. Try to identify as many as you can. Draw those you cannot identify. Do plants grow in this zone? Why or why not?

Zone C
Bottom of Log

Examine the bark and wood here carefully, as suggested for Zones A and B. Record all plant and animal life that you see.

Does life on the bottom of the log differ from that in Zones A and B? Why? Carefully roll the log away, if possible, and examine any life on the ground under it. You might find snakes, salamanders, toads, insects, or evidence of other animals (tunnels, pathways, food).

Zone D
Soil Beneath Log

How does the vegetation growing next to the log differ from that growing on the log? Push a trowel down 6 cm into the soil beneath the log. Place some of this soil on a white plastic tray and carefully observe any life. List the different kinds of organisms and types of movement you see.

Some Things to Think About

1. A tree that falls may have a mass of many hundred kilograms. The tree will eventually disappear, leaving no visible trace. What will happen to the organisms that once lived on or in the tree? the log?

2. It takes many years for a tree to rot away completely. What changes in plant and animal life might occur during this time? Can you identify fallen trees in different stages of decay?

3. If snow falls and covers the log, what will happen to the life in and on the log? What effect do you think winter has on the decaying process?

4. Compare the tolerance ranges of the organisms living in the different zones of a rotting log. Do all the organisms on the log need to tolerate the same range of conditions? Explain.

5. In nature, destructive processes such as the rotting away of trees are balanced by constructive processes. In other words, forces that renew life are constantly at work in the rotting log. How many plants and animals can you identify that have benefited from the rotting away of the log?

6. Write an article for a class newspaper explaining the niche of an animal that is associated with the rotting-log community.

In your study of the rotting-log community, you saw that the log was home to many organisms. These plants and animals used the log for shelter as well as for food. Other organisms that you didn't see—decomposers—were also present. Think back to what you read on page 19 about compost piles. Decomposers were present there, too, using the compost for food and breaking it down so that other organisms could use it for food. The decomposers in the rotting log were doing the same thing.

When the remains of living things are used in this way to help new life develop, the materials are being *recycled*. People, too, are learning to recycle many of the things they need and use. You can see many examples of recycling going on constantly in the natural world.

The matter from this decomposing animal is being recycled by the action of decomposers.

Population Changes

You know that a natural community contains a large number of niches for producers and consumers. Environmental changes that upset the producer-consumer relationship often change the natural balance within a community. Sometimes this can create serious problems.

Many settlers who came to the United States wanted to farm the land. Before they could plant crops, though, they had to cut down tracts of forest. What do you think happened to the many animals and other living things that were dependent on the forest? What would have happened if all the forests had been cut down? Do you think the settlers' fields provided a home for any living things?

Today, in many parts of the United States, forests and fields are being destroyed to make way for large real estate developments. As human communities grow larger, the communities of other living things get smaller.

However, some changes in a community occur because of what seem to be unrelated events. Consider this statement:

The number of mice in a community indirectly affects the number of rabbits in the same community.

Putting It All Together

How can this be? Do rabbits eat mice, or vice versa? Mice and rabbits are both herbivores. That is, they eat green plants. Suppose that the type of food that mice like to eat starts growing faster than usual in the area. Soon there is more than plenty of food for the mice. There is no increase in the type of food that rabbits prefer.

What immediate effect will this have on the mouse and rabbit populations?

Foxes and hawks eat both mice and rabbits. Mice are the easier prey, so with the increase in the mouse population, the foxes and hawks will eat more mice than rabbits. What is the reason for this change of diet? What will happen to the rabbit population now that the foxes and hawks have changed their diets? Judging from what you've learned, does the mouse population in this community affect the rabbit population?

What do you think will eventually happen in this community? Why do think this will happen?

The changes in population of mice (—) and rabbits (- -) are shown in the graph.

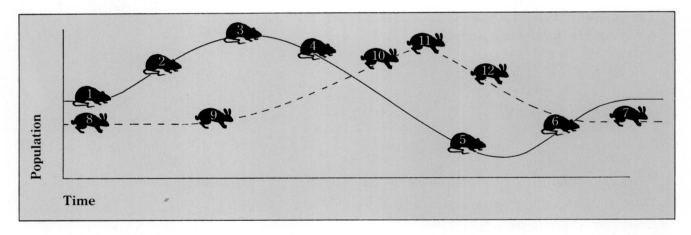

Each of the following events is represented by a number on the graph. Can you match the number to the event?

(a) The original population of rabbits; the largest population of rabbits; the largest population of mice; the smallest population of mice

(b) Mouse and rabbit populations when the increase in plants that mice like is first noticed

(c) The mouse population when they have exhausted their food supply and when they are still being preyed on by foxes and hawks more than rabbits are

(d) The rabbit population when the foxes and hawks begin preying on mice more than on rabbits

(e) The rabbit population when the mice run out of food, and foxes and hawks are preying on rabbits more than on mice

(f) The mouse population when foxes and hawks are preying more on the rabbits

(g) A return to original population counts of mice and rabbits

Population Changes to Think About and Research

1. The lemming is a small rodent that lives in cold, northern parts of the world. Every few years, according to legend, lemmings leave their habitats and drown themselves. Scientists, however, now know that they do not deliberately kill themselves. What might cause them to leave their homes and die in large numbers?

2. Study these three changes brought about by humans, and then answer the question that follows.

    In 1859, 24 European rabbits were introduced into one area of Australia. These rabbits had no natural enemies in their new environment: there were no predators, parasites, or diseases to affect them. The rabbit quickly became a pest, destroying all plant life in the area. Within a mere six years, over 20,000 of the rabbits had to be killed in an effort to control their growth.

    In 1890 an Englishman brought starlings to North America from Europe. Now the starling, a noisy and destructive bird, is found all over North America. It has displaced many of the native songbirds from their niches.

    In 1869 gypsy moth caterpillars were brought to North America from France by a man who wanted to start a cloth-making industry by breeding the gypsy moth with the silkworm. These caterpillars now kill shade trees by eating their leaves.

    How do some animals get out of control and become pests? Search for other examples of "out-of-place" animals or plants that have become pests.

3. In 1534 Jacques Cartier landed on the east coast of North America and found thousands of flightless birds that were clumsy on land but were powerful swimmers. These great auks soon became regarded as a source of profit. They were used for food; their feathers were used for making feather-beds; and they were even boiled for their oil! As a result, no one has seen a great auk alive since about 1844. Find out more about why and how this happened.

    What other animals have become extinct in the past century? Which are in danger of extinction? Are there also plants that are in danger?

A lemming

How is a "killer" bee different from a regular bee?

The last of these birds was seen in 1844.

EXPLORATION 6

A Community Study

You have already examined a rotting-log community. What other communities exist in your area? Is there a nearby woodland, pond, or stream? These examples and others offer interesting communities for study. With your classmates, choose a community that you wish to explore. Here are three possible communities for study:

- A woodland community
- A forest-floor community
- A pond community

You Will Need

- field guides for identifying organisms
- your Journal and a pencil
- a ruler or meter stick for making measurements

A Woodland Community

Woodlands provide people with many things: quiet hiking trails, beautiful spots for camping and picnicking, lakes for fishing, and so on. But too often we overlook much of the life a forest contains.

One visit is not enough to investigate all of a forest's mysteries, so this activity might make you want to return for another look.

Always keep in mind that the area you study should be left in at least the same condition as when you arrived, if not better. (Carry out those aluminum cans you find among the bushes!) If you turn over stones, be sure to return them to their original positions. Instead of taking samples of organisms back with you, try sketching them as best you can. If this isn't possible, take one organism, not a handful. Try to obtain samples in a way that won't harm the organism. For example, instead of pulling a leaf from a tree, take a leaf that has already fallen from the tree.

A Tree Study

Decide which group of trees makes up the main population of the area. If the trees have needle-like leaves, produce cones, and keep their leaves all winter, then they are *coniferous* trees. If they have flat, broad leaves, produce flowers, and shed their leaves in autumn, then the trees are *deciduous*. Choose a tree you'd like to study.

Look up the group and the name of the tree you chose in a field guide. When using a reference, pay particular attention to leaf size and shape, to how the leaves are arranged on a stem, and to how many leaves grow from a single stem. The bark can also be useful for identification purposes. Sketch a leaf of your tree, and if you can find fallen leaves, take one with you to help you learn more about the tree later.

Each type of tree within the two main groups has its own bark color and markings. You are probably familiar with the white, papery bark of paper birch trees, for instance. In your Journal, describe and sketch the bark of the tree you chose to study.

If you do your study in spring and are studying deciduous trees, check for tree flowers. Some are colorful and obvious, like cherry or apple blossoms. Others, such as the flowers of the birch and oak, are simple, small, and not at all showy. Sketch the flowers of the tree you've been studying, if they are present. Next look for its seeds on the tree or on the ground beneath. Sketch these too.

If you are studying a conifer, check for cones on the tree or on the ground nearby. Cones are made up of scales, and seeds are attached to these scales. Remove some scales and try to find the seeds or the scars showing where the seeds were once attached. Sketch a cone, a scale, and a seed or scar. Why are squirrels helpful to coniferous trees?

Observe the tree for other kinds of life. Check the bark, the leaves, and the area at the roots. List the organisms you find. Try to discover the relationship between the tree and each organism. Then construct a food chain or food web that shows how the organisms interact.

What influence do you think your tree has on other plants and animals nearby? What other organisms and abiotic conditions does it depend on? Consider its range of tolerance for each of the abiotic factors you studied earlier. Can you discover any examples of mutualism or commensalism between your tree and its immediate environment?

Putting It All Together

EXPLORATION 6—CONTINUED

A Forest-Floor Community

A whole world exists on the forest floor. Small flowering plants, ferns, mosses, lichens, and fungi can be seen on the ground. Underneath the litter of the forest floor lies a multitude of organisms.

You Will Need

- a meter stick
- your Journal and a pencil
- a newspaper or a white plastic tray
- a magnifying glass

Smaller Plants

Use a meter stick to mark off a 1- × 1-m plot. Examine the area closely and make a map of the plant and animal life you find. Use a scale of 1:5 to draw your map. In other words, your map will be 20 × 20 cm. You can use symbols to indicate each plant species, but be sure to give a key for your symbols. Also note the soil conditions of your sample area.

If the area you examined did not contain any lichens, make a special effort to search for some nearby. Some lichens that are easy to recognize include "British soldiers," with their red tips, or "old man's beard," a pale green, stringy lichen that hangs from tree branches. Other lichens, which grow on rocks, are scaly-looking and gray or orange-red in color. Sketch them, and use a field guide to identify them when you return.

Also look for the many interesting types of mosses. One type has brownish spore cases growing out of the top.

What's Under the Forest Litter?

Mark off a 50- × 50-cm plot on the forest floor. Carefully move and look under the rocks, stones, twigs, leaves, and other debris in the area. You will be surprised by what you find. Record all traces of plant and animal life. Try to construct food chains and food webs for the list of organisms you discover.

Gather some decomposing leaves and spread them on a piece of newspaper or on a white plastic tray. You might find animal life such as millipedes, spiders, centipedes, and salamanders.

Look around you for examples of commensalism, mutualism, and parasitism. Can you see signs of different types of living things cooperating with each other? Keep a record of your observations.

Sit down and look up. What do the plants and animals that live in the top layers of a forest add to the forest floor? What do they take away? How do the large trees affect the abiotic conditions on the forest floor?

Describe the tolerance ranges of the organisms that live in the leaf litter. How do these compare with the tolerance ranges of the organisms that live in the upper levels of the forest? Explain your answer.

Putting It All Together

EXPLORATION 6—CONTINUED

A Pond Community

You Will Need

- field guides for identifying organisms
- plastic bottles, containers, and bags for taking samples
- your Journal and a pencil
- a longhandled "muck scoop" (kitchen strainer)
- a white plastic tray
- a glass jar or an aquarium (optional)
- a pump (optional)

What to Do

Wear suitable clothing—preferably rubber boots. **Be Careful:** *Use caution around the pond.* Before you start looking for pond organisms, look at the pond itself. Is it a natural or an artificial pond? Where does the water come from? Where does it go? Make a sketch of the pond. Show the location of any trees or bushes that surround it. Do any streams flow into or away from the pond? Are there any rocks? Also be sure to include in your sketch any human-made features such as buildings or roads. All of these can influence the life in the pond.

One of the most interesting features of a pond is its great variety of animal life. The illustration at the right shows some of the pond life you might see.

The hardest animals to find are the ones that live in the muck at the bottom. For these you will need a "muck scoop." A kitchen strainer with a long handle should work. Use the scoop to bring up some of the decaying material on the bottom. Place a scoopful on a white plastic tray and gently poke through it, looking for anything that moves. You will likely find larvae of the dragonfly, damselfly, stonefly, and other insects. There may also be different types of worms.

As you find these animals, carefully move them to a container that has only water in it. This will allow you to look at them more closely. Use a field guide to identify some of them. Can you determine which of the animals are herbivores? carnivores? scavengers? What do you think the mouth of a carnivore would be like?

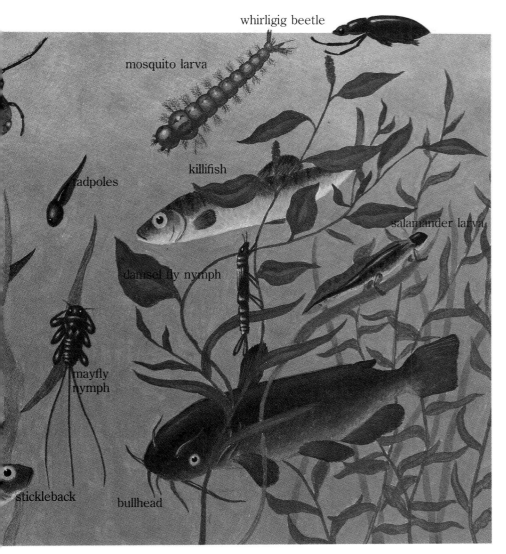

Before you leave the pond, take some of the muck from the bottom, some of the plants, a variety of animals, and pond water. You can use these materials to set up a small pond aquarium at home or in your classroom. Use a large glass jar or an aquarium. Whichever you use, try to make it as natural as possible. The water in your aquarium will have to be *aerated* (have oxygen added). You can do this with a small pump.

Watch the animals closely. Describe how each one moves. How does each feed? How do they react to each other?

Construct one large food web for the pond community, including all the organisms you either found or saw signs of. Are there some that you didn't see but that you think might visit the pond occasionally? How do they fit into your web? Classify the organisms as herbivores, carnivores, omnivores, or producers. Did you find any decomposers in or around the pond?

There may be some larger animals around your pond. You may see the animals themselves or perhaps just signs of them. In spring, frogs' eggs may be found. Can you see signs of muskrats or beavers? Make a list of any animals or birds you see.

Look for examples of commensalism, mutualism, and parasitism. Can you find signs of different organisms living or working together?

Describe the tolerance ranges of the organisms that live in the bottom muck. How are the abiotic factors different for the organisms that live in the area surrounding the pond?

When you finish your study of the pond, leave behind as few human signs as possible. Your footprints will remain, but take your plastic bottles, containers, bags, and other equipment with you. They are not a natural part of the pond.

Putting It All Together

7 Humans: Boss or Co-Worker?

Which of the diagrams below do you think better shows the interactions between humans and their environment? In your Journal, write down all the reasons for your choice.

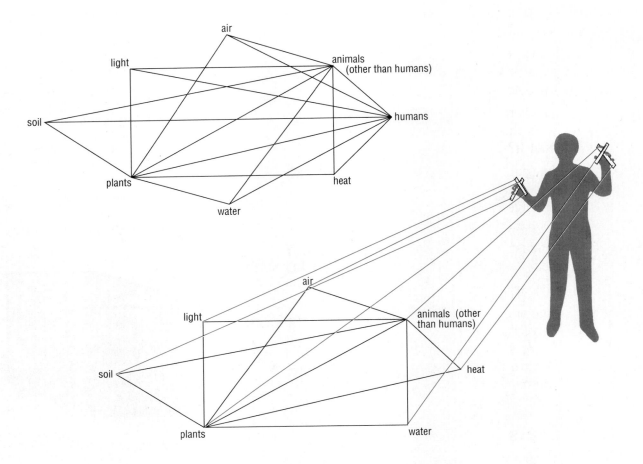

Some people consider themselves to be like other animals. They know that humans influence both the living and the nonliving environments. But they believe that humans are, in turn, equally influenced by the environment and all that it contains. Take a look at the two webs above. Which one supports this viewpoint?

Other people see themselves as separate from animals. They feel that humans have a special power or authority over both living and nonliving things in the environment. They know they influence their environment, but they often behave as if the environment has little influence on them. What you read about the great auks earlier in the book is a good example of this viewpoint. As you can see, this attitude has created many serious problems. Which of the two webs above fits this viewpoint?

Humans can manipulate and use the environment in ways that no other living thing on earth can. This ability carries with it the duty to consider the effects of our actions.

Facing the Problems

When humans think they have absolute control over nature, they often find that they have created dangerous situations. Here are some issues to consider. Think about the issues, research them, and discuss them with your fellow students.

1. Can humans bring harm upon themselves in their unwise use of the environment? How? Can you give any local examples?
2. What kind of harm can result from the careless use of pesticides? Is there a problem in the area where you live? What evidence of a problem is there? What is being done about it?
3. What are the problems created by smog in areas such as southern California? Is there air pollution where you live? What is being done about it?
4. What have industrial wastes done to the rivers and lakes? What are "hazardous wastes"? How do these kinds of wastes affect the way humans live?
5. What happens to forests and farmland when human communities, including industries and highways, expand? Is this a problem in your area?

We have come to realize that when we damage living things or thoughtlessly change nonliving things, we hurt ourselves. Because we are part of the web of life, we are affected by changes in our surroundings. Are we smart enough to survive? Survival means thoughtful, careful interactions between human beings and all parts of the environment—living, once living, and nonliving.

Can Individuals Make a Difference?

In this lesson, you have thought about and researched some environmental problems. Government and industrial agencies exist to help solve such problems. There are also many concerned individuals who encourage awareness of environmental issues. They often take an active role in solving and preventing environmental problems.

One group of students, called Kids Against Pollution (KAP), has testified before local and state governments, served as witnesses before the Environmental Protection Agency, proposed an amendment to the Constitution, and even made presentations at the United Nations. What prompted these students to do so much? One eighth-grade KAP member gave this answer: "I want my generation to be known as the one that did something about the environment." Here is the story of a group of students who decided to get involved with solving one environmental problem—acid rain.

Spraying pesticides

How can smog be reduced?

Members of Kids Against Pollution have testified before the Environmental Protection Agency about their environmental concerns.

The Story of Mr. Ripley's Class

Mr. Ripley's class in environmental science is no ordinary class. They study environmental problems that affect their community and beyond. They have expressed their concerns in such practical ways as writing a page on environmental issues in the local newspaper and participating in energy conservation activities. They have also conducted debates before the whole school on controversial environmental problems.

The students are quick to take the opportunity to initiate action. So when the city council announced the formation of a committee to study acid rain, Mr. Ripley's class decided to get involved. Here are some of the questions they investigated:

- What is acid rain?
- What causes rain to become acidic?
- Is it a natural problem, or is it caused by people?
- What harm does it do?
- Is it a widespread problem?
- Does it do any harm locally?
- If acid rain is harmful, what can we do as a group of people and as a government to control the problem?

How well informed are you? How many of these questions can you answer? What other questions would you have asked?

What the Students Found Out in Their Research

Air pollution has led to the creation of acid rain. **Acid rain** occurs when pollution from burning gas, oil, and coal mixes with water vapor in the air to form *acid*. This acid then falls to earth with snow and rain. Acid rain kills trees and makes ponds and lakes unable to support life. Some of the lakes in the eastern United States are now so polluted that they can barely support life.

The acid content of a substance is measured on a **pH scale** of 0 to 14. Look at the scale at the right. It shows the acid content of many different liquids and foods. A pH of 0 indicates the most acidic solution, and a pH of 14 indicates the most basic. Anything that has a pH value below 7 is acidic. The lower the number, the greater the acid concentration. The middle point, pH = 7, is neutral.

A pH value greater than 7 indicates the presence of a *base*. Basic solutions are often referred to as *alkaline*. Alkaline substances can *counteract,* or neutralize, acids; that is, they lower the acid content of a substance.

Check the scale for yourself.

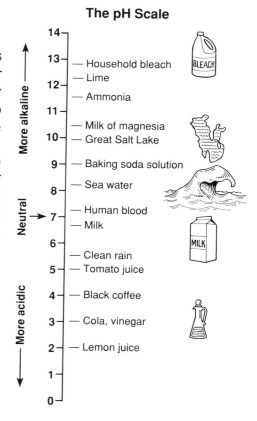

1. Is ordinary rainwater acidic?
2. At what pH does ordinary rain become acid rain?
3. Which has more acid in it—tomato juice or lemon juice?
4. Have you heard of people taking milk of magnesia for an acid stomach? Why would they do this?
5. Homeowners often put lime on their lawns in the spring. Why?
6. When the pH drops by one point, the acid content increases by 10 times. For example, a solution with a pH of 4 is ten times as acidic as a solution with a pH of 5. If the pH dropped by two points, what would the increase in acid content be?

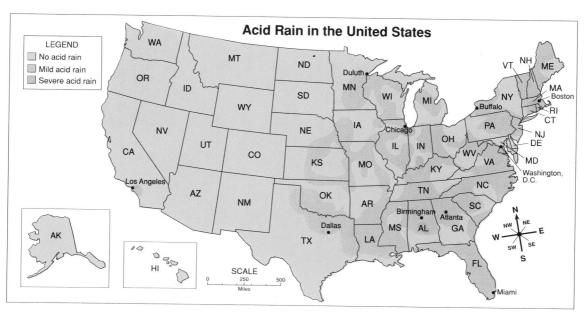

Now read what the students wrote for their report to the city council concerning the pH of rainwater in their city.

> *Acid rain has become a major problem in our city. The trees in our parks are not as healthy as they used to be, and every day we see dead fish wash up on the shores of our rivers and lakes. The thunderstorm we had last week produced rain that had a pH value of 4.0. This is far from the worst recorded case, which was a pH value of 2.4 in Scotland. But we cannot wait for our pH values to worsen to that level before we do something about it. We must act now.*

Look at the pH chart again. What liquids have a pH near 2.4? Imagine rain made of any of these liquids!

What do you think causes the pH of some rainwater to be low? Mr. Ripley's class easily discovered the answer. Industries, cars, and trucks release gases (sulfur oxides and nitrogen oxides) into the air. When rain or snow falls, these gases dissolve in the water vapor to form sulfuric acid and nitric acid. Here is more information from the students' report.

> *In North America, air pollution is a major concern. Every day, at least 165,000 metric tons of sulfur and nitrogen oxides are released into the atmosphere. This pollution remains in the air for up to 10 days, traveling thousands of kilometers from its source, depending on the wind currents. In other words, sulfuric acid clouds rising from our factories in and around the Pittsburgh area might be over the Great Lakes region in two days. So our pollution affects more than this community. Likewise, another region's pollution is making our situation worse. If every community reduced the amount of sulfur and nitrogen oxides that are released by cars and industries, we would all be better off.*

The Effects of Acid Rain

Mr. Ripley's class found plenty of information about the effects of acid rain on the environment. They learned that acid rain lowers fish populations and seriously harms aquatic life. Acid rain also destroys soil and trees. The students realized that when the soil is affected, crops will not grow as well as they would in soil with normal acidity. When acid rain kills trees, the economy suffers. That's because forests supply us with lumber for construction, paper, and pulp.

Do some research yourself and find out how acid rain actually affects the life processes in lake fish. Why do lakes become more acidic over time?

Some Controls

With your classmates, discuss what can be done about the acid rain problem. Make a class list of possible solutions.

Here are three of the eight specific recommendations that Mr. Ripley's class made in their report to the city council. Did you include any of these among your own suggestions?

1. Place more regulatory controls on factories and industries that release sulfur and nitrogen oxides into the air.
2. Penalize companies or individuals who do not follow the regulations.
3. Provide more funding for research on the effects of acid rain.

What could be done to reduce this source of pollution?

More Things to Do

1. Prepare posters for the school and community alerting people to the acid rain problem. Use the information on acid rain provided in these pages.

2. What occupations in your state are likely to be affected by acid rain? What effect do you think the increasing problem of acid rain will have on the general public? How does the public contribute to the problem?

3. Rank these factors in terms of how they contribute to the acid rain problem today: weather conditions, industry, cars and trucks, an uninformed public, a government that is preoccupied with other matters, and inappropriate education.

4. Call your local weather service and ask if they monitor the pH level of local rain. If so, chart the pH values during the next couple of months to see how they fluctuate. Would you expect the same values in the fall as in the spring? Why or why not? What local factors might influence the pH values?

5. In 1991, during the war in the Persian Gulf, many oil wells in Kuwait were set on fire. Smoke from burning oil typically contains high concentrations of sulfur dioxide. With your classmates, research the effects the burning oil wells may have on the region and on surrounding countries.

Challenge Your Thinking

Sketch a graph of the population changes in John's community to show what happened. For an example of a population graph, remember the mice and rabbits on page 42.

Putting It All Together

Unit 1 — Making Connections

The Big Ideas

In your Journal, write a summary of this unit, using the following questions as a guide.

- What are "interactions"? Give examples.
- What do the biotic and abiotic parts of the environment include?
- How is an organism's "niche" different from its "habitat"?
- Can you illustrate the differences between commensalism, mutualism, and parasitism?
- How would you describe the roles of producers, consumers, scavengers, and decomposers in a community?
- What is the difference between a food chain and a food web?
- How do each of the abiotic parts of the environment contribute to the welfare of a community?
- What kinds of changes take place in a community over time?
- Why are there continuous changes in plant and animal populations?

Checking Your Understanding

1. The mystery word tells what it's all about!

- I am a vegetarian!
- I'm not fussy. I eat *anything!*
- I stalk my food.
- I can make my own food.
- I try to keep from being eaten.
- I prefer animals for my food.
- I change dead plants and animals into soil.
- I am generous! I allow others to live on me.
- I live on the source of my food!
- I never make my own food.
- I feed on dead animals.

58 Interactions

2. Arrange these organisms into a food chain: grass, grasshopper, bear, fish, and frog.
3. Suppose that during the drought last summer, a forest fire burned a large portion of a forest near your town or city. Predict the kinds of changes that will occur in the plant and animal communities there over the next 100 years.
4. Draw a concept map that shows the energy relationships between the following items: *producer, sunlight, carnivore,* and *herbivore.*

Reading *Plus*

You have been introduced to interactions among organisms and their environment. By reading pages S1–S22 in the *SourceBook,* you will take a closer look at the role of organisms in ecosystems. You will learn more about how energy and materials are used in these ecosystems and how pollution affects our biosphere.

Updating Your Journal

Reread the paragraphs you wrote for this unit's "For Your Journal" questions. Then rewrite them to reflect what you have learned in studying this unit.

Science in Action

Spotlight on Veterinary Science

The healing and care of animals is the job of veterinarians like James Love. Most veterinarians work in clinics, but James is in charge of an animal care station for a medical research center. We talked with James in the primate section of the station.

James: We have a small colony of baboons here that are used for research purposes. However, since they have become an endangered species, the use of primates like baboons in research has become limited. We consider them to be important for research since they are very similar to human beings in many ways. Rats, mice, and guinea pigs are not enough like humans to be used for certain types of research.

One of the most exciting events we've had was the first birth of a baboon here. Immediately after birth, a newborn baboon is able to cling to its mother with all four limbs. This makes it possible for the mother to walk, run, and move about in trees without holding on to her young.

Q: What else does a veterinarian do at a medical research center?

James: Scientists come to the animal care station when they want to do experiments using animals. They discuss with me the kind of care the animal should have, the special equipment that might be needed, and any special diets that might be required. We also advise them about surgical procedures.

Q: Are there alternatives to using animals for medical research?

James: There is a movement all over the world to reduce or abolish the use of animals in research. Research on tissue cultures instead will develop very rapidly, I think, and will replace many of the animals being used in research at present. Also, future technological advances may be able to simulate the responses of animals without actually using the animals in the tests.

Q: Is the main emphasis in veterinary medicine on prevention, or on cure?

James: In caring for small farm animals, such as poultry and pigs, it is becoming uneconomical to treat individual animals for disease. It is just too expensive to call for the vet. The emphasis now is on preventing disease by vaccination or by adding antibiotics to feed or drinking water. Similar preventive medicine programs have been developed in the cattle industry.

Q: Do you have any thoughts about junior high and high school students doing research using animals?

James: Animals should not be harmed unnecessarily. Nevertheless, there is much research involving the observation of animals that can be done by young people. Those interested should first write to the Humane Society for information on the kinds of research projects that can be done. (2100 L St., N.W., Washington, D.C. 20037).

Q: Can you suggest some research projects that students might do?

James: One project that might be interesting would be to place two rats in separate cages and give only one an exercise wheel. Then over a period of time, you could compare their food intake, water intake, and weight change.

There are any number of projects of this sort, but you should write to the Humane Society or consult a veterinarian before you start your project.

A Project Idea

Some groups of people oppose the use of animals for research or medical testing. Others feel that it is acceptable to use animals for this purpose. This could be the topic of a class debate. To prepare for the debate, investigate this issue, being sure to consider both sides fairly and carefully. Why are animals used for testing? What alternatives are there to using animals in medical research?

Science in Action 61

Science and Technology

Oceanography

It has been said that the ocean is the final frontier on the earth. While many people are excited about going *up*, into space, some people have turned their attention *downward*, into the ocean. One such person is Dr. Sylvia Earle, an oceanographer who, in 1979, made the deepest untethered solo dive in history—381 meters. In her record-breaking dive off the coast of Hawaii, she planted a flag on the ocean floor in imitation of the first astronauts on the moon.

Dr. Sylvia Earle on ocean floor

The Blue Unknown

The ocean is largely unknown. Maps of the earth show different kinds of geographical regions, like deserts and rain forests. Such regions support their own kinds of **ecosystems,** or communities of plants, animals, and bacteria. The ocean, however, is still represented on many maps by a uniform blue color. This suggests that it is all the same under the surface—which certainly isn't true. There are many different geographic regions and ecosystems in the ocean waiting to be explored and studied.

Freedom Underwater

Going down to study the ocean depths has always been difficult. But this is changing. New underwater vehicles, or *submersibles*, are making it possible for single individuals to "fly" under the sea with a freedom that has never been possible before—to do "barrel rolls with the whales," according to Dr. Earle. Older submersibles were like dirigibles, with very little mobility. The new vehicles are more like airplanes.

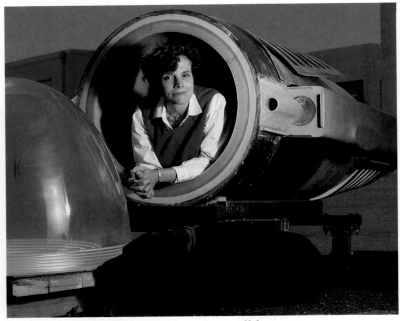

Dr. Earle in the *Deep Flight* submersible.

Touching Down

New submersibles may soon make it easy for individuals to go down to the very deepest part of the ocean, a depth of nearly 11 kilometers. This trip has been made only once before, in 1960. The trip was made in the *Trieste,* a Swiss-built bathyscaph consisting of a steel observation compartment attached to a submarine-shaped float. This particular vehicle is no longer in use.

A Sunless World

The new technology may even open up the depths of the ocean to recreational divers. If so, the world the divers see will be unlike anything they have encountered before. The deeper down you go, the dimmer the light from above becomes. As you dive below 100 meters there is not enough light for plants to produce food by photosynthesis. Only animals and bacteria live below this level.

Deep-sea swallower

Viperfish chasing a hatchetfish

People who go down to the sunless depths bring back reports of a world that is at once beautiful and bizarre. Sightless fish, luminescent creatures that glow green or blue, and fish with enormous mouths and exaggerated teeth—these are some of the fascinating sights that await the underwater explorer, perhaps even the "tourist" of the future.

Protecting the Ocean

One of the things that people see when they go down into the ocean is trash. Things like tangled fishing lines, for example, may do significant damage to life undersea. Do you think that a greater public awareness of such dangers would help bring about reforms? Do you think that undersea exploration is itself a danger to the environment?

Find Out for Yourself

Do some research to find out about the different classifications of life in the ocean. What are some of the animals that live at great depths? Where do they get their food? How are they adapted to their world? Why do you think that so many of them have such large mouths and teeth? What is the function of the *bioluminescence* of certain deep-water organisms?

Science and Technology

Unit 2: Diversity of Living Things

SUCH VARIETY!
1. Living Things in All Shapes and Sizes, p. 66
2. Why Is There Diversity? p. 74

BEGINNINGS AND ENDINGS
3. The Origins of Diversity, p. 86
4. The Loss of Diversity, p. 90

KEEPING TRACK
5. Making Sense of Diversity, p. 94
6. Ordering Diversity, p. 112
7. Life Cycles: Diverse Stages, p. 116

For Your Journal

Write a paragraph summarizing your thoughts about each of the following:
1. Why are there so many different kinds of plants and animals on earth?
2. Why can't a fish live both on land and in water like a frog does?
3. What are some of the problems that develop when species become extinct?
4. What features of the panther chameleon shown here do you think have enabled it to survive in its habitat?

SUCH VARIETY!

1 Living Things in All Shapes and Sizes

There are more than 30 million kinds of living things in the world. These living things come in all shapes, sizes, and structures. **Diversity** is the term that biologists use to describe differences in living things. In this unit you will look at the diversity of living things.

On this page and the next are illustrations of four living things, or **organisms.** Look at each one carefully. How do they differ from one another?

lizard

mushroom

66   Diversity of Living Things

With a classmate, discuss the diversity you see. Here are some questions to help you begin.

1. What is the relative size of each organism?
2. What features does each organism have? In other words, does it have roots, stems, leaves, flowers, eyes, legs, hair, scales, skin, or any other notable features?
3. What one characteristic does each organism have that distinguishes it from the other three?

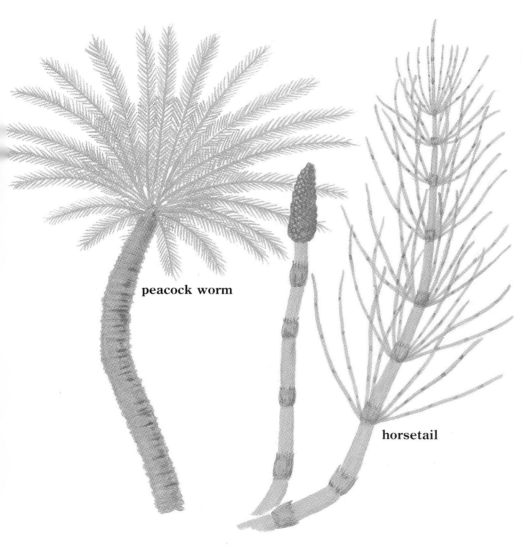

You have just described four living things that illustrate the great diversity of organisms. Living things can differ in size, habitat, appearance, eating habits, and methods of self-protection, to name just a few characteristics. Now you will investigate diversity a little further. Do either Activity 1 or Activity 2 of Exploration 1.

EXPLORATION 1

Looking for Diversity

Activity 1

Diversity in a Lawn

While playing ball on the grass or stretching out in your lawn chair to enjoy the sunshine, have you ever wondered about what is happening beneath you? There may be a food-hunting expedition or a ferocious battle going on there. You can learn a lot by getting down on your knees and carefully observing a small area of lawn. In this activity you will mark off one square meter of ground, preferably the day before the Exploration. Then you will study the diversity of animal and plant life in this square meter of lawn.

You Will Need

- sticks (4 per group)
- measuring tape
- heavy string and scissors
- a magnifying glass
- your Journal and a pencil

What to Do

Find a convenient grassy area. Using the measuring tape, measure out 1 m of ground and place a stick at both ends of the tape.

Place one end of the tape at a 90-degree angle from one of the sticks. Measure out 1 m and push a stick into the ground. Repeat this step twice to make a square. Then tie string to the sticks to outline your square.

Such Variety! **67**

EXPLORATION 1—CONTINUED

Observe the plant life in your study site. In your Journal, fill in a table like the one shown here. Then analyze the information you collected by considering the questions below.

Quick Sketch of Each Plant	Number of Same Kinds (Species) of Plants	Description of Each Plant (including size, appearance, color, and any other features you observe)	Common Name

Questions

1. How many types of plants did you find?
2. Which is the smallest plant you found? The biggest?
3. Which plant did you find in the greatest numbers? The fewest?
4. Do all the plants you found have some feature in common? If so, what is it?
5. How does the plant life affect the animal life in your study site?

The next day, carefully approach your staked area to see whether any larger animals, such as butterflies, are present. If you come to the area too quickly or noisily, you may scare away some animals. Using a magnifying glass, look closely for any smaller animals that might be hidden among the plants. In your Journal, fill in a table similar to the one shown here. When you have completed the table, analyze this information by considering the questions below the table. Then share your findings with your classmates.

Quick Sketch of Each Animal	Number of Same Kinds (Species) of Animals	Description of Each Animal (including size, appearance, color, and any other features you observe)	Common Name

Questions

1. How many types of animals did you find?
2. Did you hear animal sounds but not actually see the animal?
3. Did you find any evidence of animal life but not see the animals themselves?
4. Which is the smallest animal you found? The biggest?
5. Which animal did you find in the greatest numbers? The fewest?
6. How does animal life affect plant life in your study site?

Activity 2

Diversity Around You

What to Do

For your Journal, prepare two tables like those in Activity 1. Next, walk around your school, your neighborhood, or a park and fill in your tables. Then, answer the questions that follow each table.

In Exploration 1 you discovered the diversity that can be found in the living things around you. Now here are some riddles showing how much more diversity exists. See how many you can solve!

EXPLORATION 2

The Puzzling Diversity of Living Things

Activity 1

Who Am I?

In the following riddles, you'll find information about the structures, habits, and habitats of some organisms. Read each riddle carefully and think about what it tells you. Which living thing does each riddle describe? If you're having trouble figuring out what a riddle is describing, look at the pictures for some hints. (The pictures are not in order.) If you are really stuck, the answers follow in code. All you have to do is break the code!

Riddle 1

I move slowly when I am young but very quickly when I'm an adult. I eat flying insects, which I hunt near water. I have to be a strong flier to catch my food. When I stretch my four wings, I look like a helicopter. I have two more legs than a dog, and I have very large eyes. I am cold-blooded and have an external (outside) skeleton. Sometimes I'm very colorful. Who am I?

Riddle 2

I can walk, run, and swim. I can see well, but my sense of smell is not very sharp. I am warm-blooded. I am very adaptable and can live in many different environments. I really enjoy changing my environment. I care for my young for many years. I stand upright. Who am I?

Riddle 3

I must live in damp or wet places, avoiding the dry heat of summer and the cold of winter. If I'm living in a cold climate, I become dormant in the winter. If I'm a female, I produce young by laying eggs in water. I survive by eating anything that moves and that I can swallow. I can sing very well. Some of my close relatives can secrete a sticky white poison that can kill or paralyze dogs or other enemies who may try to eat them. Who am I?

EXPLORATION 2—CONTINUED

Activity 2
You Be the Writer!

Read the riddles again carefully, this time noticing which characteristics of the organisms are used to describe their diversity.

Then try writing your own riddles for some of the living things pictured here. Before you start, do some research to find out about the ones you chose. After writing your riddles, see whether your friends can solve them.

Here are the coded answers to the riddles:

1. CQZRNMEKX
2. GTLZM
3. SNZC
4. VGHSD ODKHBZM
5. VNNCODBJDQ
6. MMHUM
7. QZHMANV SQNTS
8. ZOOKD
9. AQNVM ADZQ
10. DKJ

Hint for decoding:
ZMS = ANT

Such Variety! 71

Riddle 4

I live in a lake, marsh, salt bay, or on a beach. I eat mostly fish and crustaceans. Although I can fly, I catch fish only by swimming. My great throat pouch is handy for scooping up fish. I fly by alternating several flaps of my wings with a glide. I always fly with my head hunched behind my shoulders. I nest on the ground in colonies. I have a wingspread of 2.5 to 3.0 m. My close cousins live only by the ocean, while I can venture inland. I am happy to report that these cousins are growing in number, even though they suffered from DDT poisoning a few years ago. Who am I?

Riddle 5

I have a very high body temperature. It is usually 7 degrees warmer than a human's. My feet are well adapted for grasping things. I have four toes on each foot: two point forward, and two point backward. I have stiff, spiny tail feathers that act as a prop when I hunt food. I eat tree-boring insects, ants, acorns, flying insects, berries, and sap. My home, which I make myself, is a hole in a tree. I use my bill to chisel the wood. Who am I?

Riddle 6

I have pointed green stalks above the ground and a rounded brown bulb below. People must pull me out of the soil before I can be useful to them. Cooks use me to improve the taste of food. If people bite me, I can bite back, making their eyes water. Who am I? (This riddle is based on an Anglo-Saxon riddle that's over a thousand years old!)

Riddle 7

I live in cold, well-oxygenated water, and I'm a fast, strong swimmer. I am slim, sleek, and colorful. I'm a carnivore; I eat mostly insects and smaller members of my own kind. I spawn my eggs during the spring in small, clear streams. I'm cold-blooded. Who am I?

Riddle 8

I undergo wondrous changes during my life. At the beginning, I am a sweet-smelling pink and white blossom. Later I'm a hard, green ball that makes your eyes and mouth pucker if you try to eat me. Finally, I become a sweet, juicy red or yellowish fruit. People say I keep physicians away. Who am I?

Riddle 9

I am a big animal. My mass is about 225 kg, but my tail is only about 15 cm long. I am dark in color. Generally, I live on forest floors and in thickets. When it begins to get really cold, I enter my shelter for the winter. I don't have very good sight, but my senses of hearing and smell are keen. Using these senses, I find lots of food: small animals, insects, any flesh, garbage, leaves, grasses, berries, nuts, and fruits. I am a good climber, and if I'm disturbed, I may retreat to the upper branches of a tree. Who am I?

Riddle 10

I am warmblooded and hairy. I feed milk to my young. I have no upper teeth, but my lower teeth tell you what kind of food I eat. I chew my cud, and I have a complex stomach. The males of my kind have huge, branching antlers. I have a heavily maned neck. Humans, wolves, and mountain lions are my only enemies, but mountain lions usually won't attack me when I'm fully grown. My young are not camouflaged from these enemies until their winter hair grows out. Sometimes you can hear the males of my kind give a high-pitched bugle call. If this call is answered by another male, a battle may follow. Who am I?

Such Variety! 73

② Why Is There Diversity?

Why are there so many different types of living things? Why isn't there a single "all-purpose" organism or at least one all-purpose plant and one all-purpose animal? These are not easy questions. This lesson will help you find some answers.

You already know that there are many different environments on our planet. There are deserts and oceans, grasslands and mountains, forests and tundra. In each of these environments, there are many different habitats for living things.

Since there is such a diversity of habitats, no single organism can survive in all of them. Obviously an organism that lives on the ocean floor would have great difficulty surviving in a desert!

You learned in Unit 1 that each organism plays a particular role in its habitat. Just as no single organism can survive in all habitats, no single organism can play all roles in one habitat.

All this leads to another question. How is each organism suited to play a certain role in a certain habitat? Keep on reading!

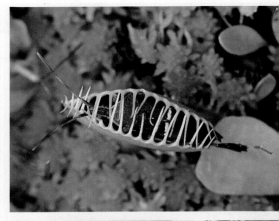

The Fine Art of Survival

Adaptations are the features organisms have that enable them to survive and produce young. Both animals and plants have many different kinds of structures and behaviors that help them survive in their environments. Below are three categories of adaptations to discuss: adaptations for obtaining food, for protection, and for locomotion. The organisms on this page and the next will give you some hints.

1. Think about the great variety of structures and behaviors (adaptations) that animals and plants have for obtaining their food.
2. Can you think of animal and plant adaptations that could be used for protection? Give some examples.
3. Consider an animal's locomotion—its movement from place to place. Name some adaptations that various animals have for locomotion. Is each adaptation you thought of related in some way to the organism's habitat? Explain.

Such Variety! 75

Disappearing Acts

Some animals in this world have spectacular survival techniques. But you have to be a very good observer to see them. Like escape artists or magicians, some animal species have developed an amazing variety of illusions. To escape the ferocious jaws, beaks, and fangs of hungry predators, they simply "disappear" into their environment! Other animals do the same trick, not to avoid being prey, but in order to catch their own prey.

Living things pull off their "disappearing acts" by a method called **camouflage** that allows them to blend in with their environment. Some organisms resemble twigs or leaves. Others resemble nonliving things like stones. Many organisms blend in perfectly with the color of their surroundings. The following examples illustrate how the shape and color of organisms can protect them in their environments.

The Katydid

Some types of katydids look like green leaves. If caterpillars accidentally nibble at the edges of the katydid's wings, spots of brown surrounded by yellow rings appear. These markings are just like the spots of decay that would appear on a leaf.

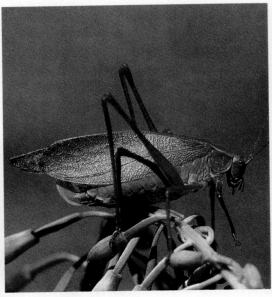

The Chameleon

Chameleons usually appear green or brown to blend in with their surroundings. However, when attracting a mate, defending their territory, or responding to light or temperature changes, chameleons may change to black, red, yellow, white, or orange.

The Stick Insect

Stick insects look like twigs. Their shape and color protect them from being eaten by birds.

Diversity of Living Things

The Polar Bear

Polar bears blend in with the snow and ice that are usually part of their surroundings.

The Flower Mantis

The flower mantis looks like a flower. Its appearance attracts insects to feed on what they think is flower nectar. The unsuspecting insects are then devoured by the deceptive mantis.

The Grasshopper

Some young grasshoppers can change their color to match their environment. For instance, nymphs that eat green leaves become green.

The Snowshoe Hare

Snowshoe hares are white in winter to blend in with the snow, and brown in summer to blend in with trees, grasses, and weeds.

Who Am I?

*We each have our special color and shape;
From our enemies it's a way to escape.*

*"I'm an extremely changeable fellow;
Any color will do—black, white, or yellow!"*

*"As for me, I specialize in white—
So I can match my habitat site."*

*"I am white too, for part of the year;
But when summer comes, then brown will appear."*

*"Colors—green and brown—are important for me too
But it's a shape like a leaf that really sees me through."*

*"Yes, shape is important—I look like a flower.
But that's to bring insects for me to devour."*

*"I could change color when I was small.
Then I was a tasty treat for all."*

*"I have the best trick that you've ever heard
I look like a twig—it'll fool any old bird."*

Such Variety!

Attention, Please!

Some animals are adapted to attract attention. Which animals shown here have an adaptation that

- Attracts members of their own kind for mating purposes?
- Warns possible enemies (predators) of their bitter or smelly secretions or powerful stings?

cardinal

peacock

lionfish

butterfly

skunk

wasp

Diversity of Living Things

Mimicry: Looking Dangerous

Some animals, such as the Mexican milk snake, have warning features that protect them from being eaten by predators. The Mexican milk snake is not poisonous, but it looks like the Texas coral snake, which predators avoid because a coral snake's bite can be serious. The Mexican milk snake *mimics* the Texas coral snake. Through **mimicry,** the harmless Mexican milk snake escapes being eaten.

Mexican milk snake

Texas coral snake

There are some very strange examples of mimicry in the animal kingdom. The larvae of the hawkmoth resemble certain poisonous snakes. Another type of moth mimics hornets. Its wings and coloring look like a hornet's, and this moth even has what looks like a stinger on its abdomen. When threatened, the moth bends and twists like a hornet that is searching for a place to inject its venom.

Another interesting form of mimicry is found in insects that have evolved large eyespots that resemble a vertebrate's eyes. These are often on the upper side of the wings, and they are usually covered when the insect is at rest. When the insect is disturbed, however, it spreads its wings, exposing the eyespots. Through experiments, scientists have shown that many birds are frightened by these eyespots.

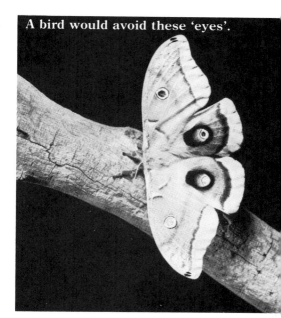

A bird would avoid these 'eyes'.

Something to Research

Find out about the mimicry of the drone fly. What does it mimic? From which animals is the drone fly protected by its mimicry?

Plants Also Adapt to Their Habitats

Plants, like animals, are uniquely suited to their habitats. Although plants cannot move about like animals, they do have certain adaptions that allow them to survive and reproduce in particular habitats.

Observe the plants shown on these two pages. Then decide which plant best matches each habitat adaptation listed below.

1. A plant that is able to hold on to rocks in swift rivers and streams
2. A plant that can store water
3. A plant that can catch insects
4. A plant that can compete with other plants for sunlight
5. A lawn mower-proof plant
6. A tree that can withstand high winds
7. A plant that grazing animals would not eat because of the plant's protective surface features
8. A plant that is capable of living on the surface of a pond

80   Diversity of Living Things

Perhaps you were able to identify some of these plants because you have seen them before. Their common names are:

   Barnyard grass
   Barrel cactus
   Birch tree
   Broadleaf plantain
   Coconut palm
   Dandelion
   Duckweed
   Pitcher plant
   Pondweed
   Scotch thistle
   Sundew
   White waterlily
   Wild rose

Match the names with the pictures.

Such Variety!

Adaptations for Seed Dispersal

Most plants produce seeds that grow into new plants. But seeds cannot always survive if they simply drop directly beneath the plant that produced them. They may need to be moved to another area. Many types of seeds, therefore, are adapted to travel. This helps ensure that new plants will grow and survive.

Look at a pea seed. Do you think it could travel very far on its own? How might you change it so that it has a greater chance to travel and survive? Working with some classmates, design changes in the pea seed that will enable it to do the following:

- Float on water
- Be thrown a distance by the parent plant
- Attract an animal that would carry it to another location
- Hitchhike on an animal for some distance
- Be carried by the wind

A Picture Puzzle

Examine each seed closely. Identify the adaptation that is being used to help each seed travel away from the parent plant. Do these adaptations resemble the suggestions you made for the pea seed?

82 Diversity of Living Things

Such Variety! 83

EXPLORATION 3

Designing an Animal

Here's another riddle for you: How is your thumb like a monkey's tail? Give up? They're both **appendages,** or protruding parts of the body. In fact, your entire hand and arm is an appendage. So are an insect's antennae, or feelers.

Appendages are adaptations designed to help an animal perform various specialized tasks. Appendages help an animal live successfully in its environment.

What to Do

Design an animal with appendages that will enable it to have the following characteristics:

1. The animal lives in water.
2. Its heavy body needs a lot of support as the animal walks.
3. It can walk on the bottom of a body of water for kilometers without stopping.
4. It can dart away suddenly from its enemies by swimming.
5. It can create water currents to bring food in the water to its mouth.

6. It tests its food before eating it.
7. It has appendages to hold larger pieces of food.
8. It can break apart hard bits of food.
9. Its diet includes shelled animals.
10. It has appendages to hold its young.
11. It has formidable defensive weapons.

In your Journal, draw the creature you designed. Does it look like any animal you have seen before?

84 Diversity of Living Things

Challenge Your Thinking

1. In letters from Tim Lawson, your pen pal in Australia, you've learned about the Australian "bush," which includes giant grain fields and tropical forests. You've also learned about the "outback," an area of great space, scattered inhabitants, and desert regions where beautiful wild flowers grow when it rains. Tim has told you about vegetation such as the gum tree (grown nowhere else) and the golden-flowered wattle. And he's written about strange animals like kangaroos, koalas, and wombats.

   Write Tim a letter telling him about the many kinds of plants and animals he would see during a visit to your area of the United States. Tell him about how these organisms are adapted for survival in your environment.

2. Sketch an adaptation that
   (a) Would enable an animal to cut down a tree.
   (b) A water plant might use to float on the surface.
   (c) Might be used as a defense against predators.
   (d) Would enable a plant to live with very little water.
   (e) Would enable an organism to blend in with its environment.

Such Variety!

BEGINNINGS AND ENDINGS

3 The Origins of Diversity

Natural Selection

Sometimes, over many generations, the characteristics of plants and animals change. The changes enable the plants and animals to live successfully in a new habitat. For example, the porpoise is an animal that lives, eats, and reproduces in the water. But unlike other aquatic organisms, the porpoise breathes air through lungs, just like land animals do. The porpoise's ancestors were land animals that, over time, changed their habitat from land to water. They probably adapted to the oceans because they found more food or more easily escaped predators there than when they lived on land. The modern porpoise, which is a strong and fast swimmer, shows just how successful these changes were.

Changes can also permit a species to live more successfully in its own habitat. Polar bears, for example, have not always been white. But the polar bears that were white were more successful in finding food. Blending in with their environment enabled them to sneak up on their prey. Hence white polar bears survived to reproduce. Over many years, the white polar bears grew in number, while the dark ones gradually disappeared.

In the 1800s, Charles Darwin developed a theory that organisms had *evolved,* or changed over many generations, by a process he called **natural selection.** Darwin called the process natural selection because the environment favors, or selects, organisms that are fit for survival. Darwin thought that organisms with traits that are well suited to the environment survive and reproduce at a greater rate than organisms that are less suited to the environment. The well-suited organisms thus pass desirable traits on to their offspring. Natural selection explains how dark polar bears were replaced by white ones and how porpoises became aquatic animals.

Sometimes the habitat of a species will change in some way. This change may force the species to move to a new habitat, adapt over many generations to the changed habitat, or become extinct. One example of a species that adapted to a changed area is the peppered moth, which you are about to examine in detail. Study the research data given on page 87. Do these facts support the theory of natural selection?

Which polar bear is harder to find?

The Case of the Peppered Moth

Date:
The 1800s, during the Industrial Revolution

Place:
England, near the industrial city of Birmingham

Researcher:
Professor Kettlewell and his assistants

White moth on a light trunk

Dark moth on a dark trunk

Research Subject:
The peppered moth. Peppered moths have lived in the forests of England for thousands of years. They rest on the trunks of trees during the day and are the diet of many birds. Peppered moths vary in color, from light-colored to dark-colored.

Research Problem:
Did air pollution, which covered tree trunks with black soot from the many new industries, affect the survival of the peppered moth?

Conditions:
Before the Industrial Revolution, the tree trunks were light-colored. The trunks and branches were also covered with silvery-white lichen.

As the Industrial Revolution progressed, pollution killed the lichen and blackened the tree bark.

Hypothesis:
Kettlewell formed a hypothesis about the peppered moth. Form your own hypothesis. It should include the effect that you think the Industrial Revolution had on the peppered moth's survival.

Procedure:
With his assistants, Kettlewell then experimented to test his hypothesis. They used the following procedure:

1. They located two areas. One was a wooded area with lichen-covered oak trees. The other was a wooded area that had been subjected to pollution for many years.
2. They released a known number of light-colored and dark-colored peppered moths into each area.
3. After a given amount of time, they recaptured as many moths as they could.

Results:
In the unpolluted area, more light-colored moths survived. In the polluted area, more dark-colored moths survived.

Discussing the Results:

1. Do these results support your hypothesis? (They supported Kettlewell's hypothesis.)
2. How can you explain the results?
3. Why didn't Kettlewell release the moths only into the polluted area?
4. Do Kettlewell's findings support the theory of natural selection? Explain your answer.

Beginnings and Endings

Darwin's Finches

In 1835 Charles Darwin spent five weeks visiting the Galápagos Islands as the naturalist on H.M.S. *Beagle*. He observed, recorded, collected, and preserved everything he could of the islands' natural history. There were many strange and colorful animals. But what excited him most were drab little birds that made unmusical sounds—finches. These birds all resembled one another closely, except for one set of features—the size and shape of their beaks. Separated for thousands of years on the different islands of the Galápagos, the finches of each island had adapted in a unique way to their own environment. The differences among the finches of the various islands are shown in this table.

Charles Darwin

Differences Among Some of Darwin's Finches

Name of the Finch	Feeding Habit	Form of Beak
small tree finch	Uses delicate bill to eat aphids and small berries.	
large tree finch	Grinds fruit and insects with parrot-like bill.	
small ground finch	Uses pointed bill to eat tiny seeds and pick ticks from iguanas.	
large ground finch	Conical bill enables it to eat large, hard seeds.	
cactus finch	Long bill probes for nectar in cactus flowers.	

88 Diversity of Living Things

1. How is the structure of the beak well suited to the diet of each group of finches?
2. Do Darwin's finches support the theory of natural selection? Give reasons for your answer.
3. It has been speculated that Darwin's finches reached the Galápagos from the mainland of South America as a single flock perhaps a million or more years ago. Think about the following questions and explain your answers.
   (a) What do you think the original finches looked like? Why?
   (b) Is it possible that the original birds were various species that arrived on the islands at different times?
   (c) Assume that one flock of finches gave rise to the 14 different species now existing on the islands. For this to occur, would it be significant that the Galápagos consist of many small islands rather than one large one?
   (d) What advantages would the finches have had in arriving on the islands under the following conditions?
      (i) When no other species with exactly the same diet existed
      (ii) When there were no predators
      (iii) When there were no parasites to live on the finches and weaken them
4. How has diversity helped Darwin's finches to survive?

H.M.S. *Beagle* arrives in South America. Can you find the Galápagos Islands on a map?

Adaptation Adventure

Would you like to live in a new, vastly different environment? How about the moon? Underwater? A planet in outer space? What adaptations (including technological ones) would help you survive? Think about the adaptations that were required for the *Apollo* astronauts to live and travel on the moon for several days.

Create a tale about how you and a group of your friends survived in a very different environment. Describe the conditions of the new environment, the difficulties you would encounter, and the adaptations you would need to survive. You might be able to develop technological solutions (adaptations) to these new problems quite quickly. But changes in the structure or function of organisms' bodies (natural adaptations) often take many, many generations to occur—or they may never occur at all. What might happen to your species if the necessary technology is not available?

Beginnings and Endings

4 The Loss of Diversity

Have you ever thought about what would happen if there were very little or no diversity among living things? Imagine what the results would be if:

- All bears were black.
- All rabbits were white.
- All plants were 5 cm tall.
- No insects could fly.
- The only living things in the oceans were seals.

Getting Started

Make 20 cards each for the following animals: whooping crane, bowhead whale, wood bison, sea otter, Peary caribou, eastern cougar, white pelican, Eskimo curlew. Have 20 blank animal cards on hand in case some of the animal populations grow beyond 20 members. Make two copies of each extinction card shown along the bottom of this page and the next.

Also make one cardboard token for each animal to travel on the gameboard pathway.

The Extinction Game

You have seen that animals may become *extinct* when their habitat changes significantly. The Extinction Game will help you discover the kinds of changes that may affect the survival of certain animals. Two to eight people can play at one time.

To play the game, choose a certain animal from the list below. Begin with a population of 20 of your animal. To be a winner in the game, you need to finish before your animal becomes extinct. You will be using two sets of cards for this game: "animal cards," which represent each of the eight animals, and "extinction cards," which tell you how many members to add or subtract from your animal's population.

Place extinction cards face down in a pile on the table. (After drawing an extinction card, place it on the bottom of the pile.) Roll the die to determine who begins—the highest goes first. Then, in turn, each player throws the die and makes a move. Remove or add animal cards from your pile as required by the extinction cards. If your animal becomes extinct, you are out of the game.

Extinction is a process that generally happens over a long period of time. This period may be represented by traveling the path a second time with your surviving population.

Animal Cards

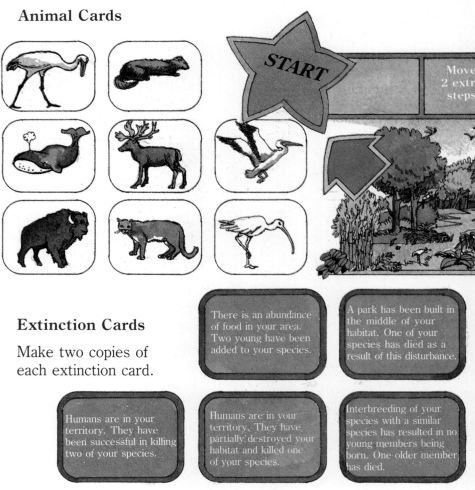

Extinction Cards

Make two copies of each extinction card.

There is an abundance of food in your area. Two young have been added to your species.

A park has been built in the middle of your habitat. One of your species has died as a result of this disturbance.

Humans are in your territory. They have been successful in killing two of your species.

Humans are in your territory. They have partially destroyed your habitat and killed one of your species.

Interbreeding of your species with a similar species has resulted in no young members being born. One older member has died.

Diversity of Living Things

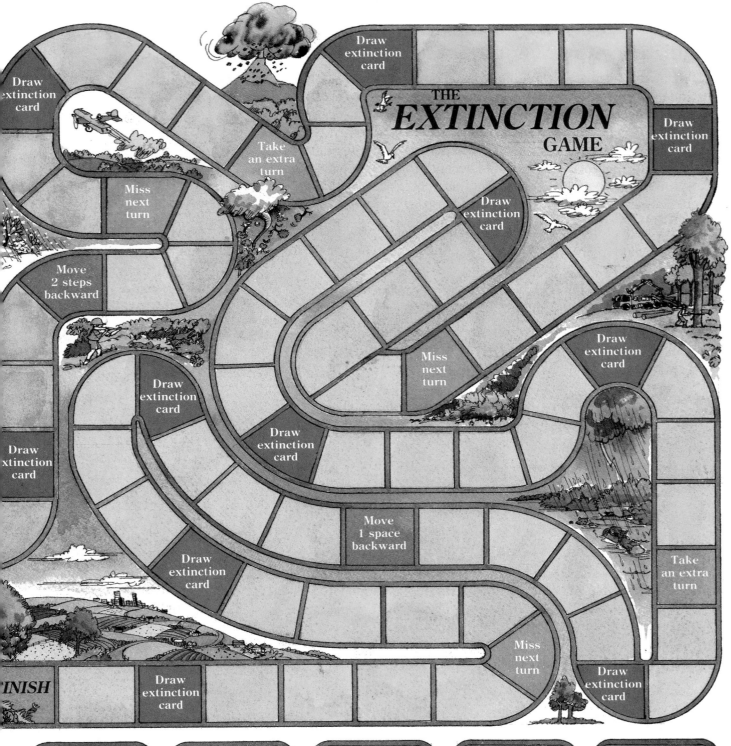

Back from the Brink

The brown pelican is a fish-eating coastal bird that nests along the Atlantic, Pacific, and Gulf shores of the Americas. Below are some facts about brown pelicans on West Anacapa Island, a major breeding colony for brown pelicans in California. The facts trace the brown pelican's story from the days when it lived in large numbers on West Anacapa Island, through its decline, to its comeback from the brink of extinction. But the facts are not placed in the right sequence. Can you put them in a better order?

(a) Investigators discovered that a chemical company had been dumping waste DDT into the Los Angeles sewer system for some time.

(b) An average of 5000 pairs of brown pelicans nested on West Anacapa Island from 1985 to 1989.

(c) In 1973 the brown pelican was placed on the United States' list of *endangered species*—those species that may not survive in the wild unless they are protected.

(d) When brown pelicans ate DDT-contaminated fish, the DDT accumulated in their bodies. This caused the shells of their eggs to be so thin and fragile that they often broke while the eggs were being laid or incubated.

(e) In 1970 only one brown pelican hatched on West Anacapa Island.

(f) Brown pelicans nested in large numbers on West Anacapa Island.

(g) In 1972 the use of DDT was banned.

(h) DDT worked well in killing mosquitoes and other insects, but biologists discovered that DDT had contaminated the Pacific Ocean and the fish in it. This caused considerable harm to the food chain.

Another Close Call

Examine the following diagram closely. What does it tell you about the survival of the brown pelican in Texas and Louisiana?

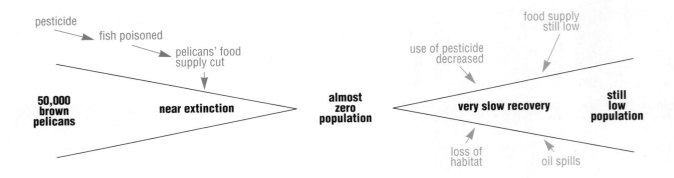

92 Diversity of Living Things

Challenge Your Thinking

1. Natural selection is sometimes referred to as *survival of the fittest*. Do the examples of the peppered moth and Darwin's finches illustrate the idea of survival of the fittest? If so, how? Think of some other examples that either support or refute this idea.

2. It is believed that the ancestors of modern giraffes did not all have long necks. Some had long necks, while some had short ones. Use the idea of *survival of the fittest* to explain why giraffes no longer have short necks.

3. An article in the local newspaper stated: "Everything that humans do to the environment causes animals to become extinct." Write a letter to the editor of the newspaper to disagree with the statement. In your letter, describe some of the ways that humans help prevent animals from becoming extinct.

Beginnings and Endings

KEEPING TRACK

5 Making Sense of Diversity

Grouping Living Things

As you may well imagine, it's difficult to keep track of over 30 million kinds of living things! The easiest way to do this is to group the organisms. You already have lots of experience with grouping things. For example, in your kitchen, how is the silverware grouped? Are the dishes kept in one place and the dish towels somewhere else? Are the pots and pans separated from the drinking glasses? Here are some more situations to think about in which items are sorted into groups, or *classified*:

- How are the books classified in the school library?
- How does the telephone company use grouping to keep track of everyone with a phone?
- There are over 240 million people in the United States. How does the postal service group all these people so that mail can be delivered to every person?

Name some other situations in which items are divided into groups.

Now think about some ways in which you might classify living things.

Try to place all the living things in the illustration on page 94 into two groups. Then further divide each group into two more subgroups. Invent a name for the living things in each group and subgroup. Your classification system should look like this:

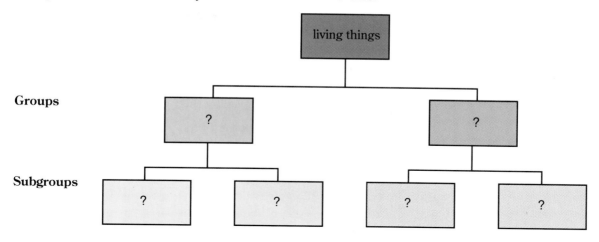

Paulette devised this way to classify the organisms:

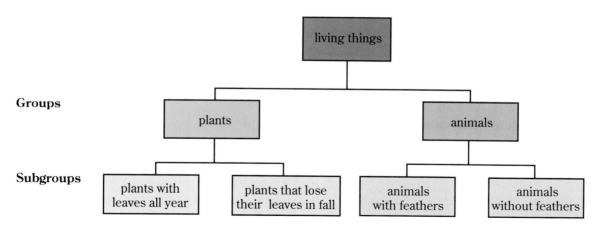

Paulette's system is only one among many ways of classifying organisms. Compare your system with Paulette's. Try classifying the living things in the picture according to Paulette's system. How useful do you think her system is?

Now let's look more closely at the differences between organisms—differences that help you make a classification.

1. Choose one of the living things in the illustration on page 94. In your Journal, list the distinguishing features that you could use to identify it. Read your list to some of your classmates. Ask them to suggest other organisms that would also match the features on your list. How many different organisms did your classmates suggest? Do you always need to talk about *each single living thing,* or can you talk about *a group of living things?*

2. Try doing the same thing again with another organism.

3. Look at the living things in this illustration. Make a group and subgroup diagram that can be used to identify them.

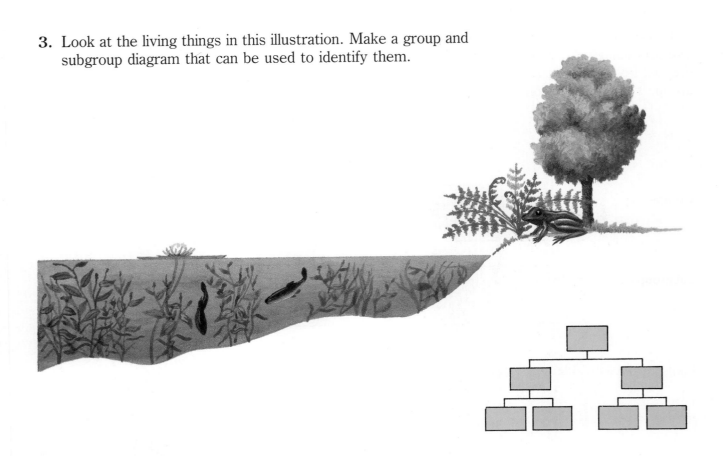

4. Try a slightly different classification system for these four plants.

96 Diversity of Living Things

Paulette divided the living things in the picture into plants and animals. Could you use her grouping for the following organisms?

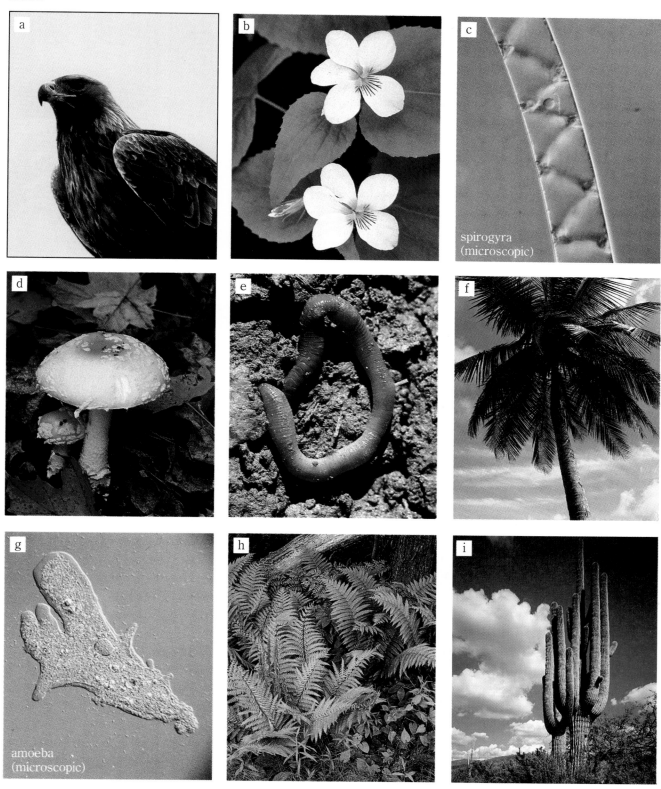

Biologists, the scientists who study living things, realized the need for a consistent system of classification. For this reason, about one hundred years ago, they divided all living things into two kingdoms: the animal kingdom and the plant kingdom. However, throughout the past century, scientists have been gaining more knowledge about the differences among living things. This has created the need to devise more precise systems for grouping all organisms. Most biologists now group living things into five kingdoms, and many biologists have even proposed a six-kingdom system. Either way, there are many living things that are regarded as neither plants nor animals.

Here is how a modern biologist would group the living things shown in the photos on page 97. Notice that the organisms that are neither plant nor animal have been placed in the "Other Kingdoms" category.

moray eel

penguins

snail

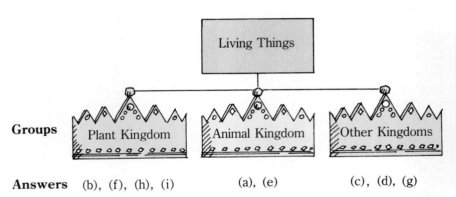

Groups: Plant Kingdom | Animal Kingdom | Other Kingdoms

Answers   (b), (f), (h), (i)   (a), (e)   (c), (d), (g)

The Animal Kingdom

Pictured here are some members of the animal kingdom. Imagine that you are a biologist, and try dividing these animals into two groups. (Hint: Think about what supports their bodies.)

snake

mussel

98   Diversity of Living Things

Keeping Track 99

The more you know about different animals, the easier it is to make useful groupings. To classify, you need to carefully observe the animals' features—not only their outside features, but their inside features too.

What characteristics did you use to classify the animals in the photos on pages 98 and 99? As you probably realize, your job was difficult because you had to depend mostly on what you could observe in the pictures. Biologists, though, study the actual animals, examining them on both the outside and the inside. The type of body support an animal has is one feature that biologists use when classifying animals. Many animals have internal (inside) skeletons to support themselves. These animals are placed in a group called **vertebrates**. Vertebrates are animals with a *vertebral column,* or backbone.

Other animals depend on a hard, external support system. For example, clams, lobsters, and grasshoppers each have a hard shell. Earthworms have tough muscles that give them support. Other animals don't really need a means of support. Their surroundings are their support. This is true for many of the animals that live and drift about in the water. Clams, lobsters, grasshoppers, earthworms, and animals that drift about in the water all can be placed in a group called **invertebrates**—animals without backbones. Is this the way you divided the animals on pages 98 and 99 into two groups? If not, do so now.

Invertebrates

Here are pictures of some invertebrates. Remember that none of these animals has a backbone for support. How might you group these animals? What features would you use as a basis for making subgroups? What names might you use to label your groups?

100   Diversity of Living Things

Keeping Track 101

The Classification System Used by Biologists

By studying invertebrates in detail, biologists have developed a classification system for them. To help you understand their system, study these two tables. The first table identifies the features that biologists use to classify invertebrates into different subgroups. The second table divides one subgroup, the arthropods, into even more subgroups.

After studying the tables, see whether you can place all the animals pictured on pages 100 and 101 into the classification scheme on page 103. Your task will not be easy; using pictures of animals to observe identifying features is more difficult than using the actual animals.

The Identifying Features of Invertebrates

Distinguishing Features Biologists Use to Group Animals	Name of the Subgroup	Examples of Animals in the Subgroups
worm-like, round bodies with many segments	annelids (means "arranged in rings")	earthworm
very soft bodies, which in most cases are protected by a shell	mollusks (means "soft-bodied")	clam (example with a shell) octopus (example without a shell)
covered with a spiny skin	echinoderms (means "spiny-skinned")	starfish
having many jointed appendages (attachments such as legs, feelers, tail)	arthropods (means "jointed legs")	lobster

The Identifying Features of Arthropods

Distinguishing Features Biologists Use to Make Subgroups of Arthropods	Name of the Subgroup	Examples of Arthropods in the Subgroup
six legs, three main body parts, wings	insects	grasshopper
eight legs, two main body parts	arachnids	spider
more than eight legs, many other kinds of appendages, crusty covering	crustaceans	lobster
numerous legs	millipedes centipedes	millipede centipede

A Simplified Classification System for Invertebrates

Look at the animals pictured on pages 100 and 101. Next, copy the classification system shown below into your Journal. Then, for each blank, fill in both the letter that corresponds to each animal and the animal's name if you know it.

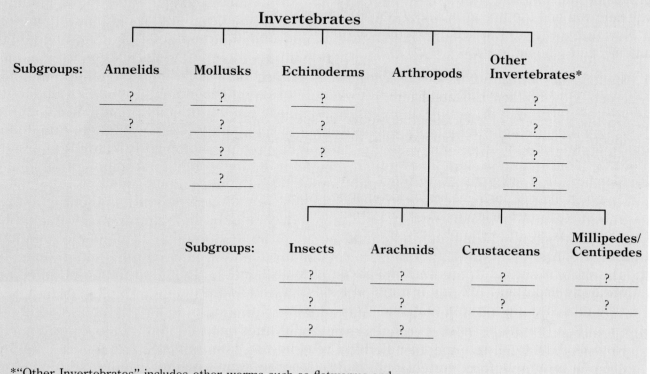

*"Other Invertebrates" includes other worms such as flatworms and unsegmented roundworms, along with many water animals such as jellyfish, sea anemones, corals, and sponges.

Now think some more about the subgroups.

1. Did you have any problems deciding which subgroup each invertebrate belonged to?

2. It's interesting to think about where invertebrates live. How many are found in water? in moist places? on dry land?

3. Does it appear that their structure enables them to live successfully in various places? Why do you think this is the case?

Keeping Track 103

Vertebrates

Because you are so well acquainted with the vertebrates around you (including yourself!), you will be able to generate vertebrate subgroups easily. Vertebrates all have backbones. Biologists classify vertebrates into five well-known subgroups. Can you classify the 15 vertebrates pictured on pages 104 and 105 into five subgroups of three animals each? (Hint: the covering of each animal will help you find the five groupings.) Name each subgroup.

Now that you have classified the vertebrates into five subgroups and named each one, discuss the following:

1. Identify the normal habitat of the living things in each subgroup. How do their habitats differ?
2. Do the members of each subgroup move in specific ways? Identify the method of locomotion that the members of each subgroup use.
3. Members of a subgroup share other features besides body covering, habitat, and means of locomotion. Keep in mind, though, that such features may vary from one subgroup to another. What are some of these features for the subgroups you have identified?
4. The ability of animals to survive in the place where they live is obviously important. Are any features of a subgroup specifically related to the survival of its members? Explain.
5. The names of the subgroups of vertebrates commonly used by biologists are found on page 106. Perhaps you suggested a different way to group vertebrates. If so, describe your method of grouping and the reasons for your classification.

104 Diversity of Living Things

Keeping Track 105

Putting It All Together

Many animals are illustrated on the next three pages. Your task is to classify them in your Journal using a classification table like the one below. Fill in both the letter that corresponds to each animal and the name of the animal. This table is an overall classification system for animals. It brings together everything you have learned about classification for the animal kingdom.

If your classroom contains any living or preserved specimens that are not represented in the pictures, classify them as well. Remember to write only in your Journal.

Animal Kingdom Classification System

Invertebrate Subgroups		Examples
Annelids		
Mollusks		
Echinoderms		
Arthropods	Insects	
	Crustaceans	
	Arachnids	
	Millipedes/Centipedes	
Other Invertebrates		

Vertebrate Subgroups	Examples
Fish	
Amphibians	
Reptiles	
Birds	
Mammals	

Sometimes the outward appearance of an animal can fool you, so be sure to examine each animal carefully. For example, the turtle does have a hard covering, but it is not a mollusk. Why not?

Keeping Track 107

108 Diversity of Living Things

Keeping Track 109

The Plant Kingdom

The great variety of plants in the world led scientists to develop a classification scheme for them, just as was done for animals. At this stage, it may not be clear to you why biologists use the particular groupings shown below. Nonetheless, you can use them to classify plants around you. Practice by using the pictures of plants on this page and the next.

cedar

roses

fiddlehead fern

oak

moss

Keeping Track 111

6 Ordering Diversity

The Swedish scientist Carolus Linnaeus, who lived in the eighteenth century, can be called the "master of order." He devised a scientific system for classifying living things according to their similarities. It is essentially his system that we still use today. You can see Linnaeus's classification system at the right.

You might be wondering why some of the groupings in his classification system have Latin names. In Linnaeus' time, Latin was the language used by all educated people. In fact, "Carolus Linnaeus" is a latinized form of "Karl von Linné"—his real name!

It was Linnaeus' ambition to name all the living things in the world. He did not quite achieve this goal, but he did name over 770 plants and 4400 animals.

Carolus Linnaeus

A Name You Can Claim

Where do you fit into Linnaeus' system? First, you are clearly a member of All Living Things. Next, you belong to the animal kingdom. Because you are a vertebrate (you have a backbone), you belong to phylum Chordata. Then, because you are warm-blooded, have hair (not feathers), and were born alive (instead of hatching from an egg), you fit into class Mammalia. There are certain similarities between you and animals such as gorillas and chimpanzees, so the next grouping you belong to is order Primates. You belong to the family Hominidae and the genus *Homo*. The most specific grouping is species. Your specific name is *sapiens*. Combine the names for our genus and our species, and you get *Homo sapiens*—the name for all human beings. (You might find it interesting to look in a dictionary to see what our name means.)

	All Living Things
(Group)	Kingdoms
(Subgroup)	Phyla (singular: Phylum)
(Further Subgroupings)	Classes
	Orders
	Families
	Genera (singular: Genus)
	Species

You have more in common with other humans *(Homo sapiens)* than you do with other primates, such as chimpanzees. But you have more in common with chimpanzees than you do with other mammals, such as cats, for instance. However, there are still more similarities between you and cats than between you and maple trees, which are also living things.

Every single type of living thing, from a fungus to a frog, has its own **scientific name.** A scientific name is a combination of its genus and its species names. It's important to remember that a scientific name contains both of these parts. You wouldn't refer to a wolf, for instance, simply as *lupus*. A wolf's scientific name is *Canis lupus*. Don't get him confused with *Canis familiaris*—your pet dog!

Now do some research. What are some organisms that belong to the animal kingdom? to phylum Chordata? to class Mammalia? to order Primates? to family Hominidae? to genus *Homo*?

112 Diversity of Living Things

A Legend: Wisakedjak Names All Creatures

A Cree legend says that the fox was the first animal to be given a name. This happened because one animal outsmarted Wisakedjak, the great spirit of the Western Indians, who said the animal was very "foxy." Wisakedjak named all the animals, fishes, birds, and insects, and he told each of them where and how to live.

He told the fish they must live in the water and swim. He told the ducks that they could live on the water, on land, and even fly in the air to escape their enemies. Wisakedjak gave the deer a white tail and said he could run fast and jump far. Squirrels could climb trees and hide in the tallest branches. To the slow porcupines, he gave a special suit of quills, which all animals learned to respect. Rabbits were trained to sit very still so they could not be seen.

Dr. Ivan H. Crowell

Do you recall any animals' names that are quite descriptive of what the animals are like? What are they?

There are many stories about how animals and plants got their characteristics. This painting illustrates the Ojibwa Indian folktale of how they received their colors.

Keeping Track 113

The Same But Different—Diversity Within Species

Do you recall your scientific name? Right, you're a *Homo sapiens*. This is the name of the genus and species to which you belong. Every person you know is also a *Homo sapiens*. All the organisms within a species are alike. Thus you share certain characteristics with all human beings—characteristics such as the general arrangement of your facial features, the number of your fingers and toes, and the ability to reason, to name just a few. People belonging to the same family tend to resemble each other even more closely. This is known as a family resemblance. It occurs because every person inherits certain characteristics from his or her parents, grandparents, and so on.

Despite family resemblances, however, everyone is unique. You are "one of a kind." Not one of your friends or acquaintances is exactly the same as you. Nor are you a duplicate of your parents or grandparents or brothers or sisters (even if you are a twin).

EXPLORATION 4

Tracing Similarities and Differences

In this Exploration, you will investigate some of your own inherited characteristics to determine how much you are like your family. You will do the same for some of your classmates.

What to Do

In the classroom, work in groups of four. Examine six inherited characteristics for each member of the group. In your Journal, construct a table like the one on page 115 and record each member's characteristics. Repeat this procedure with family members at home.

Every human is unique. How many examples of the diversity of features in *Homo sapiens* can you identify?

114 Diversity of Living Things

Characteristic	Student				Family	
	You	(2)	(3)	(4)	(1)	(2)
Hand-folding, thumb position (1) Left over right (2) Right over left						
Earlobes (3) Attached (4) Free						
Hairline (5) Pointed hairline (6) Straight hairline						
Tongue-rolling (7) Can roll (8) Can't roll						
Digit next to little finger (9) Hair (10) No hair						
Toe next to big toe (11) Same length or longer than big toe (12) Shorter than big toe						

Analyze Your Data

1. Study your table and then compare your results with the other groups in your class.
2. Calculate the percentage of individuals in your group who have each characteristic. What does this tell you?
3. Does anyone have all of characteristics 1, 3, 5, 7, 9, and 11? Does anyone have all of characteristics 2, 4, 6, 8, 10, and 12? Are there any two people in the class who have identical characteristics? Did you know that there are 64 possible combinations of these six characteristics?
4. You have looked at only six pairs of characteristics. There are hundreds of other characteristics you might have considered. Do you agree or disagree with these statements?

   "You're *just* like (name)."

   "Why, you are just the spitting image of your mother/father!"

Keeping Track   **115**

7 Life Cycles: Diverse Stages

You have seen how much diversity there is among different groups of living things. And you know that there is diversity even within a single group. There is also diversity within the different stages of an organism's life. Most organisms go through a series of stages from birth to adulthood to the birth of offspring in the next generation. This series of stages in an organism's life is called its **life cycle**. In the life cycles of many animals, the young animals live freely (on their own) and look quite different from the adults they will become. How many animals like this can you think of? Let's look at the life cycles of three living things.

From Eggs to Frogs

A small, round frog egg looks very different from an adult frog. Between these two stages in the life cycle of a frog, diverse forms can be observed.

EXPLORATION 5

Observing the Changes

You Will Need
- frog eggs
- a large jar of pond water
- algae or cornmeal
- gravel
- a pond plant
- a tall stick
- a terrarium (or a cardboard box fitted with clear plastic windows)
- plastic wrap for windows
- scissors and tape
- soil
- a small log
- a green plant
- a dish of water
- flies, worms, and small insects
- a magnifying glass
- a spoon

In the early spring, down by the marsh,
The croaking of frogs is deep and harsh.
When the male sends out his mating song,
The female responds—and before long,
She deposits her eggs in a shallow pond.
Then the male secretes sperm to complete the bond.
The egg divides into millions of cells—
"It's a tadpole!" the young boy yells.
But watch for a few weeks, and what do you see?
No longer a streamlined shape, swimming so free.
Its gills vanish and its tail disappears—
"It's a frog!" the young boy cheers.

What to Do

To study the diverse forms in the life cycle of a frog, you must construct two homes. The tadpole "nursery" is simply a large jar with a layer of gravel on the bottom. Add a pond plant, a tall stick, and pond water to complete the home.

A small terrarium or a cardboard box fitted with clear plastic windows is the beginning of an ideal home for a frog. Soil, a small log, a green plant, and a dish of water are all you need to meet the frog's needs.

Be sure to construct the home so that the frog can obtain plenty of air. Remember to feed it flies, bits of worm, and small insects.

After you have completed the homes, the tenants may move in. At this point, keep accurate notes and diagrams in your Journal so that you can complete your study with a detailed account of the life cycle of the frog.

Some Suggested Procedures

1. Obtain some frog eggs (or *spawn,* as they are called) from your teacher.
2. Put the eggs in the tadpole nursery. Using a spoon, take a few eggs out of the nursery each day and examine them with a magnifying glass.
3. Write and draw any changes you observe taking place in the jelly covering.
4. When the eggs have hatched, carefully examine a tadpole to find the outside gills it uses to take in oxygen.
5. Watch the growth of the leg buds and legs. Do the back or the front legs appear first?
6. Observe the tadpoles' tails daily. What happens to the tails?
7. Look for signs of lung development. The tadpoles will begin to use their nostrils to breathe at the surface of the water. When the tadpoles climb up the stick and remain out of water, it is time to change residences.
8. Observe the frog(s) carefully for several weeks. Make daily records and drawings of limbs, size, color, breathing, and locomotion.
9. To summarize your findings, make a poster-size diagram showing the life cycle of a frog. Include your detailed observations on the diagram or in an accompanying table.

From Eggs to Adult Insects

Insects are more abundant than any other group of animals. They inhabit every part of the earth except mid-ocean. They have been found in the atmosphere several kilometers above the earth. The life cycles of different species vary, but all insects undergo changes as they develop from young forms to adult forms. These changes are easy to identify because each stage generally looks quite different from the one before it.

Here are the life cycles of two insects. Carefully examine the two cycles and the "Label Hints" on page 120. Then copy the cycles into your Journal and use the "Label Hints" to label the diagrams.

The Life Cycle of a ___?___

An example of $\underset{\textit{(this word means "total")}}{\text{___?___}}$ $\underset{\textit{(this word means "change")}}{\text{___?___}}$

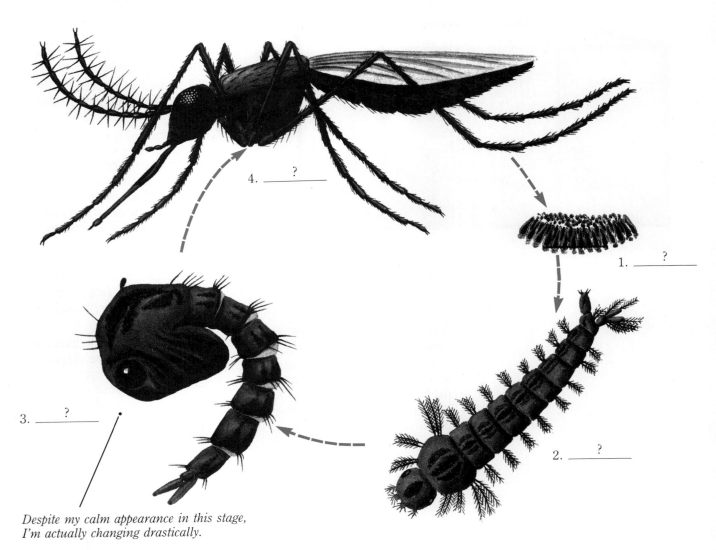

Despite my calm appearance in this stage, I'm actually changing drastically.

118 Diversity of Living Things

The Life Cycle of a ___?___

An example of ___?___ ___?___
(this word means "partial") (this word means "change")

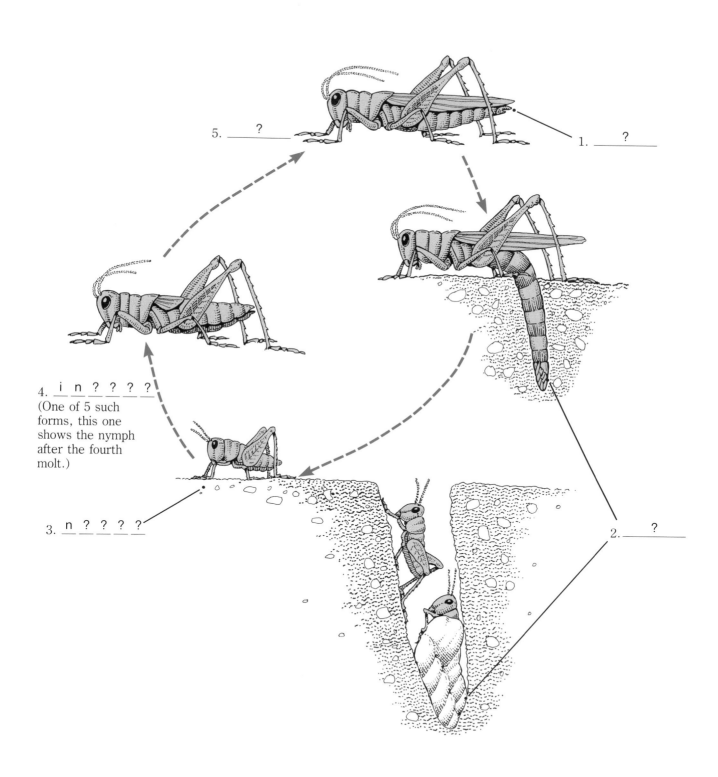

5. ___?___

1. ___?___

2. ___?___

3. n ? ? ? ?

4. i n ? ? ? ?
(One of 5 such forms, this one shows the nymph after the fourth molt.)

Keeping Track 119

Label Hints

Adult	The last stage in the life cycle of an insect (and other animals).
Ovipositor	The pointed structure at the end of the body of a female grasshopper. The female uses it to lay her eggs.
Egg	The beginning stage in the life cycle of an insect (and many other animals).
Larva	In some kinds of insects, this is the stage that looks like a worm, moves a great deal, and eats a large amount of food.
Pupa	Little motion can be seen during this stage, but great changes are taking place inside an outer case. Inside the case, the larva changes into its adult form. Not all insects go through this stage.
Metamorphosis	The word that describes the changes which occur from the egg stage to the adult stage in insects (and in many other animals, such as the frog).
Complete metamorphosis	In this kind of metamorphosis, the "in-between" stages are quite different in appearance and structure from the adult.
Incomplete metamorphosis	In this kind of metamorphosis, the "in-between" stages have the general appearance of the adult.
Nymph	A grasshopper's young, having a large head relative to the rest of its body.
Molting	The breaking and shedding of the outer skeleton of animals, such as insects, as their soft body grows inside. Grasshoppers molt six times. The wings appear on the fourth molt.
Instar	The form of an insect between moltings.

What Is It Really Like?

Your class has been invited to explain life cycles to a group of 10-year-olds who really like puppets. Write the script for a skit in which one puppet is an elderly mosquito and another is an elderly grasshopper. They're talking about their youth and the changes they went through when they grew up. Toward the end of the skit, another puppet appears—a frog. The frog compares his life cycle with the other two. What would the frog say?

Diversity in Plant Cycles

Do you think there is diversity among the stages of plant life cycles? Could there be "young plants" that live and look different from the adults they will become? Research the life cycle of a fern, and you may make some valuable discoveries.

One Form with Many Functions

Help Wanted

Insects are versatile creatures. They perform many tasks, just as humans do. However, insects have practiced their jobs for millions of years. In this time, insects have perfected their tasks so well that individuals now have specific jobs to perform for the benefit of their community.

The following want ads are taken from *The Worker's Journal* of Insectville. Study each one closely and then write your own ads for the insects shown below. Do some research to discover what each insect can do. Work in groups of two or three. Each group should prepare one ad for a class edition of *The Worker's Journal*.

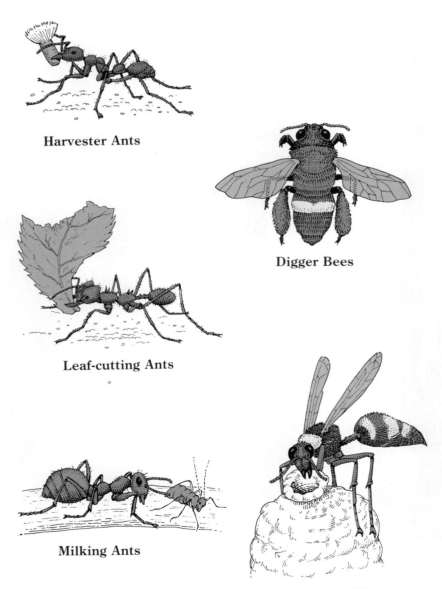

Harvester Ants

Leaf-cutting Ants

Milking Ants

Digger Bees

Potter (Mason) Wasps

The Worker's Journal
CLASSIFIED

Help Wanted: Papermaking Wasps
Two papermaking wasps needed to construct 10 nests of paper. Workers must obtain wood fibers from local dead trees. Must have good chewing ability to break down the fiber and mix it with their saliva to form pulp. Ability to shape the pulp into six-sided cells is also essential. **Apply at Hives, Inc., Box 999**

Help Wanted: 5 Female Pepsis Wasps
Must have armored bodies to hunt tarantulas and a lethal sting to paralyze the tarantula's body. Must have strength to drag the huge tarantulas to evacuated tunnels. Need strong mandibles and legs to dig new tunnels. Must be able to scrape the hair off a section of the tarantula's abdomen and deposit an egg on the cleared spot. Must be able to fill in the tunnel with sand and soil before leaving. **Apply at SPPW Tunnel 28 (Society for the Preservation of Pepsis Wasps, Inc.)**

Diversity of Living Things

Challenge Your Thinking

1. **The Animal Kingdom Pie**
   Can you explain the meaning of this pie chart? Which animals would you place in the missing piece of pie?

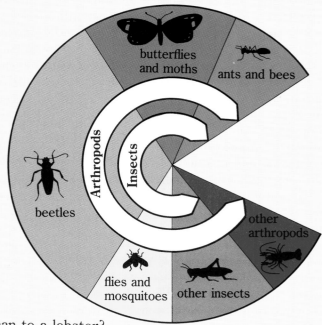

2. According to the classification systems used by biologists, why is:
   (a) a turtle more closely related to an alligator than to an eel?
   (b) an earthworm more closely related to a grasshopper than to a snake?
   (c) an octopus more closely related to a clam than to a lobster?
   (d) a whale more closely related to a tiger than to a shark?
   (e) a maple tree more closely related to a rosebush than to a pine tree?

3.

   Classify the birds above according to their characteristics. Divide them into two groups and write the characteristic each group shares in the boxes in the first row. Then divide each of these groups into two subgroups and write the characteristic each subgroup shares in the boxes in the second row. Under each of the four boxes in the second row, write the names of the birds that fit into that subgroup. Remember to write only in your Journal.

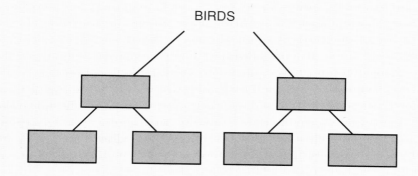

Keeping Track 123

Unit 2 Making Connections

The Big Ideas

In your Journal, write a summary of this unit, using the following questions as a guide.

- What is diversity? Why is there diversity among living things?
- What are some adaptations animals have for protection? for obtaining food? for attracting a mate? for locomotion?
- How are plants adapted to survive and reproduce?
- What is meant by natural selection?
- What evidence did Darwin use to develop his theory of how life evolved?
- What kinds of conditions cause species to become extinct?
- Why do we classify things?
- How, in a general way, are living things classified?
- How would you classify yourself, according to Linnaeus's system? What is your scientific name?
- How is diversity shown in the life cycles of living things?

Checking Your Understanding

1. For each group of words below, write one or two sentences that show how the words are related to each other.
   (a) adaptation, predator, peppered moths, habitat
   (b) organisms, diversity, environment, survival
   (c) diversity, species, inherited characteristics, unique
2. Classify the living things in these photos according to some consistent system.

124 Diversity of Living Things

3. Draw a concept map that shows how the following words are related to each other: *Texas coral snake; camouflage; mimicry; snowshoe hare; adaptations; snow; grasses, trees, and weeds; brown in summer; white in winter;* and *Mexican milk snake.*

Reading *Plus*

You have been introduced to the great diversity of life and to how this diversity arises. You also learned how to classify animals and plants into groups and subgroups. By reading pages S23–S44 in the *SourceBook,* you will take a closer look at classification, at some of the ways diversity came about, and at the origin of life.

Updating Your Journal

Reread the paragraphs you wrote for this unit's "For Your Journal" questions. Then rewrite them to reflect what you have learned in studying this unit.

Making Connections 125

Science in Action

Spotlight on Medicine

Medical Practice

Harold Still has been a family physician for nearly 40 years. Physicians diagnose and treat diseases and disorders of the human body. If you wish to enter this profession, you must spend four years in undergraduate study at a university. After this, four years of training in medical school is required. Upon graduating from medical school, you would receive an MD (Doctor of Medicine) degree. Then you would spend between three and seven years as a resident in a hospital, preparing for a specialty. Your specialty might be in family medicine, neurology, or surgery, just to name a few.

Q: Why did you become a medical doctor, and what do you find most interesting about your work?

Harold: Perhaps it was my mother who encouraged me in this direction. In any case, I wanted to be a physician from the time I was about 12 years old.

I really enjoy patient contact. I guess that's why I like being a family doctor. I like taking care of people. In particular, I like looking into the whole life of a person, not just an illness. I am interested in the relationship between the illness and the person's social environment, including work and family.

Over the past 14 years, I have also enjoyed teaching family medicine to future doctors.

Q: What kind of background and education should a person who wants to enter medicine have?

Harold: Concern for people, confidence, and the ability to interview and relate to people are all necessary personal characteristics. An understanding of the scientific basis of medicine is also very important for a family physician.

Doctors should have extensive knowledge of medicine, of course, but they should also have a good understanding of English, history, and nonscientific subjects. I feel strongly that students should not cut themselves off from the humanities and social studies when planning to become a doctor. I like to see the development of a well-rounded person. I think such physicians are the most effective in dealing with patients.

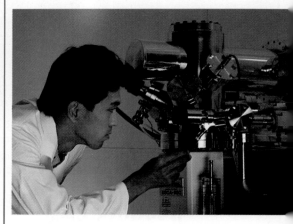

Medical Research

Barbara Pope is a medical researcher who studies the immune system of the body and the causes of cancer. Barbara has a medical degree, but instead of seeing patients, she prefers to do research.

126 Diversity of Living Things

Q: What do you find most stimulating about your work?

Barbara: The joy of analyzing experimental data is one very stimulating part of scientific work. It's exciting when you bring the data together, come up with a pattern of what you think is happening in a given field, and make your own hypothesis. There's a lot of work, but it's very rewarding work.

Q: Will there ever be a "grand" cure for all forms of cancer?

Barbara: Cancer is a very complex problem, and there is probably no single answer to it. There are a lot of partial answers, however, and eventually all the partial answers will cure cancer. But it's going to be a while yet before we understand enough basic biology. There are many systems of the body involved, including the immune system. Only when we understand how all of these systems interact will we be close to solving the problem of cancer.

Some Project Ideas

1. In many countries, medical care is basically free to all citizens, regardless of their level of income. In such countries, the government either owns and operates the health-care system, or it finances health care through national health insurance.
   Do some research to learn more about health-care systems. Do you think our private health-care system is adequate? What are its advantages and disadvantages? What are the advantages and disadvantages of a national health-care system?

2. Do some research to find out how AIDS has changed the practice of medicine. Check references such as *Magazine Index* or *Reader's Guide to Periodical Literature* for magazine articles on the subject. In addition, AIDS service organizations, your city or county health department, and local hospitals can direct you to further information. Share the information you gather with your classmates.

3. Various forms of cancer are associated with smoking. What is the connection between smoking and cancer? Present your information in a dramatic and convincing manner to your classmates.
   For assistance, contact your family physician and/or the American Cancer Society in your area. Your local library will also have valuable information on this topic.

Science in Action 127

Science and Technology

Extinction in the Modern World

In the year 1990 alone, it is estimated that *10,000 different species* of living things became completely extinct! In the year 2000, the studies say, the number of species that become extinct will be 20,000. These numbers may seem even more astonishing when you realize that in 1975 the number of extinctions was "only" 100.

It probably won't surprise you that humans are the cause of this massive destruction of life. The human population of the earth is rapidly increasing, and people are encroaching on life all around them.

Of course, there have always been extinctions from purely natural causes. Dinosaurs died off long before humans appeared on the earth. But in the past, species that died off were replaced by new species. Today the rate of extinction is so great that there is not enough time for new species to evolve to replace the ones that are lost.

The Technology Factor

As the human population increases, people move into new territory. They clear forests, drain swamps, and build dams. Technology, with innovations such as chain saws and bulldozers, has made humans effective at altering the environment very quickly. As a result, the habitats of many plants and animals have been destroyed.

Technology has also made humans effective at killing animals. Inventions such as guns and explosive harpoons have made it possible for people to kill animals on very large scales. The passenger pigeon, now extinct, was once so numerous that clouds of them would darken the sky for hours. It was killed off completely, mostly for food and fertilizer. And the American Bison would be extinct today if conservationists had not stopped their slaughter.

Certainly not all human destruction of life is deliberate. A lot of it comes as a result of the pollution that technology so often brings. Chemical wastes, pesticides, accidental oil spills, and the like, have poisoned the natural habitats of many plants and animals. Often there is a *chain reaction*. For example, poisonous wastes in the ocean might kill a plant or fish that is the food supply of a certain bird. So, because of the pollution, the bird dies too.

The American Bison has been saved from extinction.

Can Technology Help?

The work of conservationists has many sides, including proposing laws to protect endangered species. Often it involves working directly with the plants or animals. At times modern technology proves helpful. For example, researchers may attach radio-tracking devices to animals in order to learn the animals' habitats and migratory routes. In this way, they can determine if a species is being threatened by environmental hazards.

Saving the Condor

In some cases, animals are caught and raised in captivity in order to keep them safe from dangers in the environment. For example, conservationists had captured all living California condors by the end of 1987. It is hoped that they can be released back into nature when sufficient numbers have been bred, and when the environment has been made safe. By the time you read this, some of the great birds may be back in the wild.

Conservationists are working to save the California condor.

Controversy

Money is very often the reason that people deliberately kill living things in large numbers. There is a lot of money to be made from, say, killing elephants just for their tusks. But it may not be just a matter of greed. For some people, such activities might be their only source of income. Therefore, their lives would seem to depend on doing such things.

Do you think there should be laws protecting endangered species? How important is it, after all, to save an animal like the blue whale or the whooping crane? How would you feel if your personal livelihood were threatened by a law that protected some endangered species, perhaps a "minor" one that you had never even heard of before?

African elephants are being killed in large numbers.

Find Out for Yourself

Do some research to find out why the California condor was in danger of becoming extinct. What laws have been enacted by the state of California to make the environment safe for these birds? What are some other animals or plants that have become extinct in recent years? What were the reasons for their extinction?

Unit 3 Solutions

A SOLUTION OR NOT?
1. Solutions and Non-Solutions, p. 132
2. Parts of a Solution—Solutes and Solvents, p. 140
3. Soluble, Not Very Soluble, or Insoluble? p. 148

SEPARATING SOLUTE AND SOLVENT
4. The Salt Industry, p. 152
5. Desalination and Distillation, p. 156
6. Another Solution, p. 160
7. Sugar Solutions, p. 164

CONCENTRATION OF SOLUTIONS
8. Getting Colder, p. 168
9. How Much Solute? p. 170
10. Saturation, p. 176
11. Solutions in the Environment, p. 179
12. Growing Crystals, p. 183

For Your Journal

Write a paragraph summarizing your thoughts about each of the following:
1. Why can't you see a sugar cube after it has been stirred in water?
2. How could you separate the sugar from the water later?
3. What difference would it make if water were not such a good solvent?
4. Is sea water a solution or a non-solution? Explain your answer.

A SOLUTION OR NOT?

1 Solutions and Non-Solutions

The detective solves the crime. That's one kind of solution.

The math student solves the problem. That's another kind of solution.

The chemist does these things and more. The chemist also mixes solutions!

In this unit you are going to be a chemist most of the time, but you will also be a mathematician and a bit of a detective! So after much detecting, calculating, and mixing, you will be able to tell or show anyone the answers to these questions:

- How is a solution different from a non-solution?
- What is a solute? a solvent?
- What solutes are very soluble in water? not very soluble? insoluble?
- How can you separate a solute from a solvent?
- How does a solute affect the boiling and freezing points of a solvent?
- What is a "dilute" solution? a "concentrated" solution?
- How can you test a solution to see whether it is dilute or concentrated?
- What is meant by "saturated" and "solubility"?
- Does temperature affect the amount of solute that can be dissolved?

A Case of Foul Play

At first, it appeared that the carnival's Muscle Man had passed out from trying to lift too much weight, or possibly from the heat.

Binti Sirba, a paramedic, examined the Muscle Man and told Sergeant Malikia Noble, the carnival's security guard, "You're not going to believe it, but this man isn't unconscious. He's just fast asleep!"

Sergeant Noble frowned. "But that's ridiculous. No one falls asleep in the middle of lifting weights. Something's fishy here." Looking out at the audience, she asked, "Would someone tell me what the Muscle Man did right before lifting the weights?"

A boy spoke up. "Sure. He bragged that he was going to lift more weight than we'd ever seen anyone lift before. He gulped down a bunch of whatever's in that cup there, took a few deep breaths, lifted the weights to shoulder level, and then boom! Down he went." Others nodded. "Yep. Out like a light."

After hearing this, Ms. Sirba carefully examined the Muscle Man's drink. It appeared faintly yellow and transparent (clear), and it smelled like lemons.

"That's lemonade from the stand across the way," Sergeant Noble told Ms. Sirba. "He drinks it all the time, although it's not very tasty. It's that powdered lemonade mix you add water to."

"Hmm. That helps, but I still need to make one more test." Ms. Sirba put some of the lemonade into a clear glass, shook it, and held it up to the light. "This isn't a pure lemonade solution," she concluded. She held it up for the security guard to see. "My guess is that someone's added sleeping powder to the lemonade in that cup."

Chemical tests at a forensic lab showed that a potent dose of insoluble sleeping powder had been added to the Muscle Man's lemonade. Thanks to the paramedic's astute observations, further investigations were carried out, and the man who operated the lemonade stand was later convicted of the crime.

A Solution or Not?

EXPLORATION 1

Solution or Non-Solution?

Part 1

How was the paramedic, Ms. Sirba, able to distinguish the non-solution in the Muscle Man's cup from a pure solution? To tell a solution from a non-solution, try this experiment.

You Will Need

- 6 beakers
- a 100-mL graduated cylinder
- a 10-mL graduated cylinder
- the substances shown below
- water
- a stirring rod

What to Do

In separate beakers, prepare the six mixtures listed at the right. Stir each one well to ensure that it is completely mixed.

Mixture #1
Add 1 mL of powdered juice mix to 100 mL of water.

Mixture #2
Add 10 mL of rubbing alcohol to 100 mL of water.

Mixture #3
Add 2 or 3 drops of milk to 100 mL of water.

Mixture #4
Add 1 mL of sugar to 100 mL of water.

Mixture #5
Add 1 mL of oil to 100 mL of water.

Mixture #6
Add 10 mL of sand to 10 mL of salt.

Now separate the mixtures into two groups: those mixtures that are solutions (Mixtures #1, #2, and #4) and those that are not solutions (Mixtures #3, #5, and #6). Examine the two groups carefully. What differences do you see? Based on your observations, how will you know a solution when you see one?

Suppose you have been asked to appear in court as the scientific expert for the trial of the lemonade stand man. In giving evidence before the court, you are asked how to tell a solution from a non-solution. To illustrate your answer, you decide to show the court the six mixtures and explain the differences between the three solutions and the three non-solutions. Prepare your explanation in writing.

On the next page is a memorandum from Sergeant Noble to the forensic laboratory. How does your description of what a solution is compare with Sergeant Noble's?

134   Solutions

Police Department
Criminal Investigation Branch

Memorandum

To: ACME Forensic Laboratory
From: Sergeant M. Noble
Re: Carnival Lemonade

After a preliminary examination of the scene of the crime, it appeared that the victim had merely fallen asleep on the job. However, further investigation showed that the lemonade the carnival employee had consumed was not a pure solution of lemonade powder and water. This led us to suspect criminal activity.

Anyone with even a little background in chemistry knows that a solution is a mixture of two or more things, but that not all mixtures are solutions. In a solution, one of the materials becomes hidden, so that the mixture appears to be only one thing. You might say that one part becomes camouflaged. Sometimes, as in the case of food coloring and water, a solution can be colored, but it is still transparent. Any liquid mixture that is transparent must be a solution.

When we held the lemonade up to the light, it showed the Tyndall Effect. This test provided further evidence that the lemonade was not a pure solution. I suspect foul play and recommend that you test the enclosed sample to find out what was added to the lemonade drunk by the carnival employee.

M. Noble

A Solution or Not?

EXPLORATION 1—CONTINUED

Laboratory File W/P 106 (#17 and #18) in the forensic lab deals with the Tyndall Effect. Card #17 describes and explains it, and Card #18 tells how to test for it. Study these cards and then test Mixture #3—milk and water—with the solution tester shown on the bottom of the page.

File W/P 106   #17

Topic
TYNDALL EFFECT

Description:

Some mixtures that first appear to be solutions may prove not to be solutions after all. If the path of a bright light shining through a mixture can easily be seen, then that mixture is not a solution. In other words, the path of light passing through solutions is not visible. No matter what color they are, true solutions made of liquids are transparent.

Explanation:

In a non-solution, there are particles large enough to scatter or reflect light, allowing us to see the path of the light as it passes through. This scattering is called the Tyndall Effect. True solutions do not exhibit the Tyndall Effect.

A Non-Solution

136   Solutions

File W/P 106   #18

Topic

TESTING FOR THE TYNDALL EFFECT

Materials

- a bright light, lamp, or flashlight
- a piece of cardboard
- mixture(s) to be tested
- a pencil or knife
- a metric ruler

Making the Solution Tester

Using a pencil or knife, make a small, neat hole in the center of the piece of cardboard. The hole should be no more than 0.5 cm in diameter. Place the cardboard over the light source so that the only light you can see comes through the hole.

The Test

Press the light and the cardboard tightly against the beaker holding the mixture. If you can see the path of the light through the mixture, then it is not a true solution. The test should be done in a place with limited light.

A True Solution

EXPLORATION 1—CONTINUED

Part 2
What to Do

Test Mixture #3 (milk and water) with the solution tester. Can you see the light path through it? Now test Mixture #4 (sugar and water).

Next, repeat the test with other mixtures. Some that you can use are instant coffee and water, tea and water, chalk dust and water, soft drinks, and salt and water. Which ones are true solutions?

Detective Work

Are the mixtures pictured below all solutions? Discuss with a classmate. (Hint: There is water present in all beakers.)

Solution Facts

You will come across many interesting facts about solutions and non-solutions. Perhaps your class could set aside some space on a bulletin board for "Solution Facts." Use index cards to record any interesting facts about solutions or non-solutions, and put them up on the bulletin board.

> Seeing Dust in the Air
> Anyone who has ever played in a hay barn may remember seeing dust particles dancing in a beam of light that has entered through a crack or a knot hole. This is an example of the Tyndall Effect. The next time you watch a film, try to see the dust crossing the path of the projector's beam of light.

138 Solutions

Why Is the Sky Blue?

Did you know that air without the presence of dust, soot, and other such particles is a true solution? It is light from the sun being scattered by these particles that causes the sky to appear blue. If there were no particles in the air, the sky would be black.

Can Solutions Be Solid?

Yes they can! The Liberty Bell in Philadelphia, Pennsylvania is made of bronze — a solution of copper and tin. Solid solutions are called alloys. Copper and tin ore are melted and then mixed together. When this solution cools, it forms the alloy bronze. Unlike liquid or gaseous solutions, solid solutions aren't transparent.

Do you have a Solution Fact to contribute?

2 Parts of a Solution— Solutes and Solvents

Earlier you prepared a solution of sugar and water. Sugar is a solid, and water is a liquid. You also prepared a solution of rubbing alcohol and water—both liquids. Other possible combinations are listed in the data table below. For each combination, an example is given. Can you provide another example of each combination?

Some Combinations	Example	Your Example
solid and liquid	salt water	?
gas and liquid	soft drinks	?
liquid and liquid	rubbing alcohol and water	?
solid and solid	bronze (a solution of copper and tin)	?
gas and gas	air	?

Naming the Parts of a Solution

What happens when you mix things to make a solution? When a solid and liquid are mixed together, the solid separates into particles that spread evenly throughout the liquid. This process is called **dissolving**. The particles of both parts of the solution, solid and liquid, are too small to be seen. This explains why the solution is transparent. Since the particles of the solid spread evenly to every part of the liquid, the solution is the *same* throughout.

Dissolving in Action

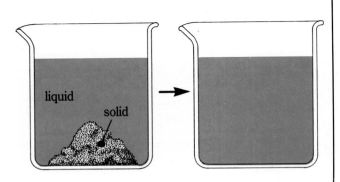

This is what you might observe as a solid dissolves into a liquid. Where did the solid go?

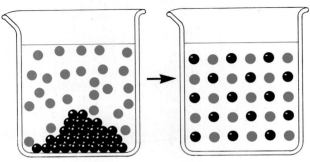

This is what you might observe if you could actually see the particles. Check to see if the solution is the same throughout.

If a solid or a gas is dissolved in a liquid, the liquid is usually the **solvent**. The solid or gas is the **solute**. What happens when you have other combinations, such as liquid–liquid, gas–gas, or solid–solid? In these cases, the substance of which there is a *greater* quantity is the solvent. The substance of which there is a *lesser* quantity is the solute.

> The *solvent* dissolves the *solute*.
> The *solute* dissolves *in* the *solvent*.

In these examples, which part is the *solute*, and which part is the *solvent*?

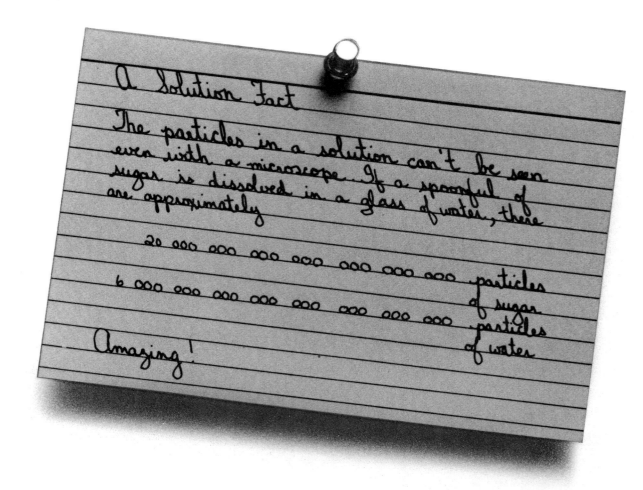

A Solution Fact

The particles in a solution can't be seen even with a microscope. If a spoonful of sugar is dissolved in a glass of water, there are approximately

20 000 000 000 000 000 000 000 particles of sugar

6 000 000 000 000 000 000 000 000 particles of water

Amazing!

A Solution or Not? 141

Solutions from A to Z

Here are situations involving real solutions. Read through them with a classmate. For each situation:

1. *Identify the type of solution* (solid in a liquid, liquid in a liquid, etc.). In your Journal, enter this information in a table similar to Table 1.
2. *Identify the solvent and solute in each solution.* Enter this information in your Journal in a table similar to Table 2 on page 143. Again, remember to write only in your Journal.

Table 1

Type of Solution	Situations (by letter)
solid and liquid	c, e, etc.
gas and liquid	
liquid and liquid	
solid and solid	
gas and gas	
other (specify)	

(a) Fish breathe oxygen, which is part of water.

(b) Vinegar is made up of 5 g of liquid acetic acid and 100 g of water.

(c) A clear mixture of baking soda and water often helps an upset stomach.

(d) Brass is an alloy that can be made by mixing 65 parts of molten copper with 35 parts of molten zinc.

(e) Mineral spirits will easily remove paint from your hands.

(f) Antifreeze used in a car radiator is a mixture of the liquid ethylene glycol and water. The ethylene glycol is present in the greater quantity.

(g) On a windy day, snow and ice disappear into thin air without melting.

(h) Sugar, lemon juice, and water make up lemonade.

(i) Rubbing alcohol is a good lock de-icer because it can dissolve water drops frozen in the keyhole.

(j) When you open a bottle of sparkling water, gas suddenly appears everywhere and rises to the top.

(k) A few drops of food coloring can change the color of water.

(l) Pulp mills discharge a gas that spreads through the air. The gas can be smelled kilometers away.

(m) Sulfur dioxide in the air mixes with water droplets in the atmosphere to form acid rain.

(n) Minerals in the ground can be dissolved by acid rain.

(o) Rain falling on the earth mixes with carbon dioxide from decaying plants and the roots of plants. The result is the formation of a weak acid.

(p) In some places, this weak acid dissolves limestone to make limestone caves.

(q) Water from lakes and streams is continually evaporating into the air.

(r) Clear ocean water contains a considerable amount of sodium chloride and potassium chloride (both are types of salt).

(s) Air is made up of almost four parts nitrogen gas to one part oxygen gas.

(t) A bit of table salt in water helps heal cuts.

(u) A seltzer tablet added to water fizzes and then becomes clear. The solution tastes a little salty.

(v) Runoff water from some farms and forests contains insecticides.

(w) There are traces of DDT in ocean water, even in the Arctic and Antarctic.

(x) If iodine crystals are dropped into a dry, warm container, a purple vapor suddenly appears and spreads evenly through the air in the container.

(y) When foods are boiled, some of their valuable vitamins are lost in the process.

(z) Gold jewelry usually has some copper in it to make it stronger and less expensive.

Table 2

Situation	Solute	Solvent
a		
b		
c	baking soda	water
d		
e		
f		
g		
h		
i		
j		
k		
l		
m		
n		
o		
p		
q		
r		
s		
t		
u		
v		
w		
x		
y		
z		

Waste Not, Want Not

Did you notice that water was involved in many of the solutions on pages 142 and 143? In the Middle Ages, people who were interested in science searched for a "universal solvent"—a liquid that would dissolve everything. Water comes closest to being this universal solvent.

Think of the many things water is used for each day, such as cooking, washing, drinking, cleaning, swimming, and industry. And after it is used, water often contains many wastes. Some waste materials are easily seen when we inspect waste water. But many other wastes dissolve in water and are therefore invisible.

Make a list of about 10 or 12 substances that can end up polluting the water supply. Look around your school, home, and neighborhood. Decide which of the materials can be dissolved in water.

Much of the used water ends up at a sewage plant, where it is partially cleaned before being released into the environment. Where does waste water go in your community? If there is a sewage treatment facility nearby, visit it and find out how the waste is removed from the sewage. Can any of these wastes be *recycled?*

Substance	Does It Dissolve in Water?
grease	No
sugar	Yes

Other Solvents Besides Water

Water is a great solvent, but as you have just seen, it does not dissolve everything. For example, it does not do a good job on cleaning grass stains from clothes. Nor does it work very well on greasy hands.

There are other solvents besides water. Many of them are useful for doing things that water cannot. The table below lists only a few of these solvents. Perhaps you know of others to add to the table. You can also ask a chemist, a pharmacist, or a hardware store clerk for further suggestions. Add about six more items to the table—more if you can!

Solvent	Solute	Use
alcohol	iodine	tincture of iodine
methyl alcohol	grease	windshield wiper fluid
toluene	rubber	rubber cement
turpentine	oil paint	paint thinner
ethyl acetate	nail polish	nail polish remover
gasoline	engine oil	fuel for two-cycle engines
alcohol	ice	de-icer for keyholes

Water is a safe solvent. It does not burn and is not poisonous. However, many other solvents are dangerous. Use your library to find out which solvents in your table can be dangerous.

144   Solutions

Waterproof Birds

Water does roll off a duck's back! When a duck or other water bird preens itself, it is actually replacing natural oils that cover its feathers. Since oil does not dissolve in water, the natural oils act as a raincoat that makes the bird waterproof. Without this coating of oil, its wings would become waterlogged, and the bird would not be able to fly or even float. The water bird would be in danger of drowning.

The grebe in the photograph is covered with too much oil. It was caught in an oil slick—a danger to wildlife that is becoming more common every year. How could the grebe be cleaned? What will happen to the bird's natural oils in the cleaning process?

Grebe covered by oil

EXPLORATION 2

A Project: Oil Spills

Our coastlines and wildlife are sometimes harmed by oil slicks. The oil is released into the water when oil tankers clean their bilges or when these ships are involved in accidents, such as the one in Prince William Sound, Alaska, in 1989.

Find out how environmental agencies get rid of oil slicks on water surfaces and on beaches. How are birds and other animals that have been in contact with oil treated?

Oil slick

A Solution or Not?

Hard Water

Hard water is *not* frozen water! It is water that contains certain types of solutes. **Soft water,** such as rainwater or melted snow, is mostly free of dissolved matter. Hard water will not form as many suds when mixed with soap as soft water will. Does your drinking water contain solutes? Here is an easy test to determine whether your drinking water is hard or soft.

EXPLORATION 3

Tap Water—Hard or Soft?

You Will Need

- a 25-mL graduated cylinder
- 10 mL soft water (distilled water works best)
- 10 mL tap water
- 25 mL soapy water, made by adding flakes of basic, unscented hand soap to 25 mL soft water. Mix well, but do not make suds.
- 2 test tubes
- a test-tube rack
- an eyedropper

What to Do

In one test tube, place 10 mL of drinking water from the tap. Into the second test tube, pour 10 mL of soft water. Add 10 drops of soapy water to each, and shake the test tubes an equal number of times. Compare the amount of suds formed in each test tube. Is the tap water harder than the soft water? Do you think this is a reliable test?

Make a list of at least three things that you did in this experiment to ensure that your results would be reliable.

Do you think the results would have been the same if you had used drinking water from the hot-water tap?

Collect samples of water from different sources in your city and test each one for hardness.

What Not to Do!

Will this procedure give reliable results?

EXPLORATION 4

Solutes in Hard Water

You Will Need

- sugar
- sodium chloride
- magnesium chloride
- calcium chloride
- iron chloride
- calcium nitrate
- magnesium nitrate
- a 25-mL graduated cylinder
- distilled water
- 7 test tubes
- test-tube racks
- soapy water, made according to instructions in Exploration 3
- an eyedropper

What to Do

In test tubes, dissolve a few crystals of each of the solutes listed above in separate 10-mL amounts of distilled water. Test the solutions for hardness, as you did in Exploration 3. Make a table that records which of the solutes is able to make suds.

What are some of the solutes that may be responsible for making water hard?

Something to Solve

Does the amount of calcium chloride added to the 10-mL portion of distilled water affect the number of drops of soapy water required to make suds? What do you predict? Try adding different amounts of calcium chloride to test your prediction.

A Solution or Not? **147**

3. Soluble, Not Very Soluble, or Insoluble?

As you have seen, water comes closest to being a universal solvent because it dissolves a great many substances. However, you know that there are some substances that it cannot dissolve, such as grease and oil. Are there still other substances that water can dissolve a little but not completely?

Here are some terms that describe how well a solvent (such as water) dissolves a solute.

- A substance is **soluble** if it disappears in the solvent. This shows that it has dissolved completely.
- A substance is *not very soluble* if some of it is still visible in the solvent. This shows that only some of it has dissolved.
- A substance is **insoluble** if none of it disappears in the solvent. This shows that none of it has dissolved.

There are certain factors that affect how well a substance dissolves. One of them is time. When testing how soluble a substance is, you should allow enough time to make sure you get reliable results. Another factor, as you will see later, is temperature. There are others besides. These factors are also known as **variables.**

Whenever you are comparing different substances, you must remember to keep all the variables except one in every test *constant*—in other words, the same. When you do this, you are *controlling* the variables, and your test results will more likely be fair and reliable.

EXPLORATION 5

How Soluble Is It?

You Will Need

- a 10-mL graduated cylinder
- 21 test tubes or beakers
- test tube racks
- a stirring rod
- a variety of substances to test:
  flour
  sodium bicarbonate (baking soda)
  magnesium sulfate (Epsom salts)
  candle wax
  mothballs
  starch
  chalk dust
  bouillon cubes
  alum
  orange drink crystals
  pepper
  salt
  gelatin
  sugar
  coffee
  tea
  instant coffee
  instant tea
  chili powder
  aspirin
  seltzer tablet
  solid watercolor paint

What to Do

1. To 10 mL of water, add a pinch or a piece of one of the solutes listed above and stir. Does the substance dissolve? If it does, add a second pinch and stir again. If you observe no change, leave it for 5 minutes, stirring occasionally. Why is it necessary to leave it for 5 minutes?
2. Decide whether the substance is
   (a) soluble,
   (b) not very soluble, or
   (c) insoluble.

3. Repeat the test on the other materials you have gathered. Write down any variable(s) in your experiment. Can you think of anything you could do to make this a more controlled test?

4. In your Journal, record your findings in a table like the one below.

Substance	Soluble	Not Very Soluble	Insoluble	Was Time A Factor?	Any Other Variable(s)	Special Observations

An Experiment Out of Control

John had results that did not compare well with those of his classmates. Here's how he performed each test:

Test 1

He half-filled a test tube from the cold-water tap, dumped in some crystals, and shook the test tube until his arm got tired.

Test 2

He added 5 mL of water to a beaker, added two crystals of another solute and stirred the mixture once or twice with a stirring rod.

What would you suggest to John to help him improve his experimentation technique?

Detective Work

Soluble or not soluble— that is the question!

Dyani and Mario were having a friendly argument over the experiment they had just finished. Mario thought that baking soda should be entered in their data table under the "Not Very Soluble" heading. Dyani insisted that baking soda was insoluble. Suddenly Mario snapped his fingers and said, "I know a way we can figure out who's right!" What do you think Mario's idea was?

EXPLORATION 6

Let's Have a Sugar Cube Race!

If you drop a sugar cube into a glass of water, how long will it take to dissolve? Will it disappear in 5 minutes, 10 minutes, or even longer? What could you do to speed up the rate at which it dissolves? Before reading further, make a list of things you could try. Record the list in your Journal.

Object of the Race

To dissolve one sugar cube in 150 mL of water as rapidly as possible.

You Will Need

- 2 sugar cubes
- a 250-mL graduated cylinder
- 2 beakers
- a stopwatch

Rules

1. Each team will be given a sugar cube to add to 150 mL of water. Watch it dissolve without disturbing it. Make a note of the time it takes.
2. Each team must then decide what single thing they will do to the sugar cube or the water to speed up the rate at which the cube dissolves.
3. Each team will get a new sugar cube and another 150 mL of water.
4. Taking care to note the time, start your procedure for speeding up the dissolving of the cube. Record the time it takes the cube to dissolve completely.

What was the single variable you changed to speed up the rate of dissolving? What were the variables of the other teams in the class? Which of these variables is the most important?

Detective Work

The Case of the Missing Candy

Two identical candies are placed in beakers of water. Fifteen minutes later the experimenter returns to the room and sees that one piece of candy has disappeared.

What do you think could have happened while the experimenter was out of the room? Can you explain why one piece of candy disappeared and one did not?

Challenge Your Thinking

1. Based on your knowledge of hard water, explain the "bathtub ring." What solvents might you use to get rid of this ring?

2. Sarah keeps confusing *solute* with *solvent*. How are you able to remember the difference between these similar-sounding terms?

3. Van is trying to determine if water containing detergent will dissolve salt faster than water alone does. When you ask how he plans to perform this experiment, he tells you:

   I'm going to add a pinch of salt to 25 mL of water and time how long it will take to dissolve while stirring. Then I'll add salt to some water containing liquid dishwashing detergent, stir, and check the time again.

   What variables must you remind Van to control so that his test results will be reliable?

4. The illustration below shows two beakers of water. One beaker contains a sugar cube and the other an ice cube. After 15 minutes, both cubes will have disappeared. Will both beakers now contain a solution? Justify your answer.

A Solution or Not?

SEPARATING SOLUTE AND SOLVENT

4. The Salt Industry

Around the world, salt is obtained in several different ways. Many methods use an *evaporation* or *boiling* process to separate the solute (salt) from the solvent (water).

Evaporation is the gradual process by which a liquid changes into a gas or vapor. For example, wet clothes will dry over time by taking heat from the surrounding air. The drying, or evaporation, process will go faster if you add heat, as in a clothes dryer.

Boiling is the process by which a liquid rapidly changes into a gas. This occurs when the temperature of the liquid is raised to its boiling point. Water boils at 100 °C at sea level. At higher elevations, the boiling point of water is a bit lower. Other liquids have their own characteristic boiling points.

Read the two reports that follow, written by students from different parts of the world. They describe how salt is produced in their countries. The first report is written by students in Zambia, a country in south-central Africa. The second report is by a student in Nova Scotia, a province of Canada.

As you read the reports, count the number of times the evaporation process and the boiling process are referred to. Think about the ways in which the two processes are similar.

By:
Michael Nosenge
and
John Bavalya

Cibwa salt is found mainly in the country of Zambia. This kind of salt is found in a sort of grass.

Grass

In August, the grass is collected using hoes and then spread on the ground to dry. This grass may remain there for about two months. When it is dry, it turns brown in color.

In September and October, the Cibwa salt is extracted from the grass. People do this by burning the dried grass, which turns it into ash. The salt is in the ash.

Collecting the grass

This ash is put on a filter made of another kind of grass, and water is poured over the ash into a clay pot. The water dissolves the salt and leaves behind the insoluble impurities of the ash on the filter.

The salt solution in the clay pot is placed over a fire. During the boiling, the water evaporates away. When the water has all evaporated, the salt is in the bottom of the pot in solid form.

Drying the grass

The people in our part of Zambia still make salt the same way.

Boiling away the water, leaving the salt behind

Pouring water over the ash

Burning the dried grass

Separating Solute and Solvent   153

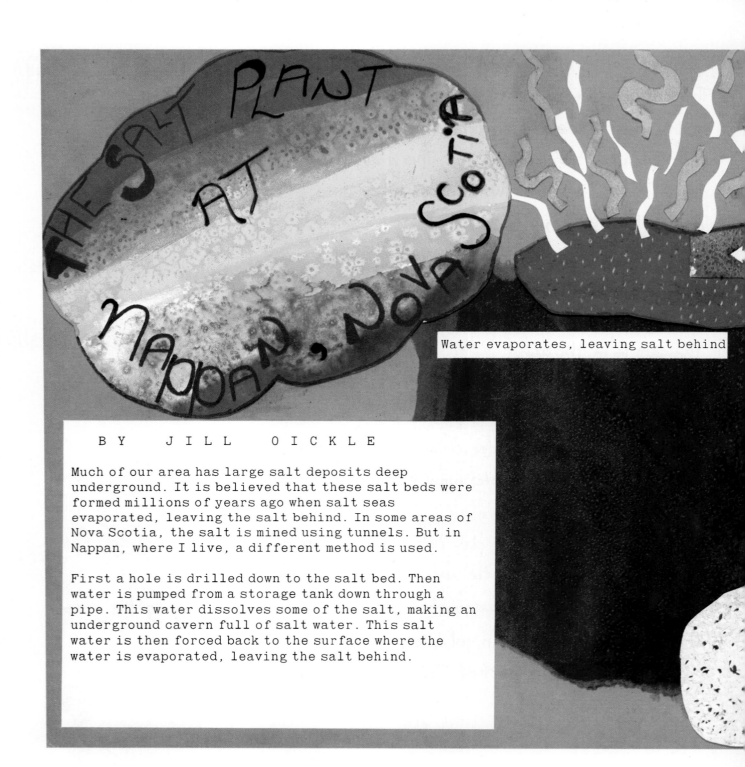

Water evaporates, leaving salt behind

The Salt Plant at Nappan, Nova Scotia

BY JILL OICKLE

Much of our area has large salt deposits deep underground. It is believed that these salt beds were formed millions of years ago when salt seas evaporated, leaving the salt behind. In some areas of Nova Scotia, the salt is mined using tunnels. But in Nappan, where I live, a different method is used.

First a hole is drilled down to the salt bed. Then water is pumped from a storage tank down through a pipe. This water dissolves some of the salt, making an underground cavern full of salt water. This salt water is then forced back to the surface where the water is evaporated, leaving the salt behind.

EXPLORATION 7

It's Your Turn to Design!

You Will Need

- sand–salt mixture
- a heat source
- beakers
- a funnel and filter paper
- stirring rods
- paper towels
- tweezers
- a magnifying glass
- a stand
- water

Ask for any other apparatus you need.

What to Do

Here's the problem! You have just found a deposit of salt, but it is quite impure. In fact, it's 50 percent sand. Work as a team with several other people to devise a scheme to extract the pure salt. Check your plan with your teacher to make sure it's safe, and then carry out your experiment. Report on the success of your design.

Comparing Methods

Compare the method of salt extraction used in Zambia with the method used in Nova Scotia. What similarities are there in the methods of gathering the salt? How was water used in each process? How was the salt separated from the water? Where did the water go?

Separating Solute and Solvent 155

Desalination and Distillation

In certain parts of the United States and the world, fresh drinking water is in short supply. One method of meeting the ever-increasing demand for fresh water is to remove the salt from ocean water. This process is called **desalination.** In this case, it is the solvent (water), not the solute (salt), that is being saved.

In some situations, it may be useful to save both the solvent and the solute. To save the solvent, as is done in desalination, or to save both the solute and the solvent, a process called **distillation** is used. Let's look more closely at both desalination and distillation.

The Future of Desalination

Should we be using desalination of the ocean as a source of fresh water? Here are some facts about desalination:

- Desalination can relieve water shortages in dry regions along seacoasts.
- One large desalination plant alone can produce about 280 million liters of fresh water daily.
- The desalination process requires large amounts of energy, regardless of how that energy is produced.
- Providing desalted water for dry regions far from seacoasts requires a transportation system.
- The world's desalination plants today produce only a small fraction of the world's daily demand for fresh water.
- Nuclear-powered desalination plants might someday produce desalted water and electricity at the same time.

Form small groups and discuss the following questions. Do you think it would be wise for us to rely on the oceans for our supply of fresh water? Why or why not? How do you think the cost of getting water from desalination compares with the cost of fresh water from other sources? What are some ways in which we could make better use of the fresh water already available to us from rain, lakes, and rivers?

Distillation

Distillation involves two steps. The first step is heating the solution to change one part of the solution into a gas or vapor. The second step involves cooling the vapor back into liquid form and collecting it.

Here is a design that Karen used for collecting both solvent and solute by distillation. What are some weaknesses in her design? Is she taking the necessary safety precautions?

With other students, discuss how Karen's design might be improved, using similar equipment. Check your suggestion with your teacher (for safety), and then test your own design. Compare your results with those of other groups.

Something to Write About!

Here's a chance to apply what you have learned about evaporation, boiling, and distillation.

During an expedition to a remote mountainous area, you learn that a person living there urgently needs distilled water to help treat a rare medical condition. You have a teakettle, a camp stove, and a few bottles. Draw a diagram to show how you could use them to obtain distilled water. Explain how your method works. See how many different ways your class can devise for distilling water with just the equipment listed.

Separating Solute and Solvent

What Is the World's Largest Distillery?

Surprisingly enough, it is the earth itself! The earth's water is constantly involved in a purification process similar to that of a giant distillery. This process is called the **water cycle.**

Each day, vast quantities of water evaporate from the oceans, lakes, rivers, and land. Eventually, this moisture condenses and falls as rain or snow. This is nature's way of purifying the world's water supply. How is nature's way similar to the method you developed to purify water?

The Water Cycle

Because of the water cycle, all the water on the earth today was here when the earth was formed. The next time you have a drink of water, stop and ask yourself where the water may have been before. Perhaps that very water, or part of it, had been in a royal bath in ancient Egypt. Or perhaps George Washington Carver might have drunk some of it. Take the two cartoon characters that appear on this page and develop a cartoon strip showing their adventures through time. How many different and unusual places can you picture? Let your imagination roam, but there is one rule: at some point, you must include the water cycle in your cartoon series.

6 Another Solution

Detective Hanamoto's Forgery Case

Scientists do many of the things detectives do. This includes using scientific equipment, detectives' methods, and detectives' ways of thinking about clues. Detectives must be good observers, and so must scientists. Detectives search for clues, just as scientists do. Detectives make **inferences** and draw conclusions from the clues; so do scientists. Detectives prepare an explanation or a **hypothesis** to explain why certain things happen the way they do. Scientists do the same. Both detectives and scientists often make educated guesses about something that has happened or that might happen. Detectives working in a forensic lab often conduct experiments and make measurements; so do scientists. In this lesson, you will be asked to do many of the things detectives and scientists do: observe, make inferences, form hypotheses, make predictions, and conduct experiments.

Detective Yori Hanamoto made the following summary of a forgery case he investigated. As you read his report, try to pick out the similarities between detective work and scientific investigation.

Detective Hanamoto's summary:

> The accused was charged after a complaint was made about an unauthorized withdrawal of money from another person's bank account. The accused claimed he was given a check for the amount. The police investigation suggested that the accused had altered the check: instead of "$7000," the check was now made out for "$70,000"!

```
                                                10/15     19 92
PAY TO THE
ORDER OF   Cash                                 $ 70,000
SUM OF   Seventy Thousand                       00/100 DOLLARS

FIRST BANK OF COMMERCE
23 Martin Luther King Blvd.
Los Angeles, California 90052
                                                G. Millar

    :01712 0101
```

On close examination with a magnifying glass, the check still looked genuine. The only way to settle the issue was to perform a test on the ink. If the check was a forgery, there would be a difference in the inks where there shouldn't be; that is, there would be a difference between the "Seven" and the "ty." There would also be a difference between the comma in "70,000" and the numbers themselves. After I made the test, there was no doubt: the check was a forgery.

What kind of test could Detective Hanamoto have used? Play detective by trying Exploration 8.

EXPLORATION 8

Analysis of Inks

You Will Need

- a metric ruler
- a strip of filter paper about 10 cm long and 6 cm wide (You can also use ordinary newspaper—but use only the edge so that there is no ink on it to confuse your results!)
- a beaker or other wide-mouthed container
- distilled water
- equipment to permit the paper to be held upright in the container
- several samples of washable (water-soluble) inks. Certain felt-tipped pens can be used.

What to Do

1. Pour water into the container to a depth of 2 cm.
2. Draw a line in pencil at a distance of 3 cm from the bottom of the filter paper or newspaper.
3. Put small crosses in pencil at least 1 cm apart on the line you have drawn.
4. Place a drop of one of the inks on one of the pencil crosses. Put one drop of each of the other inks on each of the other crosses.
5. Let the inks dry, then suspend the strip by some means so that the bottom 2 cm of the strip is immersed in the water inside the container.
6. After 15 minutes, remove the filter paper or newspaper and let it dry.

What do you observe?

EXPLORATION 8—CONTINUED

Questions

1. Now explain why Detective Hanamoto *inferred* that the check was written with two different inks.
2. Do you think that inks are a true solution? Devise an experiment to find out.
3. Read the following cartoon, which shows Keenan being a detective. Then explain the difference between observations, inferences, hypotheses, and predictions.

Observation   Inference

Hypothesis   Prediction

4. Did you notice the various ways in which Detective Hanamoto was doing the types of things that scientists do? Complete the table below in your Journal, using specific examples from Detective Hanamoto's summary.

What Scientists Do	What Detective Hanamoto Did
observes	?
hypothesizes	?
infers	?
predicts	?
experiments	?

5. "Being scientific" means doing the things scientists do. In learning about solutions, you are being scientific. The list below suggests a number of ways in which scientists learn about our world. For each one, suggest in your Journal an example of a situation in this unit in which *you* have been scientific.

Being Scientific Means	I Was Being Scientific by
observing	?
measuring	?
classifying	?
inferring	?
predicting	?
experimenting	?

6. *Detective Work at Home.* The method you used to separate the colors of ink is called "paper chromatography" (*chroma* means "color"). You can use the same method to determine whether food coloring is made up of only one color or is, instead, a solution of different colors. Try it, and report back to your class.

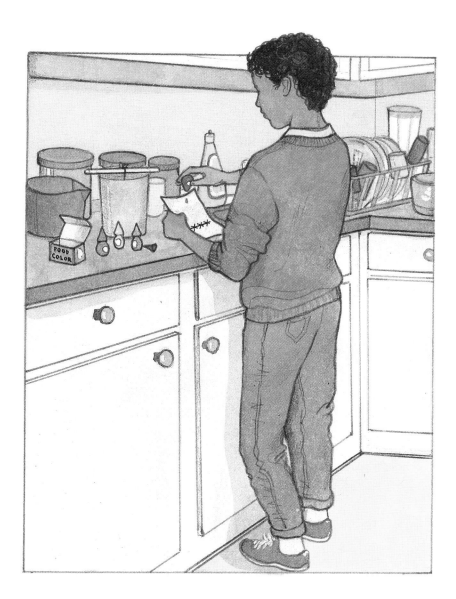

Separating Solute and Solvent 163

7 Sugar Solutions

Maple syrup and maple sugar are products made from the sap of the rock maple tree, often called the sugar maple. The sap of the sugar maple is a solution of sugar and water. Much can be learned about solutions by visiting a grove of sugar maples (a sugarbush) and seeing how maple syrup is made. That is just what Beth's class did. She visited a sugarbush in Vermont, which is the largest producer of maple syrup in the United States.

Read the report Beth presented to her English class the day after she and her class made the trip. As you read the report, write down three important discoveries Beth made about solutions. Compare your list with that of a classmate, giving reasons for your choices.

Mr. Lowther, the owner of the sugarbush I visited, starts tapping his sugar maple trees about the second week of March. If it is a warm spring, he begins sooner. If it's a large tree, Mr. Lowther can bore several holes into it. A small spout, called a spile, is pounded into each hole. In this sugarbush, all the spiles are connected to a network of plastic tubing that carries the sap to the sugar house. There the sap is boiled down to make maple syrup. Mr. Lowther said that a single tree can supply between 40 L and 100 L of sap in a season.

The sugar house is an exciting place, full of steam and delicious odors from the boiling sap. I found out that sap is only 2 percent sugar and 98 percent water! The sap is boiled in the evaporator until the water content is decreased to 34 percent and the sugar content increased to 66 percent. Mr. Lowther said that it takes 60 L of sap to make 1 L of maple syrup. Only when the syrup has the proper water and sugar content is it drawn off, strained, and bottled. I asked Mr. Lowther how he knew when the syrup was 66 percent sugar. He said he uses a thermometer! That's right—when the temperature reaches 104 °C, the solution is right for bottling.

Mr. Lowther makes other maple products as well. He prepared a whole batch of maple candy for the class by boiling some syrup in a big pot until the thermometer reached 110 °C. Then he poured the syrup onto a patch of clean snow where it hardened. It was great!

A trip to the sugarbush is an interesting way of investigating solutions. What a great time we had!

Here is a chart of the other maple products Mr. Lowther makes, and their boiling points.

Maple butter	113°C
Maple cream	114°C–115°C
Maple wax	119°C
Hard maple sugar	123°C

Questions to Ponder

With a classmate, discuss the answers to the following questions.

1. What method of removing a solute from a solvent is used in making maple products?
2. What effect does sugar have on the boiling point of water? (Pure water boils at 100 °C.)
3. What is the relationship between the number of degrees of change in the boiling point of the solution and the percentage of sugar in the solution?
4. How did Beth know that maple wax contains less water than maple cream?
5. At most, how many liters of maple syrup could one tree produce in a season?

EXPLORATION 9

Making Artificial Maple Syrup

You Will Need

- a beaker
- a balance
- 65 g white sugar
- a 50-mL graduated cylinder
- 35 mL water
- 3 drops maple flavoring
- a hot plate
- a stirring rod
- a candy thermometer (or one that reads over 100 °C)

What to Do

Simply mix the sugar and water in a beaker. While stirring, heat the mixture gently until all of the sugar dissolves. Then add 3 drops of maple flavoring and stir.

At what temperature do you predict this solution will boil, if it has the same percentage of sugar as Mr. Lowther's maple syrup?

If you have a thermometer that reads over 100 °C, test your prediction. To make a fair test of your prediction, what else should you do?

The Honey Factory

The honey made by bees comes from *nectar,* a solution that is mainly sugar and water. The bees gather nectar from flowers and then store it in open cells in the beehive. The bees fan their wings in front of the cells to help evaporate the excess water in the nectar. When the solution has reached the right thickness, they close off the open cells with wax. The result is honey. Bees can do what Mr. Lowther does without boiling.

Actually, there are many similarities between making maple syrup from sap and making honey from nectar. One is given in the table below. In your Journal, fill in the rest of the steps from Mr. Lowther's operation that correspond to the descriptions of the steps performed by the honeybees.

Mr. Lowther's Business	The Honeybees' Business
1. Mr. Lowther collects sap from many maple trees.	1. The bees collect nectar from many flowers.
2. ?	2. The bees store the nectar in a central place in the hive.
3. ?	3. The bees speed up the evaporation of water from the thin nectar solution by fanning their wings.
4. ?	4. The bees store the product in cells and seal off the cells with wax.
5. ?	5. Honey is used as a food by bees as well as humans.

Challenge Your Thinking

1. You are given a clear liquid. It does not show the Tyndall Effect when a light is shined through it. Without tasting the liquid, how would you tell whether it is just one substance or a solution? Is there any case where your method might not work?

2. Sometimes you find spots on your dishes after they have been washed. How can you explain this?

3. Li is boiling a beaker of sea water to find out how much salt remains.

   (a) Describe all of the things Li might observe.

   (b) What will happen to the temperature of the boiling liquid as time passes? Why does this happen?

4. Salim wondered what would happen to the artificial maple syrup he made if he put it in the freezer. To his surprise, it would not freeze. Why wouldn't it freeze?

Separating Solute and Solvent

CONCENTRATION OF SOLUTIONS

Getting Colder

Beth discovered that there is a relationship between the boiling point of a solution and the amount of solute (sugar) dissolved in the solvent (water). Would a solute affect the freezing point of a solvent as well? The following drawings suggest an answer to this question.

What's the message of these cartoons? Here are three experiments to help you find out: one for the laboratory (Exploration 10), one for home (Exploration 11), and one already completed for you to analyze (Exploration 12).

EXPLORATION 10

What Is the Temperature of Water When It Freezes?

You Will Need

- a 250-mL graduated cylinder
- a plastic-foam cup
- ice cubes
- a thermometer

What to Do

1. Add 150 mL of water to a plastic-foam cup.
2. Now add an ice cube and stir with a thermometer. **Be Careful:** *Stir slowly and carefully so that the thermometer does not break.* Observe the changes in temperature.
3. Keep adding ice cubes until the temperature will not go any lower. What is the lowest temperature the water can reach?
4. As soon as the temperature is constant, remove any ice that is left. How much ice did you use to cool the water to its lowest temperature? There is a way you can find out, without finding the mass of the ice.

EXPLORATION 11

Effect of a Solute on the Freezing of Water

You Will Need

- a 15-mL measuring spoon (1 Tbsp.)
- 3 juice glasses
- 45 mL of sugar
- 15 mL of salt
- water

What to Do

Pour 75 mL of water into a glass. Put 45 mL of sugar in the second glass and 15 mL of salt in the third glass. Add enough water to the second and third glasses to make their solutions even with the water level in the first glass. Stir. Put the glasses in the freezer. Examine them every half hour. In what order do they freeze? Why do you think this is so?

Water

Sugar solution

Salt solution

EXPLORATION 12

Mr. Acosta's Home Investigation

What to Do

Analyze the results of Mr. Acosta's experiment and draw some conclusions from his findings.

What He Did

Mr. Acosta surprised his family by taking over the kitchen one evening to do some experimenting. He had brought home several thermometers. He took the salt, measuring cups, and glasses out of the cupboard. If you analyze the table of results below, you will learn what he did and the information he found. What conclusions can you make about the effect of the solute on the solvent?

Trial	Solutions Mixed	Temperature When a Crust Forms on Top	Time to Freeze Completely
1	5 mL of salt in 100 mL of water	−3.6 °C	1 hour 5 minutes
2	10 mL of salt in 100 mL of water	−7.0 °C	1 hour 30 minutes
3	15 mL of salt in 100 mL of water	−10 °C	1 hour 45 minutes
4	20 mL of salt in 100 mL of water	−14 °C	2 hours
5	100 mL of water (no salt)	0 °C	48 minutes

Look again at the drawings on p. 168. For each drawing, write a two-line caption that explains in scientific terms what the drawing illustrates.

Concentration of Solutions

9 How Much Solute?

Sap is a *dilute* sugar solution. Maple syrup is a more *concentrated* sugar solution. What do these terms mean?

Both beakers in the illustration contain the same type of solution. But how might you determine which one contains more solute?

The beakers show an example of two different **concentrations** of the solute and solvent. The one with *more* solute in the solution is the **concentrated** solution. The one with *less* solute in the solution is the **dilute** solution.

What would happen to the color of solution A if you added a lot of water to it? How would adding water to solution A change its concentration? Would solution A become more concentrated or more dilute? Now suggest two ways of changing the concentration of solution B. Would each change make solution B more concentrated or more dilute?

By now you may be noticing that the concentration of a solution is a *relative* thing. It applies only when you compare solutions made of the same solute and solvent. You can compare the concentration of one solution with another having the same solute and solvent by using the terms *more concentrated* or *more dilute*.

Concentrate on These!

1. Beakers C and D contain the same solute and solvent. The amount of solute is the same in both beakers.

    (a) What is the difference between solutions C and D?

    (b) After stirring, which will be more dilute? more concentrated?

Stir to dissolve.

2. Beakers E and F contain the same solute and solvent. The amount of solvent is the same in both beakers.

    (a) What is the difference between solutions E and F?

    (b) Which will be more dilute? more concentrated?

Stir to dissolve.

3. Stir to dissolve completely. Then add more solvent to solution G. The result is a more __?__ solution.

4. Stir to dissolve completely. Then boil off some of the solvent in solution H to make a more __?__ solution.

G

Stir to dissolve.

H

Numbers for Concentration

If you and your classmates were told to prepare a dilute solution of salt and water, would all of the results be identical? Probably not. The concentration of each solution would likely differ. In order for everyone to make an identical solution, you would have to be told the amounts of both the solute and the solvent. Another way of stating this is to say that you would need to know the concentration of the solution.

The concentration of a solution is often expressed in "x grams of solute" dissolved in "100 g of solvent." Here are the concentrations of a few familiar solutions.

Solution	Solute	Solvent
maple sap	2 g sugar	100 g water
maple syrup	194 g sugar	100 g water
vinegar	5 g acetic acid	100 g water
sea water	3.5 g salt	100 g water
apple juice	not less than 0.035 g vitamin C	100 g water

Mystery Solutions

It is not always possible to identify the more concentrated solution just by looking at it. However, there are other properties or characteristics that can be used to distinguish a concentrated solution from a dilute one. For example, what property or characteristic of maple sap is used to determine the right concentration of maple syrup? Exploration 13 suggests some other ways to discover the relative concentrations of solutions.

Concentration of Solutions

EXPLORATION 13

Concentrated or Dilute?

You Will Need

- 2 salt solutions of different concentrations—solution A and solution B
- 2 beakers
- a balance
- a 10-mL graduated cylinder
- 2 hard-boiled eggs
- 2 test tubes and test-tube rack
- food coloring
- a clear plastic straw
- a plastic-foam cup and ice
- a ruler
- a thermometer

What to Do

Form small groups for this Exploration. Your teacher will give each group two salt solutions—solution A and solution B. One is concentrated; the other is dilute. Perform the following tests on each solution. Explain how each test enables you to distinguish the dilute solution from the concentrated one.

Test 1

Find the mass of a beaker. Add 10 mL of solution A to the beaker and find its new mass. What is the mass of 10 mL of solution A? Repeat with 10 mL of solution B. How does the mass of solution A compare with the mass of solution B? Which is more concentrated—solution A or solution B? Why? Return each solution to its original container.

Test 2

Drop an egg into both solutions. What happens to the egg in solution A? in solution B? Which is more concentrated, solution A or solution B? Why?

mass of the beaker = ?

A mass of 10 mL of solution A = ?

B mass of 10 mL of solution B = ?

Test 3

Fill two test tubes halfway with each solution. Add several drops of food coloring to test tube B. Using a clear plastic straw or glass tube, form two separate layers of the solution inside the straw.

Step 1
Lower the straw into Solution A by about 1 cm. Let solution A enter the straw. Now cover the top of the straw securely with your finger.

Step 2
Transfer the straw with the solution from test tube A to test tube B, and lower the straw by 3 cm or more.

Step 3
Lift your finger off the top of the straw to let solution B enter. Cover the top of the straw again. Remove from test tube B. Which solution is on top, solution A or solution B? Or did the two mix together?

Reverse the whole procedure, this time putting the straw into test tube B first. Can you form layers with the colored solution on the bottom? on the top? Which appears more concentrated, solution A or solution B? How did you decide?

Test 4

Place 25 mL of solution A in a plastic-foam cup. Add ice to the cup and stir carefully with a thermometer. What is the lowest temperature you obtain? Repeat the procedure with solution B. Which solution has the lowest freezing point? Which solution is more concentrated, solution A or solution B? Why?

Can you suggest other tests that might let you distinguish solutions of different concentrations?

Concentration of Solutions

Solution Fact

The following is an advertisement from the window of a travel agency.

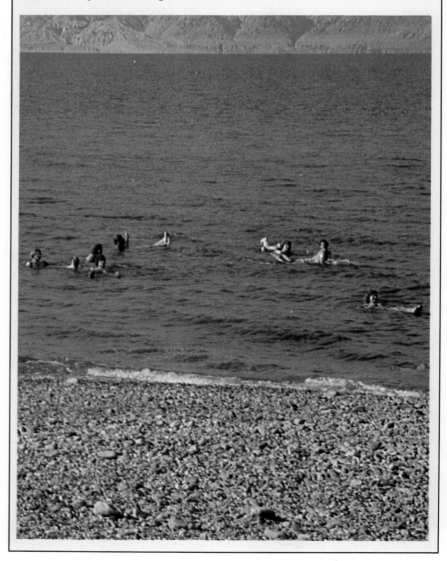

THE DEAD SEA

Enjoy life at the Dead Sea! Spend your next vacation on the shores of the Dead Sea, where you can enjoy the sunshine ... and the water. We guarantee that you will float. A great vacation spot for all you sinkers!

What property of the Dead Sea makes it easier to stay afloat in its waters?

EXPLORATION 14

Make a Concentration Tester for Salt Water

You Will Need

- 3 different concentrations of salt solutions
- 3 test tubes
- test-tube rack
- 3 used pencils about 8 cm long each (or 2 straws, each cut in half)
- 6 thumbtacks (or 6 nails and modeling clay)
- a balance
- a 25-mL graduated cylinder
- a ruler
- scissors

What to Do

1. Make a tester using one of the methods illustrated.

Tester 1

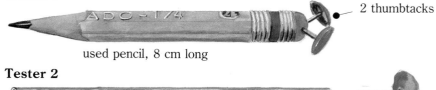

used pencil, 8 cm long

Tester 2

half a straw

2. Find the mass of 25 mL of each solution. Then pour 25 mL of each solution into its own test tube.
3. Place one tester in each test tube as shown.

In which solution does the pencil or the straw stand the highest? the lowest? in between?

Compare the masses of the solutions. Can you see any relationship between the mass of a solution and its concentration? between the concentration of a solution and how well an object floats in it?

Do these solutions have the same concentration?

Concentration of Solutions 175

10 Saturation

A sponge soaks up water easily, but there comes a point after which it is unable to soak up any more water. At that point a sponge is said to be **saturated**. When no more solute can dissolve into the solvent, a solution is said to be saturated. Earlier, you learned that water does not dissolve all things equally well. For instance, much more solute is needed to saturate the solution if the solute is sugar than if the solute is baking soda. The amount of solute needed to saturate a solution can be measured.

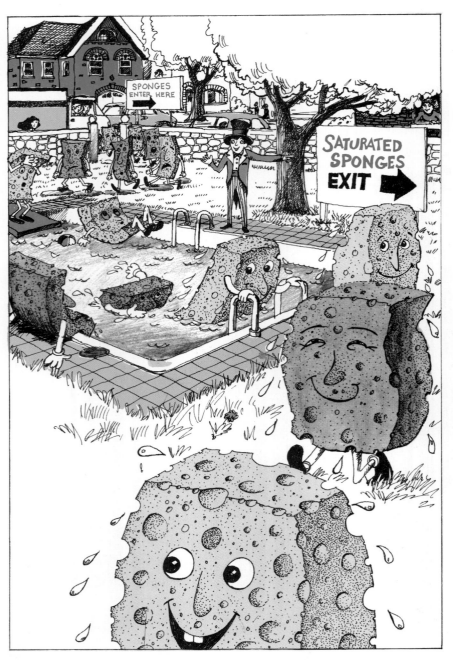

EXPLORATION 15

How Much Solute for Saturation?

You Will Need

- a salt sample
- a 10-mL graduated cylinder
- water
- a beaker
- a stirring rod

What to Do

1. Using a dry graduated cylinder, measure the volume of salt in your sample.
2. Add salt gradually to a beaker containing 10 mL of water until no more salt dissolves. Make sure you stir well after each addition of salt.
3. Measure what is left of your sample of salt. How many milliliters of salt did you use?

Comparing Solubility Figures

One milliliter of salt has a mass of about 1.2 g. What mass of salt did you use in 10 mL of water? What mass would dissolve in 100 mL of water? What mass of salt is needed to saturate 100 g of water? (Remember, 100 mL of water is about the same as 100 g of water.) How does this figure compare with the figure for salt given in the Table of Solubilities on p. 177?

176 Solutions

Saturation and Temperature

The amount of solute needed to saturate a solution is a measure of the solution's **solubility.** Solubility is often expressed as the number of grams of solute that can dissolve in 100 g of solvent. Do you think the temperature of a solvent affects its solubility? The following solubility table may provide the answer. It shows the maximum amount of different solutes (in grams) that can be dissolved in 100 g of water at different temperatures.

Table of Solubilities

Temperature (°C)	Salt (sodium chloride) (g)	Sugar (sucrose) (g)	Baking Soda (sodium bicarbonate) (g)	Alum (potassium aluminum sulfate) (g)
0	35.7	179	6.9	4.83
10	35.8	191	8.2	6.90
20	36.0	204	9.6	9.63
30	36.2	220	11.1	14.30
40	36.5	238	12.7	22.20
50	36.8	260	14.5	30.30
60	37.3	287	16.4	46.90
70	37.6	320	—	74.50
80	38.1	362	—	138.80
90	38.6	414	—	—
100	39.2	487	—	—

Making Sense of the Data

1. How does temperature affect the amount of solute that can be dissolved?
2. What substance is most soluble at 0 °C? least soluble at 0 °C?
3. What is the order of solubility of the given solutes at 0 °C? at 60 °C?
4. What is the solubility of sugar at 0 °C? At what temperature is the solubility of sugar approximately doubled?
5. Look at the solubility of other solutes at 0 °C. At what temperature would the solubilities of each be approximately doubled?
6. What substance shows the least increase in solubility as the temperature rises?
7. How much sodium chloride must you add to 100 g of water at 50 °C in order to saturate the solution?
8. Suppose you added 25 g of alum to 100 g of water and heated the solution to 80 °C. Would all of the alum dissolve? What would happen if you now cooled this solution of alum to 20 °C?

Concentration of Solutions

Picturing the Data

Another way of representing the data in the table is to show it in a graph such as the one below. The vertical axis of the graph is a scale of the solution's solubility. The horizontal axis shows the temperature. Use the information in the Table of Solubilities to graph each solute. Use different symbols for each solute (for example, X for salt, * for sugar, O for baking soda, and + for alum).

1. All the curves start at a different place on the vertical axis. What does this tell you?
2. Which line is almost horizontal? What does this tell you?
3. Certain lines are very curved. What does this tell you?
4. One curve is very steep. What does this tell you?

This is the graph of one solute. Which solute is it?

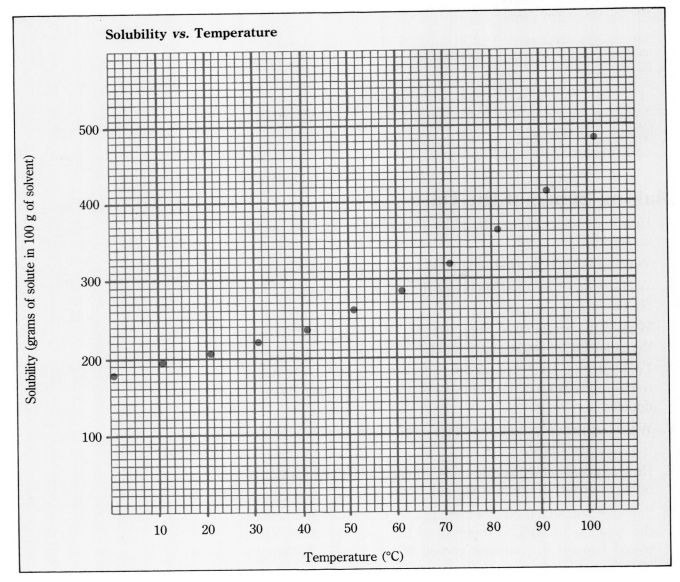

178 Solutions

11 Solutions in the Environment

While reading the following section about two important solutions, make your own list of any words referring to solutions. The list at the right will help you get started. Your list will be useful later on when you will be working on a crossword puzzle.

> concentrated, concentration, dilute, dissolve, distillation, evaporate, mixture, saturated, solute, solution, solvent, transparent

The Atmosphere

The air around you and around the earth is called the **atmosphere**. It is a solution. How do you know that it's a solution? Here's how:

(a) The atmosphere is a mixture.
(b) Like liquid solutions, the atmosphere is transparent.
(c) The atmosphere can be separated into its parts.
(d) The concentrations of the parts need not always be the same.

The atmosphere is like a huge blanket around the earth. It is several thousand meters thick. Human actions, nevertheless, can change the concentrations of some of its parts. When we drive a car, the gasoline burns. This produces a gas called *carbon dioxide* that becomes one of the solutes in the atmosphere. Burning wood, oil, and other fuels also produces carbon dioxide. Some people think that putting so much carbon dioxide into the air will warm up the "blanket" of the atmosphere, slowly warming up our climate. What effect would this have on us?

Carbon dioxide isn't the only gas produced when we burn fuel. Factories that burn coal send many metric tons of *sulfur dioxide* into the air every day. Sulfur dioxide is a very poisonous gas, and it causes changes faster than carbon dioxide. You already know from studying Unit 1 that sulfur dioxide is soluble in water, and when the two are mixed, the solution they create is acidic. Rain that falls through air with sulfur dioxide in it is a solution called "acid rain." Many lakes and rivers are becoming more and more acidic because of the acid rain falling into them. Some are so acidic that fish cannot live in them any more.

What can you do to stop the atmosphere from being "saturated" with pollution?

Concentration of Solutions

The Ocean

The atmosphere is not the only large and important solution in the world. The ocean is another. Indeed, the ocean is the largest liquid solution on earth. Parts of it are deep enough to swallow up Mount Everest—the world's highest mountain!

The solvent of this huge solution is water. The ocean contains many different solutes. What are they? This is the sort of question that can be answered by a chemist. Here is what a chemist would find when separating the water from the solutes in some sea water.

Parts of the ocean are deep enough to swallow up Mount Everest.

Mt. Everest
8,850 m

Marianas Trench
(Pacific Ocean)
11,036 m

If 100 g of sea water is evaporated to dryness, 3.5 g of dissolved solids will be left behind. Most of the dissolved solids, about 2.4 g, is sodium chloride—table salt. The remainder consists mainly of compounds of magnesium, sulfur, calcium, potassium, bromine, and iodine. Someday, the ocean may be a valuable source of these materials, but so far only sodium chloride, magnesium, and bromine are being extracted on a commercial basis.

Solids are not the only solutes found in sea water. Two important gases that are dissolved in the ocean are oxygen and carbon dioxide. Without dissolved oxygen, most water animals could not exist; without dissolved carbon dioxide, plant life in the ocean would die.

180   Solutions

Word Hunt

The answers to the clues for this crossword puzzle are probably in the list on page 179. Copy the puzzle into your Journal and complete it there.

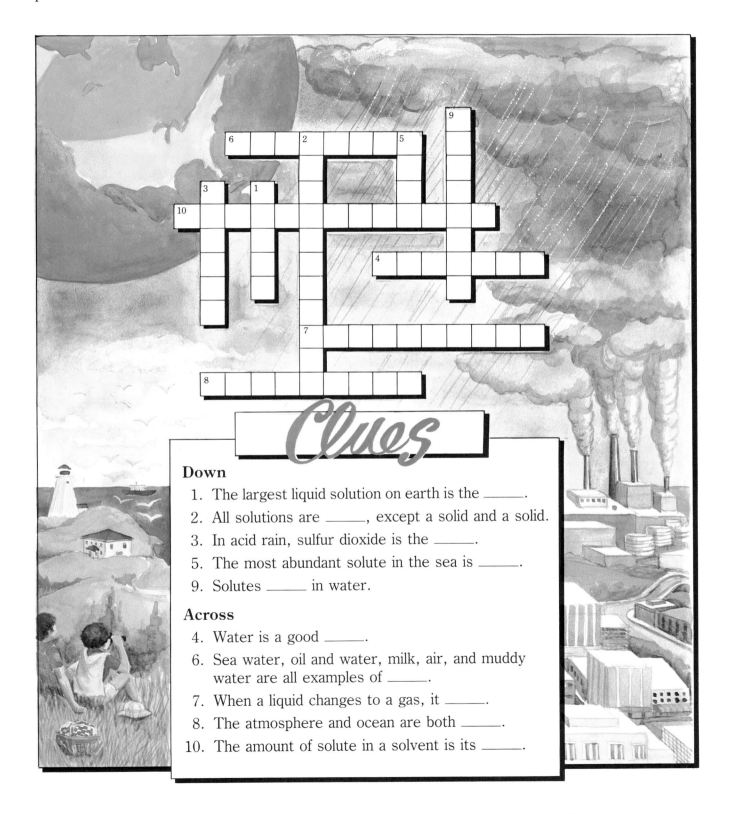

Clues

Down

1. The largest liquid solution on earth is the _____.
2. All solutions are _____, except a solid and a solid.
3. In acid rain, sulfur dioxide is the _____.
5. The most abundant solute in the sea is _____.
9. Solutes _____ in water.

Across

4. Water is a good _____.
6. Sea water, oil and water, milk, air, and muddy water are all examples of _____.
7. When a liquid changes to a gas, it _____.
8. The atmosphere and ocean are both _____.
10. The amount of solute in a solvent is its _____.

Concentration of Solutions

Scientific Talk

Take another look at the terms in the list on page 179. Write a short paragraph that contains all of these words.

Gold from the Sea

> Water is sometimes called the universal *solvent,* since so many *solutes* are found in it. Even gold is found *dissolved* in sea water. If just one cubic kilometer of this *solution* is *evaporated,* we will recover 0.004 metric tons of gold. That's a big gold mine!

1. To find out how difficult it is to talk about solutions without using any of the words listed on page 179, rewrite the paragraph above, replacing the italicized words with words or phrases that mean the same thing.
2. Listen to the national news or look in a newspaper for today's gold prices. Can you work out how many dollars' worth of gold is found in 1 cubic kilometer of sea water?

182   Solutions

12 Growing Crystals

Growing your own crystals is fun, and all it requires is a suitable solute and some patience. Many solutes are suitable for crystal growing. A few of these are copper sulfate, magnesium sulfate, and alum. Alum will be used in Exploration 16. You may try others if you wish.

> **EXPLORATION 16**
>
> ## Growing a Crystal
>
> ### You Will Need
> - a 100-mL graduated cylinder
> - water
> - alum
> - a beaker
> - a stirring rod or a straw
> - an evaporating dish
> - a 25-mL graduated cylinder
> - plastic wrap
> - a small jar
> - thread
> - a ruler
> - a thermometer

Tourmaline

Aragonite

Quartz

Gypsum

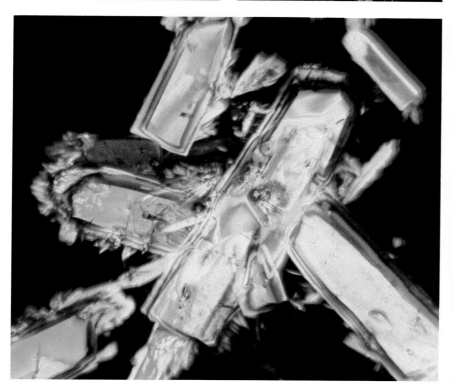

Sugar crystals (magnified 40 times)

Concentration of Solutions   **183**

EXPLORATION 16—CONTINUED

What to Do

1. **Prepare a saturated solution**
   (a) From the Table of Solubilities on page 177, determine how much alum is needed to make a saturated solution in 200 mL (or 200 g) of water at 20 °C. Measure out this amount of alum and then add about 10 g.
   (b) Dissolve the alum in a beaker by adding 200 mL of water at about 50 °C. (Water from the hot-water tap works fine.)
   (c) Let the solution cool to room temperature (2–3 hours). After cooling, there should be some undissolved solute on the bottom of the beaker. If there is, the solution is saturated.

2. **Grow a seed crystal**
   (a) Carefully pour about 20 mL of this saturated solution into an evaporating dish. *Be careful not to transfer any undissolved solute.* Cover your beaker with plastic wrap to keep the rest of your solution from evaporating.
   (b) Place the evaporating dish containing the 20 mL of solution in a quiet spot where it will not be disturbed. In a couple of days, you should have a supply of seed crystals of sufficient size to use in step 3.

3. **Suspend your seed crystal**
   (a) Remove the plastic wrap from the container holding your supply of saturated solution. Slowly pour this liquid into a clean jar. *Be careful that no undissolved alum enters the jar.* The jar should be small enough so that it will be nearly full.
   (b) Choose a well-formed crystal from your supply of seed crystals and tie a thread around it.
   (c) Attach the thread to a stirring rod or straw so that when the straw or rod is placed across the mouth of the jar, the seed will be suspended in the center of the solution. Place the jar in a quiet spot and observe what happens over the next week or two.

A Final Challenge

In millimeters (mm), measure the dimensions of your crystal every few days. Keep a record of its growth. Also make a diagram that shows its shape and size. On the average, how much mass is added to your crystal each day? How could you find out?

Challenge Your Thinking

1. Maya was given five solutions of sugar in water, in different concentrations. Four of the solutions had been dyed the following colors by adding food coloring: red, yellow, green, and blue. The fifth solution was clear. She used the layering test to determine their concentrations. The illustration below shows her results.

   Which solution is the most dilute? Which is the most concentrated? Arrange the five liquids from the least concentrated to the most concentrated.

2. Josh wants to check a claim by Brand X that their canned fruit is packed in very light syrup (dilute sugar solution). Suggest a way that he can compare the "lightness" of the syrup used by Brand X with that of the syrup used by Brand Y.

3. Salt is mixed with sand and placed on roads in the winter after snowstorms in many states. This practice does not work as well in Alaska as it does in Washington. Why not?

4. The graph shows how temperature affects the solubility of gases and most solids in water.

   (a) When can a lake hold more dissolved oxygen—in the winter or in the summer? Why?

   (b) Liter for liter, which ocean can hold more salt—the Caribbean or the North Atlantic? Why?

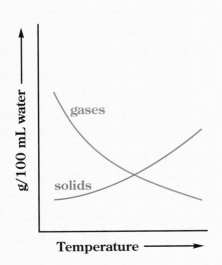

Concentration of Solutions 185

Unit 3 — Making Connections

The Big Ideas

In your Journal, write a summary of this unit, using the following questions as a guide.

- How is a solution different from a non-solution?
- What are the parts of a solution? Why are they not visible?
- How would you determine whether a substance is soluble, not very soluble, or insoluble?
- How would you separate a solute from a solvent?
- How would you make a solute dissolve faster?
- How is evaporation different from boiling? Which process occurs in the water cycle? in the making of maple syrup? in the making of honey?
- How does the presence of a solute affect the boiling point and freezing point of water? Of what practical value is this?
- How would you distinguish by experiment the difference between a concentrated solution and a dilute solution of a given solute in a given solvent?
- How are the terms *saturated, solubility,* and *temperature* related?

Checking Your Understanding

1. Suppose 6 g of salt is mixed with 50 g of water.

    (a) What is the mass of the solvent?

    (b) What is the mass of the solute?

    (c) What is the mass of the solution?

2. Jennifer discovered that when she put a little washing soda (sodium carbonate) in water, the solution became warm. She wondered if the amount of washing soda mattered. So she found 5 bottles, filled each one with 100 mL of water, and took the temperature of the water in each. Jennifer then added different amounts of soda to each and measured the temperatures again. Her results are on the next page.

Bottle	Number of Teaspoons* of Soda Added	Amount of Water (mL)	Temperature Before (°C)	Temperature After (°C)
A	1	100	22	26
B	2	100	22	30
C	3	100	22	34
D	4	100	22	37
E	5	100	22	39

*1 teaspoon of washing soda has a mass of about 7.1 g.

(a) What variables were held constant?
(b) What variable changed?
(c) What variable changed in response?
(d) What did Jennifer find out?
(e) Which solution was most dilute?
(f) If washing soda happened to make a blue solution, which solution would be the deepest blue?

3. Draw a concept map that shows how the following words are related: *solute, antifreeze, solvent, boiling point, salt,* and *freezing point.*

Reading *Plus*

Now that you have been introduced to solutions and non-solutions, let's take a closer look at these mixtures and at some of the special properties they display. By reading pages S45–S60 in the *SourceBook,* you will learn more about different solutions and about the importance of water in making many of them. You will also learn more about acids, bases, and salts.

Updating Your Journal

Reread the paragraphs you wrote for this unit's "For Your Journal" questions. Then rewrite them to reflect what you have learned in studying this unit.

Making Connections **187**

Science in Action

Spotlight on Chemistry

Biochemistry

Kathleen Mailer is a biochemical researcher and university teacher with a Ph.D. in chemistry. Her research concerns the biochemistry of living tissues, mainly the heart. Kathleen says she loves research and greatly enjoys teaching. The most stimulating part of her job, she says, is the freedom to wonder whether an idea is true and then to try to find out.

Q: Why did you choose to do heart research?

Kathleen: To me, the heart is the most fascinating part of the body. It's the one organ that must never stop. Your muscles, your liver, and other organs can stop temporarily, but not your heart.

The heart is like a super-efficient factory. Even during the sprint at the end of a marathon, only a part of its capacity is being used. It is said to be "over-engineered."

Nevertheless, the human heart can be afflicted with heart disease. Heart disease also occurs in rabbits, but very rarely in rats. It's curious that the same things that predispose human beings to heart attacks, such as absence of exercise or bad diets, also predispose rabbits to heart attacks—but not rats. It is important, therefore, to study the differences between rat hearts and rabbit hearts.

Chemistry Teaching and Research

Enrique Diaz has devoted years to teaching chemistry in high school and college. In addition, he did chemical research before becoming a teacher. The basic qualification for his field of work is a Ph.D. in chemistry. Enrique obtained his doctorate after earning a Bachelor of Arts in languages and a Master's degree in science. His many and varied responsibilities as a researcher, teacher, and administrator make his life busy and rewarding.

Q: How did you get interested in science?

Enrique: As a child, I was curious and always wanted to know how things worked and why things were as they were. My interest in chemistry was sparked by very enthusiastic high school and college chemistry teachers. Their love of chemistry and teaching helped tremendously when the studies became more difficult.

Q: What do you enjoy most about your work?

Enrique: I enjoy the opportunity to work with students and to pass on some of the enthusiasm I have for chemistry and research. Making a new compound that no one has made before is quite a thrill. Then, determining the structure of that compound makes all the effort worthwhile.

Q: What subject areas are important for students who are interested in becoming chemists?

Enrique: First of all, they must have a knowledge of basic math and chemistry and physics. Of course, this is important for everyone—not just for chemists. Scientific literacy is a must for today's world.

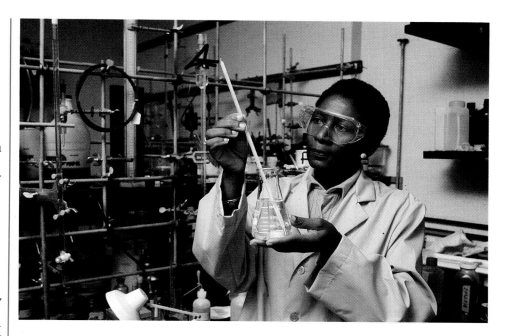

Students interested in chemistry must also be able to write clearly, because scientists share their research discoveries with other scientists by writing articles that describe their work.

Q: Are there any particular abilities that are most important to a chemical researcher?

Enrique: I think that an analytical problem-solving approach, logical thinking, attention to detail, and careful library research habits are all important abilities. One hour in the library recording what others have already discovered can save many hours in the laboratory.

A Project Idea

You've probably seen labels on laundry detergents that say "Phosphate-Free." Find out why these detergents are better for the environment than detergents containing phosphate.

Chemists have played a key role in replacing the phosphates in laundry detergents with safer substances. Find out what these safer substances are and why they won't damage the environment as much as phosphates do.

Science and the Arts

The Perfume Maker's Art

Making perfumes is an ancient art. It was practiced, for example, by the ancient Egyptians, who rubbed their bodies with a substance made from soaking fragrant woods and resins in water and oil. From certain references and formulas in the Bible, we know that the ancient Israelites also practiced the art of perfume making. Other sources indicate that the art was known to the early Chinese, Hindus, Arabs, Greeks, and Romans.

A Complex Art

Have you ever wondered why some perfumes are traditionally so expensive? To begin with, a fine perfume may contain over 100 different ingredients! And it's not just a matter of mixing things together that have pleasant odors, as you will see.

Plant, Animal, or ?

The perfume ingredients you are probably most familiar with come from fragrant plants or flowers like sandalwood, jasmine, or patchouli. These plants get their pleasant odor from their *essential oils,* which are stored by the plants in tiny baglike parts called *sacs.* The parts of the plants used for perfumes may include their flowers, roots, or leaves.

Another kind of ingredient that is used in making perfume comes from animals. These substances are useful as *fixatives,* to make other fragrances long-lasting. In some cases, the substances as they occur full-strength in nature have distinctly unpleasant odors! For example, *civet musk* is actually a foul-smelling liquid that the civet, a small cat-like animal, sprays on its enemies. Other examples of fixatives are *castor,* from the beaver, and *ambergris,* from the sperm whale.

A third class of ingredients is human-made, or synthetic. In some cases, these ingredients are chemically the same as naturally occurring materials. In others, however, they are chemicals unknown in nature. These extend the creative options of the perfume maker.

Jasmine flowers

African civet

Taking "Notes"

When you smell a perfume, the first odor you detect is called the *top note*. It is a very fragrant and refreshing odor that evaporates rather quickly. The *middle note*, or *modifier*, gives a more substantial character to the impression made by the top note. The *base note*, or *end note*, is the odor that lasts the longest.

A Catalog of Fragrances

Perfume fragrances can be grouped according to their most dominant odors. The table below shows some of the major groups, with examples.

Group	Examples
Floral	Jasmine, rose, gardenia
Spicy	Clove, cinnamon, nutmeg
Woody	Sandalwood, cedarwood
Mossy	Oak Moss
Orientals	Combinations with vanilla, balsam, etc.
Herbal	Clover, sweetgrass
Leather/tobacco	Leather, tobacco, birch tar
Aldehydic	Synthesized from aldehydes (these generally have a fruity odor)

Roses have long been admired for their fragrance.

Find Out for Yourself

Test a number of different perfumes and colognes to see if you can distinguish three different "notes" of each. Write a report on your findings. Do some research on the history of perfume making.

Perfumes have other uses than as personal fragrances. Make a list of as many of these uses as you can.

Science and the Arts

Unit 4: Force and Motion

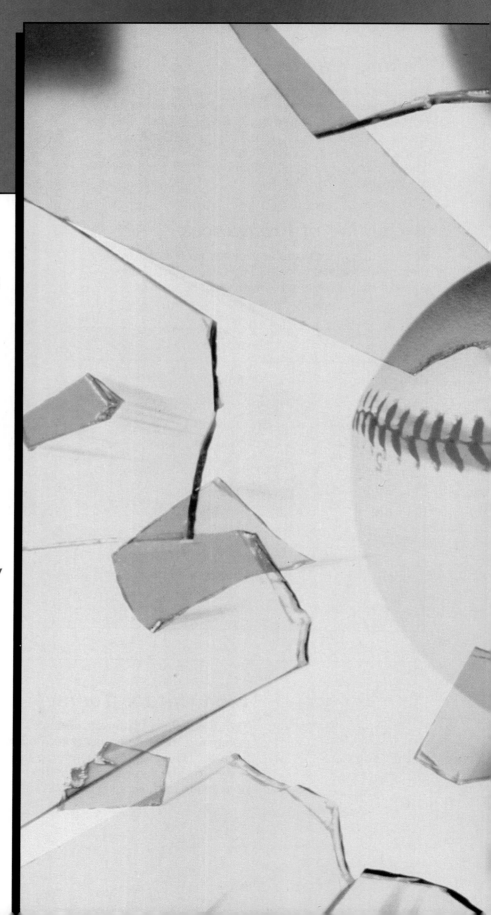

UNDERSTANDING FORCES
1. Forces at Work, p. 194
2. Different Kinds of Forces, p. 200
3. What Makes Things Fall? p. 204

MEASURING FORCES
4. The Tools to Use, p. 210
5. Estimating the Size of a Force, p. 216
6. How Strong Are You? p. 220

FRICTIONAL FORCE
7. Take a Look at Friction, p. 225
8. Causes and Consequences of Friction, p. 231

GOING THROUGH THE MOTIONS
9. Thought Experiments, p. 236
10. Puffball: A Game of Force and Motion, p. 242
11. What is Inertia? p. 246
12. Forces in Pairs, p. 251

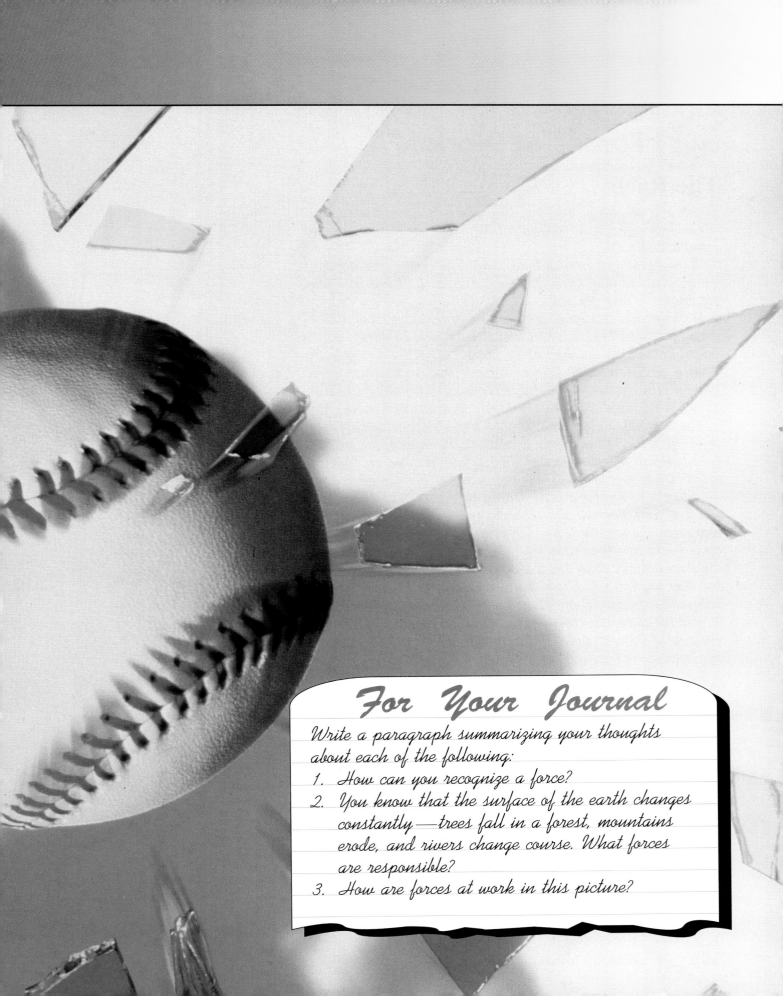

For Your Journal

Write a paragraph summarizing your thoughts about each of the following:
1. How can you recognize a force?
2. You know that the surface of the earth changes constantly—trees fall in a forest, mountains erode, and rivers change course. What forces are responsible?
3. How are forces at work in this picture?

UNDERSTANDING FORCES

1 Forces at Work

The Race

Today was the day of the race! Kathleen and Kimiko were excited! With difficulty, they dragged their sailboat across the sand into the water.

A favorable west wind was blowing. The craft was ready to go. Their hearts beating hard, they pulled on the halyards with nimble fingers to raise the sails.

With one hand, Kimiko took control of the tiller. With the other, she tugged on the sheet for the mainsail. Kathleen, sitting mid-ship, secured the sheet of the jib.

As a gentle breeze pushed on the sails, the craft moved slowly and steadily out beyond the jetty. There the full force of the wind stretched the sails. The craft suddenly sped up. They were off!

The boat shuddered as the choppy waves slapped and sprayed the hull. With the experience of a summer's sailing, the girls sailed toward the starting line. Just as they reached the line, the starter's flag was lowered. The race was on! Kimiko pushed on the tiller. The rudder twisted against the water. Their course was set. The sails strained and the mast creaked as the boat skimmed and broke the water's surface. Kathleen pushed downward on the centerboard to offset the leeway. Passing the second red buoy on the starboard, Kimiko decided to make a quick turn. She drew the tiller toward herself, hauled in the sheet at the same time, and hollered "Coming About!" The boat swung around until the wind was at the stern. The boom of the mainsail swung suddenly from one side of the craft to the other. In an instant, the boat changed its course.

Kathleen, who was watching another boat coming up very rapidly on the leeward side, responded too late to the warning. The boom smacked her arm. She fell over the side and with a splash hit the water.

Quickly, almost automatically, Kimiko brought the boat up into the wind. The sails flapped harmlessly; the boat slowed down. . .

194   Force and Motion

The sailing terms used in this story about Kathleen and Kimiko may be new to you. Look up the definitions for the terms below, or ask someone who sails. Then read the story again.

- halyards
- tiller
- mainsail
- sheet
- jib
- hull
- rudder
- centerboard
- leeway
- buoy
- starboard
- stern
- boom
- leeward
- jetty

Recognizing Forces

Sailing is an exciting, fast-paced activity. Many forces are at work as a sailboat skims across the water. Sailors put out tremendous effort and action to keep their boats sailing. Notice how many action words appear in the story above: pushed, hauled, smacked, and slapped, just to name a few. Write a list of all the action words in the story and all the forces that you recognize.

Questions

1. Can you have a force without some object (someone or something) exerting the force and some other object receiving the force?
2. Name some effects that forces have on objects.
3. Is it possible for one object to exert a force on another without ever touching it? Where are there examples of this in the story?
4. Examine the story for pairs of opposing forces, which are forces that work against each other. How many did you find?
5. How would you define or describe a force? Make sure your definition includes all the things you have discovered about forces.
6. Write an exciting conclusion to the story. Illustrate the events of your story that involve forces. Then identify the forces found in your drawing.

Describing Forces

The passages below are taken from the story of the boat race. Pay careful attention to the underlined words and the words in italics. Then answer the questions, and see what you can discover about forces.

1. What is similar about the forces described in these sentences?
   (a) "With one hand, Kimiko took control of the tiller. With the other, she *tugged* on the sheet . . ."
   (b) "As a gentle breeze *pushed* on the sails . . ."
   (c) "The boom *smacked* her arm."

2. What do these sentences tell you about the effects of forces?
   (a) ". . . the full force of the wind *stretched* the sails."
   (b) "The craft suddenly *sped* up."
   (c) "In an instant the boat *changed* its course."
   (d) "She *fell* over the side . . ."
   (e) ". . . the boat *slowed down* . . ."

3. Think of other examples of forces that could cause an object to do the following:
   (a) change its shape   (c) speed up or slow down
   (b) begin to move   (d) change direction

4. What aspect of force does this quotation describe? "Kathleen pushed *downward* on the center board . . ."

The Direction of Forces

In the story of the race, the directions of most of the forces are not described. When scientists make diagrams of forces, they often use an arrow to indicate the direction of a force. The length of the arrow gives you an idea of the size of the force. The tail of the arrow indicates the point where the force is exerted.

Which way do you think the table will move?

Understanding Forces 197

EXPLORATION 1

A Picture Puzzle

Several forces are illustrated in these pictures. For each picture, pick out the object (the agent) that is having a noticeable effect on another object (the receiver). In your Journal, sketch each situation, and place arrows indicating the direction of each force and the point where each force is being applied. Write the name of the agent and the receiver, and describe the effect of the force.

Are there cases where two forces are acting on the same object? What is their combined effect on the object?

(a)

(b)

(c)

(d)

(e)

(f)

(g)

Understanding Forces 199

Different Kinds of Forces

EXPLORATION 2

You're familiar with the many different game booths at a carnival. For this Exploration, you will test your skill at a carnival of science.

Experiencing Forces

The seven mini-experiments in this Exploration provide examples of specific forces at work. Each one is set up at its own station in the classroom. At each station you will find materials arranged as in the illustrations on pages 201 to 203. Your purpose at each station is to discover which forces are at work. As you work through the mini-experiments, think about these questions:

- What is the agent, receiver, and effect of each force?
- Are there any new types of forces?
- How are they different?
- Are there any *noncontact forces*—forces exerted by an agent that does not touch the receiver?

Record your observations and conclusions in your Journal.

You Will Need

- an empty jar
- a pail
- water
- a glass
- paper clips and rubber bands
- a magnet
- a plastic bread bag, cut in strips
- a sock
- hooked masses
- string
- books
- drinking straws
- a bottle
- newspapers and a cloth towel

200 Force and Motion

Station 1
What to Do

1. Push the empty jar slowly into a pail of water. Keep its open end facing up.
2. Submerge it; let the water fill the jar.
3. Now take the jar out of the water.

What forces do you feel at each step?

Station 2
What to Do

1. Bring a strong magnet near the glass container.
2. Try to lift the paper clips to the top of the glass.
3. Lift the container off the table.

Think about the forces involved.

Station 3
What to Do

1. With one hand, hold up two long strips cut from a plastic bread bag. Observe them as they hang freely.
2. Now rub both sides of the strips with a wool sock. Allow the strips to hang freely again. Observe any forces at work.
3. Bring one of the strips near a table or a wall. What happens?

Identify the forces acting on the plastic strips.

EXPLORATION 2—CONTINUED

Station 4
What to Do

1. How fast does paper fall? Why does it fall? Try dropping the following:
   (a) a single sheet of paper
   (b) a piece folded into quarters
   (c) a piece crumpled into a ball
2. Try two different pieces at the same time. Do some fall faster than others?

What are the forces involved here?

Station 5
What to Do

1. Hang a hooked mass on the rubber band.
2. Add more masses. How many are needed to break the band? Let the masses drop onto something soft.
3. Replace the broken rubber band with a new one so the next person can try the experiment.

Identify all forces.

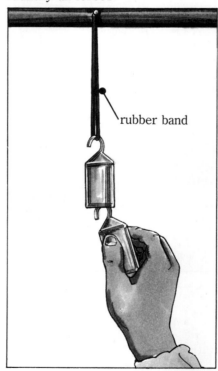

rubber band

Station 6
What to Do

1. Cut a rubber band and tie it securely to a paper clip. Attach the clip to a string tied around a book.

   How long must the band stretch before the book starts moving?

2. Try it again with another book placed on top of the first one. What do you notice?

3. Place the books on drinking straws and repeat the experiment. How has the opposing force changed?

What tends to prevent the books from moving?

Station 7

What to Do

1. Find out how far a bottle will roll on the floor before stopping.
2. Place several sheets of newspaper on the floor and try again.
3. Repeat the experiment by rolling the bottle over other surfaces, such as a cloth towel spread on the floor.

What is slowing down the bottle?

Which Forces Did You Meet in the "Carnival of Science"?

1. Water exerted an upward **buoyant force** on the empty bottle at Station 1. Where did you see this force before?
2. A **magnetic force** was present at Station 2. Did the magnet need to touch the paper clips to exert an attractive force on them?
3. An **electrical force** acted upon the charged plastic strips at Station 3. Each strip exerted a force on the other. Is this a noncontact force like the magnetic force?
4. A **gravitational force** acted upon the sheets of paper at Station 4. Is this a noncontact force? What agent is exerting the force? Why did the uncrumpled sheets fall more slowly?
5. The rubber bands at Stations 5 and 6 exerted an **elastic force.** Where else could you find elastic forces?
6. **Frictional forces** were exerted on the bottle and book at Stations 6 and 7. As an object moves in one direction, frictional forces oppose the motion. In what direction was the frictional force at Stations 6 and 7 acting?

Understanding Forces

3 What Makes Things Fall?

You are holding a ball in your hand. You let it go. It falls. What is the direction of its motion? Is there a force acting on the ball? If so, what agent exerts this force? What is the direction of the force in relation to you? to the earth? Does it matter where you are on the earth? What would happen to the ball if you dropped it into a tunnel that passed through the center of the earth?

Which way is down?

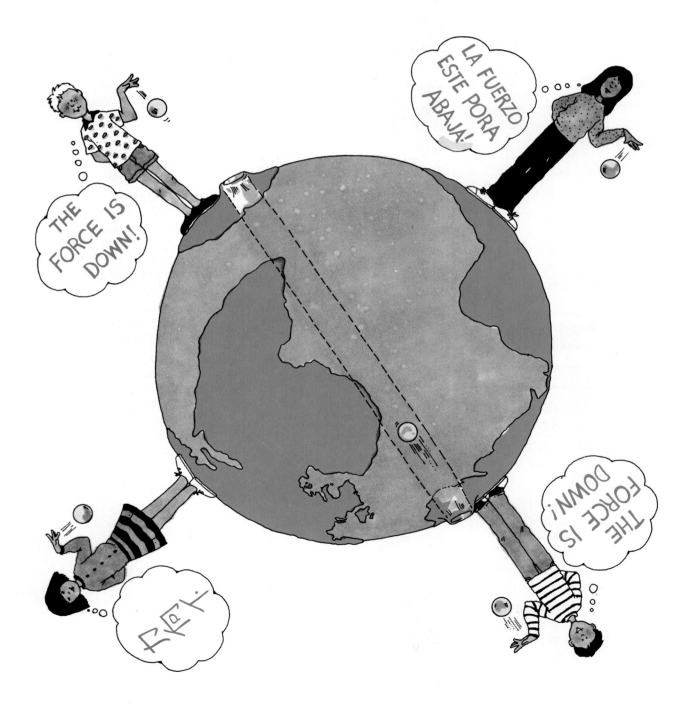

204 Force and Motion

Blow up a round balloon to use as a model of the earth. Rub the whole surface of the balloon with a cloth or against your hair. Now hold little pieces of paper near its surface at different places. Release the paper.

Does the paper fall toward the balloon? What kind of force is acting on the paper? What is the agent of this force?

The balloon (or earth model) exerted an *attractive* force on the bits of paper. It "pulled" the paper toward its surface. The earth appears to do something similar to the ball. The earth exerts an attractive force on the ball and pulls the ball toward its surface.

The forces exerted by the model and the earth itself are similar in two ways: both are attractive, and both are noncontact forces. However, there is an important difference between the earth and the model.

The attractive force exerted by the balloon is *electrical*. You gave the balloon an electrical charge by rubbing it against your hair or the cloth. The force exerted by the earth, however, is not electrical. It is an attractive force that all objects made of matter exert on each other. Because the earth is made of matter, it is able to attract other things made of matter.

The force exerted by the earth is called **gravitational force.** This force causes objects to move toward each other. The gravitational force the earth places on an object is called the **weight** of the object. To find out where the theory of gravitational force came from, read the following story.

electrical force

gravitational force

Understanding Forces

A Striking Discovery

Why do objects fall? If an object starts to move, isn't there a force causing this movement? The scientist Isaac Newton pondered these questions. The old story says that he was sitting under a tree when an apple bounced off his head and set him thinking!

Newton wanted to understand the motion of the moon. Why does it keep circling the earth? Newton already knew that if you push a ball along the floor, it moves in a straight line—even after the initial force stops pushing it. It takes a second force to change the ball from its straight path.

Newton wondered, "What if the earth exerts an attractive force on the moon? This would explain why the moon orbits the earth instead of moving in a straight line." The earth's force causes the moon to change its path just as a force exerted on a moving ball causes the ball to change its path. Then Newton thought, "If the earth exerts an attractive force on an object far away, it must also exert an attractive force on objects near its surface. It is this force that causes objects to fall toward the earth."

Thus, Newton had an explanation for both the apple falling and the moon curving from its straight path. The earth exerts a gravitational force on every object. What are the agent, receiver, direction and effect of this gravitational force?

Changing the direction of a moving ball

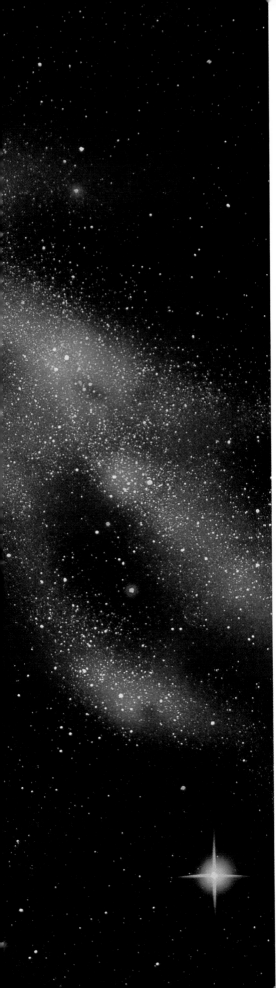

Newton's Laws

Newton's laws of motion and gravitation were extremely important discoveries. They explained many fundamental properties of our universe. For example, Newton concluded that every object in the universe exerts an attractive force on every other object in the universe. This law applies to huge bodies of matter like stars and planets: the sun attracts the earth; the earth attracts the sun. It also applies to bodies of very different size: the earth attracts the ball; the ball attracts the earth. It is true of small bodies as well: the table attracts the chair; the chair attracts the table. With small objects, the gravitational force is extremely small. With large objects, the force is much greater. The size of the attractive force depends on the size of the object. In other words, the strength of the force depends on how much matter there is in the object. The amount of matter is what scientists call **mass.**

Did Newton ever test his theory? No. That came much later when Lord Cavendish devised an experiment that proved that Newton's theory was correct. He showed that if you doubled the quantity of matter (mass) in one of the bodies, the attractive force also doubled. Cavendish also demonstrated that when bodies were separated farther apart from each other, the force of attraction became smaller.

Dr. X

Write a science fiction story about the future. Here is the scenario: Dr. X has just invented an anti-gravity spray. The harmless-looking liquid can cancel out the gravitational force exerted by the earth.

Now he wants to try it out. He chooses your school and sprays it all over the cafeteria and gym. Your principal rushes out of the office just as . . .

Understanding Forces

A Massive (Or Weighty) Problem

You know that the earth exerts a gravitational force on objects. This force is commonly called the weight of the object. If the earth were to disappear, an object would have no reason to fall. The object would become "weightless." This does not mean that it would no longer contain any matter. The object contains the same amount of matter with or without the earth's pull. That is, an object has the same mass. Obviously, mass and weight are really quite different ideas.

How much do you weigh? Suppose you were to take off in a spaceship and go far above the earth's atmosphere. What would happen to your weight? Remember that Cavendish showed that the attractive force of one object for another decreases as the distance between them increases.

At 6400 km above the earth (the distance from Washington D.C. to Paris, France), you would weigh only ¼ of your weight on earth. Of course, you would still have the same mass.

Earth
object has weight

No Earth
no weight

Force and Motion

Challenge Your Thinking

1. Your weight on the moon is the measure of the attractive force that the moon exerts on you. On the moon, you would weigh only ⅙ of your weight on earth. Why?
   Would you have any less mass on the moon?
   Why could you jump so much farther on the moon than on the earth?

   Earth 6400 km   **Moon** 1700 km

   The mass of earth is 80 times greater than the mass of moon.

2. If there were no gravitational force at a baseball game, what would happen if:
   (a) a ball were hit by a batter toward left field?
   (b) the left fielder made a jump to catch the ball?
   (c) the batter started to run for first base?
   (d) the surprised coach jumped out of the dugout?

3. How do you win a game of tug-of-war? What kinds of force must you overcome? Why must you play on even ground for the game to be fair?

4. The spring scale below is being used to find the weight of a jar with bees resting on the bottom of it. All of a sudden, the bees start flying around. What happens to the pointer on the scale? What does this tell you about weight? about force?

Understanding Forces

MEASURING FORCES

4 The Tools to Use

Forces cannot be seen. But there is a way to discover the size of a force—measure the size of its effect. In this section, you will follow a series of Explorations showing you how to make *force meters* and how to use them to measure forces.

EXPLORATION 3

Force Meters

Carefully study the following examples. In each case, what effect is brought about by the applied force? If the force were increased, how would the effect change? How could you use the effect of the force as a means of measuring the force? The first example has been completed for you. For the remaining examples, draw a diagram in your Journal showing what would happen if there was an increase in force. Your drawing should also show how you would use the effect to measure the size of the force. If you prefer, you may describe in words the increase in force, its effect, and how to measure the size of the force.

rubber band

The force of the single mass causes the rubber band to stretch to a length of 4.5 cm.

(a) Blocks hanging on a rubber band

Adding a second mass increases the force. This causes the band to stretch to a length of 8.5 cm.

210 Force and Motion

EXPLORATION 3—CONTINUED

(b) Force pushing on a plunger in a syringe

syringe

end plugged

(c) Force of a weight on the end of a ruler that is fastened to the edge of a table

ruler

(d) Hacksaw blade pulling on a book

hacksaw blade fastened solidly

(e) Weights hanging from a spring

EXPLORATION 4

Making Your Own Force Meter

Here are plans for constructing a force meter. This method uses rubber bands to measure the size of forces.

Construct the meter or design one of your own. Instead of using rubber bands, you could use a spring or some other elastic material. After completing your design, find the materials you will need and build the meter.

You Will Need

wooden dowel 1.5 cm to 2.0 cm in diameter

 cup hook

rigid cardboard tube (strengthen with masking tape, with holes on opposite sides near end)

bent paper clips

 rubber bands

A Push-Pull Force Meter

Three ways to fasten the rubber bands

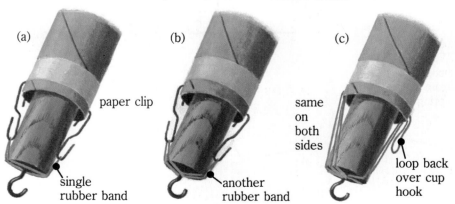

(a) paper clip / single rubber band

(b) another rubber band

(c) same on both sides / loop back over cup hook

1. Smaller Forces   **2. Medium Forces**   **3. Larger Forces**

212 Force and Motion

Every meter needs a scale. The scale should indicate the size of a force according to some standard. The **newton** (N) is the international metric unit (SI) used to measure force. The newton is approximately equal to the weight of a 100-g mass on earth. In the figure, you can see how the stretch of a rubber band can be used to indicate the size of forces in newtons.

Two ways to mark the newton scale on your force meter

Hold the cardboard tube.

commercial force meter

Use this end for pulling forces.
Use this end for pushing forces.

Each 100-g mass pulls the tube 1-N mark.

100 g

EXPLORATION 5

Using a Spring to Measure Gravitational Force (Weight)

A spring makes an extremely good force meter. The following experiment will help you discover why. Keep a record in your Journal as you carry out this experiment. You'll want to keep precise records of what you do and what you discover. After you complete the experiment, write a formal report using the information from your Journal. In your report include a title, purpose, procedures, predictions, results, discoveries, and answers to the questions in this Exploration.

You Will Need

- a coil spring
- washers
- hanging masses
- a meter stick
- a force meter

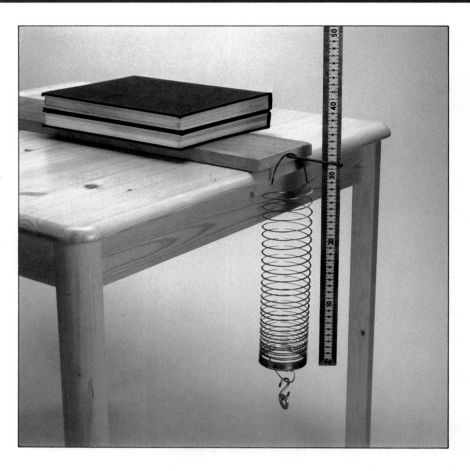

What to Do

1. Set up the apparatus as shown in the diagram.
2. Using the meter stick, read the position of the bottom of the spring. Repeat after adding one washer. What is the result? How much longer is the spring? The position indicates the size of the force that the earth exerts on one washer; that is, the weight of one washer.
3. Repeat for 2, 3, 4, 5, or more washers. Predict the effect each time before adding a washer. (Remember that the gravitational force on two washers is twice the force on one washer. How large would the size of the force be on 3, 4, or 5 washers?)
4. Enter your results in a table like the one below. This helps you see all your data clearly. What does your table tell you about the length of stretch and the number of washers? Why does a spring make a good force meter?

Number of Washers	Position of Bottom of Spring	Stretch of Spring
0		
1		
2		
etc.		

5. Another useful way to study your data is to make a graph of it like the one to the right. In your Journal plot the amount of stretch against the number of washers.

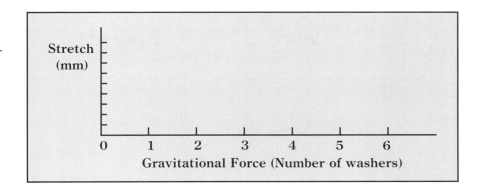

Questions

1. Suppose that an object hung on the spring causes it to stretch 22 mm. What would be the size of the gravitational force on the object (its weight)? Use your graph to find the answer.
2. Choose an object that is not too heavy to distort the spring. Predict how many washers it will weigh. Now hang it on the spring and find out if your prediction was correct.
3. Could you use the spring to measure the force needed to pull a book across a table?

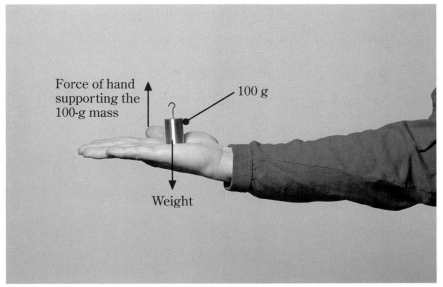

Converting Washers into Newtons (N)

1. Hold a 100-g mass in your hand. How much force are you exerting to hold it up? Find out by placing the mass on a commercial force meter marked in newtons.
2. Repeat for a 500-g mass and a 1000-g (or 1-kg) mass.
3. How many "washers of force" are equal to 1-N of force? Use the commercial force meter to find out.
4. For the original spring you used, how many millimeters would it have to stretch to indicate a 1-N force?

Measuring Forces

5 Estimating the Size of a Force

You are now familiar with the use of the newton (N). But what does a newton feel like? Try lifting and holding objects like these:

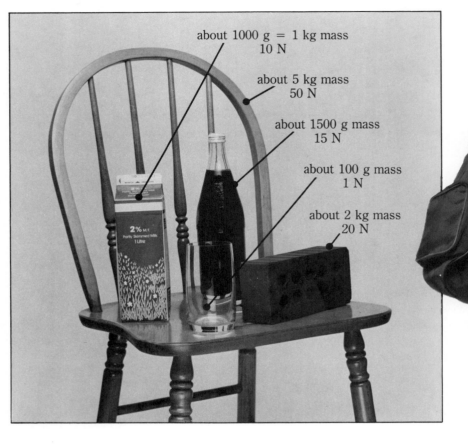

about 1000 g = 1 kg mass
10 N

about 5 kg mass
50 N

about 1500 g mass
15 N

about 100 g mass
1 N

about 2 kg mass
20 N

about 50-kg mass
500 N

Now, estimate the force needed to lift the following. Approximate answers are given in the box. But don't look now!

(a) Your science textbook
(b) Your teacher
(c) A pail of water
(d) A metal wastepaper basket
(e) A dozen eggs
(f) A basketball
(g) A hammer
(h) An iron shovel
(i) A concrete block
(j) $5 in quarters
(k) A newborn baby
(l) A compact car

Approximate Answers

(a) 6 N (e) 5 N (i) 110 N
(b) 700 N (f) 6 N (j) 1 N
(c) 60 N (g) 5 N (k) 30 N
(d) 12 N (h) 17 N (l) 8000 N

EXPLORATION 6

A Finger Exercise

Here's how to feel different sizes of forces with your finger.

You Will Need

- a meter stick
- a 1-kg mass or equivalent

What to Do

1. Balance a meter stick over the edge of the table (at the 50 cm mark).
2. Place a 1-kg mass on the ruler so that it is centered at the 45-cm mark.
3. Press down at 100 cm to raise the mass. You are using a 1-N force.
4. Press down at 60 cm for a 5-N force.
5. Now shift the 1-kg mass to 0 cm. Press at 100 cm for 10 N, at 75 cm for 20 N, at 60 cm for 50 N, and at 55 cm for 100 N.

   With the 1-kg mass at 0 cm you can calculate the force in newtons with this formula:

   $$\frac{500}{(P - 50)}$$

   where P = pressing point.

Figure 1

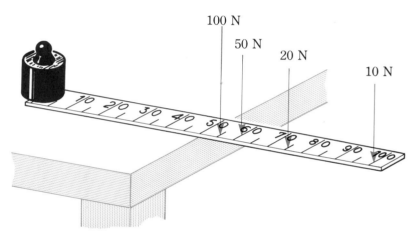

Think About It

Can you exert a 100-N force with one finger? with two fingers? What is the maximum size of force each finger can exert? Are all your fingers equal in strength? How does your left hand compare to your right hand?

Measuring Forces 217

EXPLORATION 7

Estimating and Measuring the Size of Forces

Now it's time to test your ability to estimate different sizes of forces.

You Will Need

- a commercial force meter or the one you made yourself

What to Do

1. Construct a table in your Journal like the one on page 219. As you perform this Exploration, fill in the information to complete your table.
2. Predict the size of the force exerted in each of the situations shown. Record your estimates in your Journal.
3. Select a force meter which enables you to make an accurate measurement of the force required in each case. Record these measurements in your Journal.

Situation	Estimated Force (N)	Measured Force (N)
(a) Opening a drawer		
(b) Pulling a door knob		
(c) Closing a door using the knob		
(d) Closing a door by pushing near the hinge		
(e) Lifting a brick		
(f) Causing a brick to start moving on a horizontal board		
(g) Keeping a brick moving on a horizontal board		
(h) Pushing a brick up an inclined board		
(i) Pushing a table		
(j) Lifting a sandbag using a pulley		
(k) Lifting a sandbag using two pulleys		
(l) Pulling down a screen or blind		
(m) Your choice		

6. How Strong Are You?

Here's a chance to measure your own strength. First, guess how many newtons you'll be able to exert. Then see how close your guesses were to your true strength.

1. Arm-Muscle Strength

How far can you stretch the luggage cord? Measure the length that the cord stretches in cm.

Find the force you used! Place the cord beside a meter stick. Pull the cord with a force meter until it matches the length that you were able to stretch the cord.

Now record the number of newtons shown on the force meter.

220 Force and Motion

2. Forearm Pull

Stretch the luggage cord using only your forearm muscles.

luggage cord

support held firmly

Be Careful: *Don't perform this activity in direct line of other people in case the cord slips out of your hand.*

Copy this chart into your Journal.

My Strength Record in Newtons
1. Arm muscle: _____?_____
2. Forearm pull: _____?_____
3. Hand grip: _____?_____
4. Leg thrust: _____?_____

Test the strength of other muscles; you decide which ones to try!

Measuring Forces

3. Hand Grip

Grip a bathroom scale with both hands. Be sure to use a scale that shows kg. The amount in kg multiplied by 10 gives you the force in newtons.

4. Leg Thrust

Using a chair as a backrest, push your legs against a bathroom scale braced against a wall. Calculate the force you exerted.

Measuring Mass and Weight

As you have already learned, the quantity of matter in an object is called its mass. Whether you move an object to the top of Mount Everest or to the moon, its mass remains the same. The object still has the same quantity of matter in it. Mass is usually measured in kilograms (kg).

Weight, on the other hand, is a measure of the gravitational force exerted on an object. The newton is the unit used to measure gravitational force (weight) and all other forces. A spring scale is the device used most often to measure weight.

Commonly, the terms "mass" and "weight" are interchanged and used incorrectly. This can cause some confusion—even for students of science. Carefully study the table and picture below. Account for the difference in the weight (N) of the spring scales shown in the figure.

Mass is . . .	Weight is . . .
the quantity of matter in an object.	a measure of the gravitational force exerted by the earth on an object.
constant for an object taken anywhere in the universe.	varied as you move away from the earth's surface.
measured with an equal arm balance or a pan balance.	measured with a spring scale.
measured in grams (g) and kilograms (kg).	measured in newtons (N).

How do measuring devices indicate a constant mass and a changing weight?

Challenge Your Thinking

1. Take the mass and weight quiz! People often confuse the meaning and use of the terms mass and weight. Rewrite these incorrect sentences and correct the confusion.

   (a) If I go far enough away from earth's surface, I will get to a place where my mass will be almost zero.

   (b) An object of 200 N on the surface of the earth should still measure 200 N in the orbiting space station.

   (c) In the space station, I would use a spring scale to find the mass of an object.

   (d) Rachel placed a football on one pan of an equal arm balance and got a perfect balance with a 250-g mass on the other pan. Naturally, she couldn't expect the same result on the moon.

   (e) Food is sold in the supermarket by weight—for example, $2.50 for a kg of sugar.

2. Design an innovative way to measure the force needed to:

   (a) perform a chin up

   (b) close a door

   (c) push a sports car

   (d) lift an enormous boulder

FRICTIONAL FORCE

7. Take a Look at Friction

You already saw some examples of friction in Exploration 1. Take another look at those examples. What did they tell you about friction? Now read the three case studies below and answer the questions.

Case Study A

As Jennifer reaches the level field, she gradually slows down.

1. What slows down the skis?
2. Is there a force involved?
3. What is the direction of the force?
4. What name do we give to this kind of force?

hard-packed snow

Case Study B

Driving east on a level stretch at 100 km/h, Andy shifts into neutral. The car coasts for 300 m before coming to a stop.

1. Identify the frictional force and its direction.

road surface

Frictional Force 225

Case Study C

Don dropped his book behind the large refrigerator. He pulled on one edge of the refrigerator to move it away from the wall. It did not budge. His sister, Susan, joined him. Together they just managed to move it.

1. What is the frictional force?
2. What is it doing here that is different from the frictional force in the other case studies?
3. How do the combined forces of Don and Susan compare with the size of the frictional force?

Case Study Conclusions

The following general conclusions about frictional forces can be drawn from the case studies:

1. When one surface moves against another, a frictional force results. The frictional force works against the direction of motion.
2. If one body moves over a stationary body, the stationary body tends to slow down the moving body.
3. If a force tries to move a body at rest on a surface, the surface resists this motion with an opposing force of friction.

Questions

Discuss these questions with a group of classmates. Try to reach an agreement about each one.

1. Name some other examples in which frictional forces slow down motion.
2. In the sailboat story at the beginning of this unit, were there any examples of friction? If so, identify them.
3. In case study A, what if there were no frictional forces opposing the motion of the skis?
4. For case study B, what would have happened if Andy had been driving on gravel? on ice?
5. In case study C, how could you help Don move the refrigerator by himself? In your solution have you reduced, increased, or maintained the size of the force of friction?

226 Force and Motion

Measuring Friction

How large or small are frictional forces? How can they be increased or decreased? You measured friction in Exploration 7. Review these measurements to see whether they can help you discover some new ideas about friction.

Janell had some ideas (hypotheses) about friction. She decided to check her new ideas by conducting a series of experiments.

The results of her experiments are summarized on page 228. Look at the results, then try to solve the Friction Hypothesis Puzzle below. In this puzzle, you will match Janell's experimental results with the hypotheses she began with. Do the experimental results confirm all of her hypotheses?

Friction Hypothesis Puzzle

Experiments designed to test the hypotheses:	Janell's hypotheses:
(a) Experiments 1 and 4	1. The size of the frictional force depends on the kind of surfaces that are rubbing together.
(b) Experiments 2 and 3	2. The heavier the object, the greater the force of friction.
(c) Experiments 3 and 4	3. The force of "starting" friction is greater than the force of "sliding" friction.
(d) Experiments 3 and 5	4. The force of "rolling" friction is greater than the force of "sliding" friction.
(e) Experiments 3 and 6	5. The force needed to lift an object is greater than the force needed to drag a body along a level surface.
(f) Experiments 4 and 7	6. Frictional force is constant for an object on a surface, no matter what the area of the surface contact is.

Expt. No.	Janell's Experiment	Diagram	Results
1	Janell measured the weight of a brick.	spring balance	Weight = 20 N
2	Janell measured the force needed to start the brick moving on a glass surface.	glass, spring balance	Force = 3.5 N
3	Janell measured the force needed to start the brick moving on a wooden surface.	wood	Force = 6.3 N
4	Janell measured the force needed to keep the brick moving on wood once it had started to move.		Force = 5.0 N
5	Janell measured the force needed to start two bricks moving on a wooden surface.		Force = 12.5 N
6	Janell turned the brick on its narrow side and measured the force needed to start it moving on wood.		Force = 6.3 N
7	Janell placed straws between the brick and the wooden surface and measured the force needed to keep the brick moving.		Force = 0.70 N

Force and Motion

> **EXPLORATION 8**

Your Turn to Test Friction

Do you still have doubts about Janell's hypotheses? After all, she compared the results of only two or three experiments for each of the hypotheses. Perhaps more proof is needed. It might be valuable to conduct more experiments.

Here is your opportunity to confirm or correct the results of Janell's experiments and to find out whether or not her hypotheses hold true.

What to Do

1. Choose one of Janell's hypotheses to test.
2. Design your own experiment.
3. Decide what equipment you will need.
4. Collect the equipment.
5. Perform the experiment.
6. Record your data.
7. Check your results. Did you perform the test enough times? Were the variables controlled?
8. Decide whether the hypothesis you tested is right or wrong.

More About Friction

By now, you realize that friction is present all around us. From the cars on the highway to the sneakers on a basketball court, friction is a familiar force. Summarized here are some important scientific points about friction.

Friction . . .

- opposes a starting or continuing motion.
- is greater when motion is started than when motion is continued.

Frictional Force

Friction . . .

- is greater for heavier weights.

- varies in size with the kinds of surfaces opposed to one another.
- is reduced when the opposing surfaces are separated by something round.

Now, answer the following questions about friction.

1. What causes friction?
2. What are some ways to reduce friction?
3. What harm can friction cause? Why?
4. Where is friction useful? Why?
5. What substances besides solids can cause friction?

8 Causes and Consequences of Friction

EXPLORATION 9

The Cause of Friction

Is it easy to move one piece of sandpaper over another? No—it's rough going! Turn over the two pieces of sandpaper and move their smooth surfaces over each other. Is it easier? Why is there a difference? The rougher the surfaces, the harder it is to move one over the other. The bumps and irregularities on the surfaces interlock with one another and disrupt motion. This causes friction.

Reducing Friction

If we make rough surfaces smoother, we reduce friction. Cover the rough side of two pieces of sandpaper with margarine or grease. Now rub the surfaces together. Not all the friction is gone, but it is significantly less.

Try rubbing your finger on glass. Then place a drop of oil between your finger and the glass. Glass without oil seems quite smooth. Is there any difference with oil?

Roller bearings

But there is friction even when the smooth sides are rubbed together. If you examine the smoother side with a magnifying glass, you see that it is not perfectly smooth. Test the smoothness of various surfaces listed below with your finger.
- piece of metal
- ordinary paper
- book cover
- window glass
- table top
- floor tile
- wall
- sandpaper

Which surfaces are roughest? smoothest?

Even the smoothest surfaces appear rough when viewed with a microscope. Can you identify the objects shown in the magnified photograph below?

(ballpoint pen on paper)

EXPLORATION 9—CONTINUED

It is very important to oil or grease the moving metal parts of machines. Can you suggest some examples where this is done? Often, metal surfaces which must move or turn against each other are separated by metal balls or rollers. This reduces the friction between the moving parts. Recall that rolling friction is less than sliding friction. Wheels of roller skates, bicycles, motorcycles,

Ball bearings

and cars turn on metal axles that are separated from the wheels by ball or roller bearings. Adding oil to the bearings further reduces the friction.

The Harm of Friction

What harm does friction do? Suppose you rub two pieces of sandpaper together for 30 seconds. What do you observe? Feel how warm the surfaces are. You may also notice that the surfaces have been slightly worn down.

> **Friction**
> - opposes motion
> - produces heat
> - causes surfaces to wear away

Science "Friction"

In the following pages you will read about many examples of friction—some harmful, some helpful. Before you look at these pages, try writing a little science "friction."

Your home has been invaded by a UA (unfriendly alien) armed with an AF (anti-friction) spray gun. You open the front door to enter your house and . . .

Describe what happens in the next few minutes.

Ball bearings in a bicycle wheel

Force and Motion

EXPLORATION 10

Friction Projects

Transportation by Brick

What to Do

1. Move a brick a distance of 10 cm in various ways: dragging only the brick; wrapping it up and dragging it; dragging it on rollers; dragging it on a lubricated surface; dragging it on a roller skate; placing the brick in a large plastic bowl and dragging it over water. Compare the force needed in each of the various methods.
2. Draw conclusions about how friction is decreased in vehicles used for transportation.

Reducing Friction

What to Do

1. Try rotating a book on a table.

2. Place a jar lid over enough marbles to fill the lid. Place the book on top and rotate it.

3. Try rotating one paint can over another. Separate them with glass marbles in the outer ridge of the top of one can. Add oil and observe what happens.

Comparing the Friction of Different Types of Paper

What to Do

1. Set up your equipment as shown in the illustration. Remember, paper and jars must be of equal size.
2. Attach a string to the center of the wooden dowel. Pull gently on the string. Do the two jars move along evenly? Or does one lag behind, causing the dowel to rotate? Why? Where is the friction greater?
3. Add water to one of the jars until the two move along evenly. What does adding water do? How do the frictional forces compare now?
4. How can you use your data to compare the friction between waxed paper and the table, and between the paper towel and the table?
5. Compare the friction between other surfaces. Use various kinds of paper, cardboard, and sandpaper of different grades.

4. Go to a garage, auto salvage shop, bicycle shop, or ball bearing company. Get samples of roller bearings, ball bearings, and sleeve bearings. Find out how each type works. Make a list of devices that use the different types of bearings.

Frictional Force

Friction: Friend or Foe?

Picture Puzzle

In the pictures on this page, identify the surfaces that are rubbing together to cause friction.

1. In which situations is there little friction?
2. In which situations is there a lot of friction?
3. Where is friction harmful?
4. Where is friction helpful?
5. How could you reduce friction in each of the pictures?
6. How could you increase friction in each of the pictures?

234   Force and Motion

Challenge Your Thinking

Do you know the following friction facts? Read through them, and answer the questions about them. Afterward, find two friction facts of your own.

1. A hovercraft travels faster than a boat. Why is this so?
2. Chalk marks differently when its writing end is pushed rather than pulled across the blackboard. In what way are the marks different? Why are they different?
3. Archeological discoveries reveal that the wheel was first made in Assyria in 3000 B.C. How does the wheel reduce problems with friction?
4. A crucial part of a space shuttle is the 33,000 thermal panels for re-entry into the earth's atmosphere. Why are the panels so important?
5. Sliding quickly down a rope may burn your hands. Why?
6. An ancient method of lighting fires was to rub two pieces of wood together. How could this start a fire?
7. Shown here is a road sign used internationally. What does it mean?

Frictional Force 235

GOING THROUGH THE MOTIONS

9 Thought Experiments

Everyone is familiar with the motion of objects, but not everyone has the same explanation for how objects move through space, or how forces are related to their motions. See what you can discover by conducting a survey.

Look at the questions in the survey below, and answer them on your own. Then ask the views of at least two other people. How do your views differ? How are they the same?

A Survey

1. Is a force necessary to start a ball rolling on a level surface?
2. Once the ball is moving, must a force be exerted on the ball to keep it moving?
3. Is a force required to slow the ball down?
4. In the 1500s, Galileo wondered what kept the moon circling about the earth. What explanation for the moon's orbit do you have?
5. Once the Apollo spacecraft got away from the earth on its path toward the moon, the astronauts shut off the rocket engines. Is a force necessary to keep the spacecraft going at over 3000 km/h?
6. The path of the Apollo spacecraft was straight; the path of the moon is circular. How would you explain this difference?

To find out whether or not your answers are correct, think about the following five thought experiments. Remember, scientists regularly perform thought experiments as they sort out their ideas and develop their hypotheses.

236   Force and Motion

Thought Experiment 1

Over 350 years ago, Galileo performed an experiment similar to the one below. For the purpose of his thought experiment, Galileo regarded the surfaces involved in his experiment as being so smooth that there was no friction between them. Of course, there are no surfaces like that in the real world.

(a) When a ball rolls down the slope on the left, we know that it rolls up the slope on the right to the same height—2 m.

(b) The ball does the same thing here in diagram (b), but note how much farther it goes.

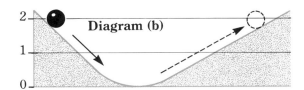

(c) How far would you predict the ball would go in diagram (c)? diagram (d)? diagram (e)? Is any force necessary to keep the ball moving along the horizontal plane (e)?

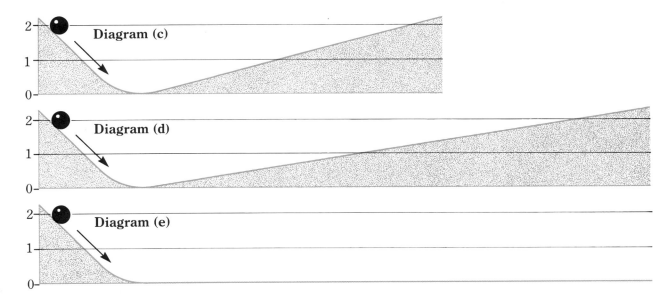

Galileo said that once an object is moving, it tends to keep on moving in a straight line without slowing down or speeding up. But, if a force were placed on this moving object, the force could slow it down, speed it up, or change its direction.

Going Through the Motions

Thought Experiment 2

Suppose Galileo lived in the twentieth century. He might think about the game of bowling. When a player places an initial force on the ball, the ball leaves his or her hand and moves away quickly. It meets an opposing force of friction. Diagrams (a) through (e) show what is likely to happen. Of course, how far the ball goes depends upon the surface on which it is moving.

1. In your Journal, match each diagram with the appropriate surface: grass, a concrete road, a frictionless surface, a bowling alley of indefinite length, living room carpet
2. If there were no opposing force on the moving ball, what would happen to its motion?

Diagram (a)   Surface: ____?____   (20 m)

Diagram (b)   Surface: ____?____   (50 m)

Diagram (c)   Surface: ____?____

Diagram (d)   Surface: ____?____

Diagram (e)   Surface: ____?____

Thought Experiment 3

Recall what you learned earlier about gravitational force.

1. In diagram (a), what stops Bob from throwing the ball higher?
2. Why can he throw the ball so much higher in diagram (b)?
3. What would happen to the ball if Bob threw it off the spacecraft in diagram (c)?

Questions, Part 1

If you have been working through these Thought Experiments carefully, some conclusions should already be clear. Record your answers in your Journal.

Conclusions	Questions
A force is needed to start an object moving.	What force starts the object moving in each Thought Experiment?
An object may keep moving when this force is no longer exerted on the object.	In each Thought Experiment, when was the force that started the object in motion removed from the object?
If there are no forces on the moving object, it will keep moving in a straight line with steady motion.	Which drawings in each Thought Experiment show that a force is needed to change the direction or the speed of motion?
Objects in motion may be slowed down by opposing forces.	What are the opposing forces in each of the Thought Experiments?

(a) Earth's surface 10 m

(b) Moon's surface 60 m

(c) Spacecraft

Going Through the Motions

Thought Experiment 4

1. You are holding a weight, as in diagram (a). Why doesn't the weight fall?
2. In diagram (b), the added weight, called weight B, takes the place of your hand. Weight A is still not moving. Why?
3. What happens to A in diagram (c)? Why?
4. What happens to A in diagram (d)? Why?

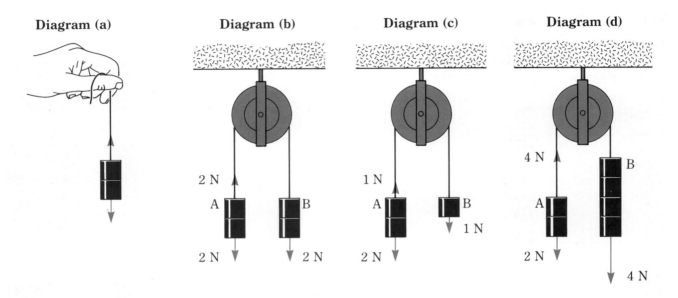

Questions, Part 2

Draw some more conclusions from these Thought Experiments and record them in your Journal.

Conclusions	Questions
If forces on an object at rest are balanced, the object stays at rest. It is just as if there were no forces on the object.	Do you remember Don's attempt to move the refrigerator? Why couldn't Don move it?
An unbalanced force causes an object at rest to start moving.	In which parts of Thought Experiment 4 is there an unbalanced force?

Thought Experiment 5

1. A car is moving at 70 km/h. If the car is allowed to coast freely in a straight line until it stops, what would slow it down?

   Diagram (a) (The solid arrows refer to forces. The dotted arrows refer to the speed of the car.)

2. Now, instead of allowing the car to coast, the motor is kept running. The gas pedal is pressed just enough to maintain the car's speed at 70 km/h. What is the size of the force that the car's motor is providing to move the wheels along the road?

   Diagram (b)

3. Suppose that the speed of the car is steadily increasing. How do you think the force of friction and the force provided by the motor compare? Are the forces on the car balanced or unbalanced?

   Diagram (c)

Questions, Part 3

Here are the final conclusions to be drawn from the Thought Experiments.

Conclusions	Questions
If the forces on an object in motion are balanced, the object will stay at rest or in steady motion in a straight line.	Which diagram in Thought Experiment 5 shows balanced forces?
An unbalanced force causes an object to speed up or slow down.	Which diagrams in Thought Experiment 5 show this? What else can an unbalanced force do to an object in motion?

Going Through the Motions

Puffball: A Game of Force and Motion

Here is a game you can play in class or on your own. As in any game, there are forces at hand. The purpose of Puffball is to blow a light ball into your opponent's goal. The straws cannot touch the ball, and your hands cannot touch the table.

You Will Need

- a playing area of approximately 1 m × 1.5 m
- a barrier about 4 cm high around the edges of the area
- 2 goal areas
- a plastic-foam or table tennis ball
- drinking straws
- 2 players (or teams of 2 or 3 players)

Play for 15 minutes and then write down all you have observed about the effects of two or more forces on the motion of a light ball.

242   Force and Motion

Next, work through the situations on the following pages. A dotted arrow beside the ball indicates the direction in which the ball is moving before Magda or Pauline begins to blow. No dotted arrow means that the ball is at rest. The direction of the straw in Magda's or Pauline's mouth indicates the direction of the force.

Your task is to show the size and direction of the forces. Copy the diagrams into your Journal. Use solid arrows on your diagrams as shown in the examples below. Also, predict whether a point is going to be scored, and if so, who scores it. In a difficult case, *you* may set the conditions and decide the outcome. Two examples have been given to help get you started.

Example 1

Ball is at rest. Pauline is blowing the ball toward Magda's goal.

It will go in. Pauline gets a point.

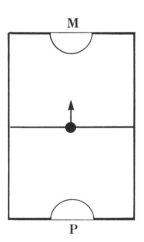

Example 2

Ball is at rest. Magda and Pauline are both exerting a force. Who is exerting the greater force? Where does the ball go? Who scores the point?

You decide that: Magda exerts a greater force. The ball goes in Pauline's goal. Magda scores.

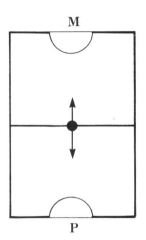

Going Through the Motions   243

The Games

(a) Only Magda is exerting a force.

(b) Both are exerting a force; Pauline's is greater.

(c) Both are exerting an equal force.

(d) Only Pauline is exerting a force.

(e) Only Magda is exerting a force.

(f) Both are exerting a force; Magda's is greater.

(g) Both are exerting an equal force.

(h) Only Pauline is exerting a force.

(i) Both are exerting a force; Magda's is greater.

(500 m)

(j) Both are exerting an equal force.

(k) Neither is exerting a force.

(l) Both are exerting an equal force.

(m) Only Magda is exerting a force.

(n) Neither is exerting a force.

Questions

Check how well you have understood the forces and motion in Puffball by answering the following questions.

In which of situations (a) to (n) . . .

1. are there essentially no forces?
2. are there single forces?
3. are there balanced forces?
4. are there unbalanced forces?
5. is the ball at rest set into motion? Why?
6. does the ball remain in constant motion? Why?
7. does the ball, already in motion, slow down? Why?
8. does the ball, already in motion, speed up? Why?
9. does the ball, already in motion, change direction? Why?
10. does the ball at rest stay at rest? Why?

11 What Is Inertia?

It takes a force to move a body.

When a body is moving, it takes a force to alter its motion.

You have seen that objects at rest tend to stay at rest. Objects in motion tend to continue in motion, maintaining a constant speed in a straight line. This tendency is called **inertia.** Inertia is not a force. It is the state in which an object remains as it is. A measure of an object's inertia is its mass. The greater the mass, the greater the inertia—that is, the greater the force needed to move the object.

Where Have You Met Inertia Before?

You are standing unsupported in a parked bus. The bus starts moving forward. What happens to you? What is inertia doing to you? Would holding onto something help?

At rest

Here you go!

246   Force and Motion

You were at rest, so you tended to stay at rest. Therefore, as the bus moved forward, you tended to stay where you were. The result was that you fell backward.

Now the bus is moving along. You are standing in the aisle unsupported and are keeping your balance quite easily. Suddenly, the bus driver applies the brakes, slowing down and stopping the bus. What effect does inertia have on you?

Stop!

You were in motion. When the bus slowed down, you tended to continue forward. As a result, you fell forward.

Suppose you are sitting in the back seat of a car being driven at 80 km/h. Without warning and with tires squealing, the driver swerves around a sharp corner. What happens to you? In what direction do you tend to go in relation to the car? in relation to the road?

Can you think of other situations in which you have experienced the effects of inertia?

What does inertia feel like? Try devising some experiments to help you "feel" inertia. You may use such devices as a four-wheeled cart, a bicycle, a skateboard, or a see-saw. For example, try sitting on a cart blindfolded. Have a couple of friends pull the cart. Can your inertia tell you when they are starting, speeding up, slowing down, or stopping?

EXPLORATION 11

Experiencing Inertia

The five mini-experiments in this Exploration provide other examples of inertia. At each station you will find materials arranged as shown in the illustrations below.

Spend as much time as you need at each station. Then explain how inertia is working in each case. Record your observations and explanations of inertia in your Journal.

Station 1

Use a ruler to hit the bottom domino sharply toward the barrier. What happens if you hit it gently? Why? Can you reduce the pile completely, one domino at a time?

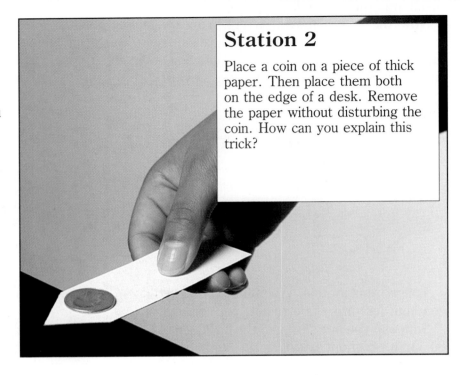

Station 2

Place a coin on a piece of thick paper. Then place them both on the edge of a desk. Remove the paper without disturbing the coin. How can you explain this trick?

Station 3

Which egg is hard-boiled? Spin each egg on a smooth surface. How does inertia help you decide which egg is boiled and which is not?

248   Force and Motion

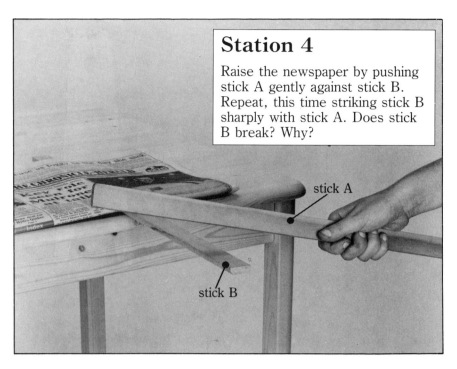

Station 4

Raise the newspaper by pushing stick A gently against stick B. Repeat, this time striking stick B sharply with stick A. Does stick B break? Why?

stick A

stick B

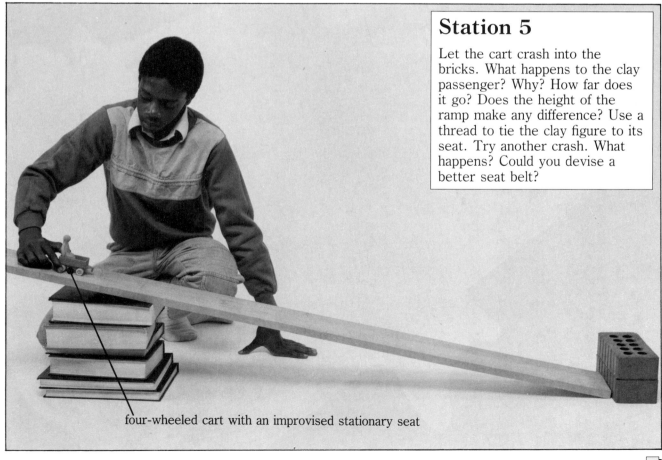

Station 5

Let the cart crash into the bricks. What happens to the clay passenger? Why? How far does it go? Does the height of the ramp make any difference? Use a thread to tie the clay figure to its seat. Try another crash. What happens? Could you devise a better seat belt?

four-wheeled cart with an improvised stationary seat

Going Through the Motions

EXPLORATION 12

Something to Try at Home

Prove this hypothesis: The greater the inertia of an object, the less effect a force has on the object's motion.

You Will Need

- a clothespin
- two pencils of different size
- string
- matches
- a cookie sheet

What to Do

1. You should do this activity with an adult present. Be sure no young children are playing nearby.
2. Tie the ends of the clothespin together with a string.
3. Set up the clothespin and the pencils on an overturned cookie sheet as shown in the drawing.
4. When the pencils are in place and everyone is safely away, burn the thread.
5. Try other objects of differing masses.

Questions

1. What do you observe?
2. Is the force the same on each pencil? Explain.
3. Can you think of some practical examples of this principle?

250 Force and Motion

12 Forces in Pairs

Until now, you have been looking at forces acting on the same object in balanced or unbalanced states. But it is also interesting to consider two forces acting on two different objects at once. Study Contests A, B, and C, and then work through the equations.

Contest A

Compare the size and direction of Deven's and Jill's forces. Do you observe that

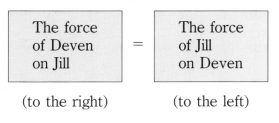

(to the right)   (to the left)

Contest B

Mark and Tomás are having a force meter tug-of-war. Compare the size and direction of their forces. Do you notice that

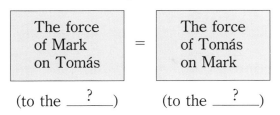

(to the ___?___)   (to the ___?___)

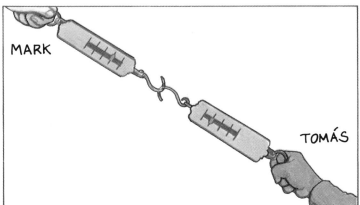

Contest C

Suppose Team Red pulls on Team Blue with a force of 800 N. With what force does Team Blue pull on Team Red?

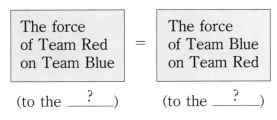

(to the ___?___)   (to the ___?___)

Going Through the Motions

Do All Forces Act in Pairs?

You may recall that in Exploration 2, two strips of plastic were rubbed with a wool sock. They pushed each other apart. P and Q are both equally distant from the dotted vertical line. What does this suggest about the size of the two forces?

Now, use what you learned from Contests A, B, and C to write the "force-pair" equation for P and Q.

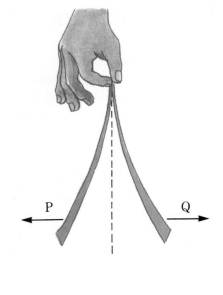

Isaac Newton's Reaction

In addition to his other important discoveries, Newton noticed that forces always occur in pairs. The forces are always equal in size to each other but opposite in direction. He called the forces **action** and **reaction**. Newton's formula states: For every action, there is an equal but opposite reaction.

| Action | = | Reaction |
| (one direction) | = | (opposite direction) |

There is a clearer way to write this:

If A puts a force on B, then

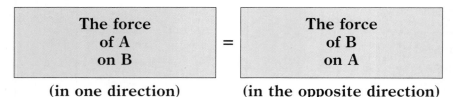

| The force of A on B | = | The force of B on A |
| (in one direction) | | (in the opposite direction) |

More Action-Reaction Pairs
Example 1

The block puts a force on the rubber band, stretching it downward. The rubber band puts a force on the block, holding it upward. Draw the force-pair equation.

Another pair of (noncontact) forces is also at work in this case:

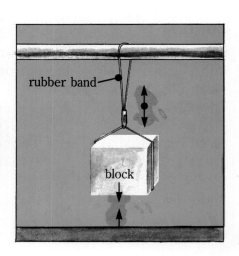

| The gravitational force of the earth on the block | = | The gravitational force of the block on the earth |
| (earth pulling the block downward) | | (the block pulling the earth upward) |

252 Force and Motion

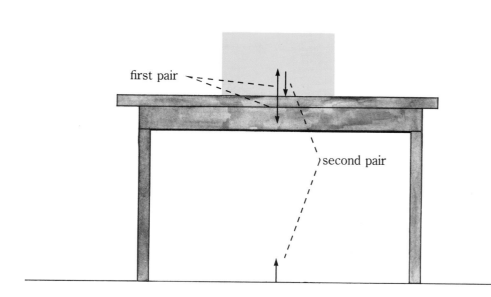

Example 2

What is the force-pair equation for each pair?

First pair: the block on the table

Second pair: the table on the earth

Example 3

Pat asked his friends to answer this question.

Here are their results. What is the force-pair equation?

Going Through the Motions

EXPLORATION 13

Making Use of Reaction Force

The four mini-experiments in this Exploration show various uses of reaction force. At each station you will find materials set up as shown in the illustrations below.

In each situation, watch what happens. Identify the action-reaction pair of forces. What is each force doing? In your Journal, make a sketch, drawing in the appropriate forces. Write the force-pair equation for each situation.

Station 1

Suspended test tube
Cork pushed firmly into test tube
candle

Station 2

Fill the water bottle one-third full with water. Break up a seltzer tablet and put it in the bottle. Insert the cork firmly. Wrap the bottle in paper towels. What happens next could be messy, so have more paper towels handy.

straws or wooden dowels

Station 3

Stand on roller skates, holding a bag of sand. Throw the bag ahead of you.

Be Careful: *Use caution when standing on roller skates.*

Station 4

Blow up a balloon. Release it. What happens? How far did it go? Where did it go? Does the shape, size, or amount of air in the balloon matter? Explain.

254 Force and Motion

Challenge Your Thinking

1. **Reaction Please!**
   Look at each situation below and identify the action and reaction forces involved.

2. Are forces, motions, and inertia found only in nonliving systems? Where do you find physical actions and reactions in your own body? Try doing a jumping-jack. Did you experience inertia? feel the force of gravity? cause an object to start moving? Draw a diagram showing direction of force in a person doing jumping-jacks.

Going Through the Motions 255

Unit 4 Making Connections

The Big Ideas

In your Journal, write a summary of this unit, using the following questions as a guide.

- What characteristics do all forces share?
- What is a gravitational force and what does it do?
- How can you distinguish the mass of an object from its weight?
- What devices may be used to measure forces?
- What units are used to measure mass? force? weight?
- What does frictional force do? How can you increase it? decrease it?
- What happens to an object (at rest or in motion) if balanced forces are at work on it? unbalanced forces?
- What does an action-reaction equation represent?
- What is inertia? How does it relate to forces?

Checking Your Understanding

1. You began to learn about forces in the story of the sailboat race on page 194. Read the story again, and identify the following:

    (a) frictional forces (Name the agent, receiver, and effect.)//
    (b) gravitational forces (Name the agent, receiver, and effect.)
    (c) buoyant forces (Name the agent, receiver, and effect.)
    (d) balanced forces
    (e) unbalanced forces
    (f) examples of inertia
    (g) pairs of action-reaction forces

2. One of the largest tug-of-war pulls in the world was performed by 1015 high school students in 1976.

    (a) If each student literally pulled his or her own weight, what force did the rope have to endure? (Estimate the average weight of a student.)
    (b) Can you identify an action-reaction force pair at work? What is its equation?

256 Force and Motion

3. Jenny, after experiencing free-fall, opens her parachute and drifts steadily downward until her feet hit the ground. She is carried along for several meters before coming to a complete stop.
   (a) In each stage of her fall, describe what forces are acting on Jenny. Give the directions of these forces.
   (b) At what points are there unbalanced forces on Jenny? balanced forces?
   (c) When is Jenny most aware of inertia?
4. Draw a concept map showing the relationship between these words: *newtons, mass, matter, weight,* and *grams.*

Reading *Plus*

Building on what you know about force and motion, you can discover more about describing motion, the laws of motion, and the relationship of force to work and power. Read pages S61–S76 in the *SourceBook* to learn about the physical laws of force, motion, work, and power.

Updating Your Journal

Reread the paragraphs you wrote for this unit's "For Your Journal." Then rewrite them to reflect what you have learned in studying this unit.

Making Connections

Science in Action

Spotlight on Electrical Engineering

Electrical Engineering Technology

Craig Gamble is an electrical engineering technologist for a company that supplies electrical energy. He earned the qualification for this position by completing a two-year program in electrical engineering technology after graduation from high school.

Q: What do you do?

Craig: I design the systems that carry electricity to homes and other buildings.

Q: What do you find most interesting about your work?

Craig: I think the best thing is seeing a job through, from the beginning stages to its completion. I follow a job through budgeting, design, and completion in the field.

Q: Is the science you studied in high school important in relation to your work now?

Craig: Most of my technical knowledge, aside from on-the-job experience, I gained while training to be a technologist. But basic ideas of mathematics and physics—particularly electricity—I learned in high school. Especially important in my present work is understanding the transmission of electrical energy and how high voltages carried by transmission lines are "stepped down" to the lower voltages required in homes and industry.

Q: Do you tend to do your work alone, or does your work involve cooperation with others?

Craig: I am very much involved with other people in my work. In fact, I would say that the ability to work with other people is very important. First of all, I usually consult colleagues in my own department about my new designs. Then having done the design work, I work with several other departments, some concerned with the technical aspects, others concerned with customer service. Then I work with the builders of the system, and then with its operators.

Research and Training

Wagih Fam is a professor of electrical engineering and the head of the Electrical Engineering Department at a university.

Q: What do you do, and what do you find most interesting about your work?

Wagih: My working life can be divided into two parts: teaching and research. As far as teaching is concerned, the joy lies in giving knowledge to my students. It is particularly rewarding to discover that they have really understood what I have said and are able to use and apply this new knowledge.

As for my research, I find it very exciting to obtain new results and new knowledge. I spent three years on my Ph.D. project. I studied and measured the flow of electrical and magnetic forces in the air gaps of electric machines.

When the method you apply in researching or in designing a project succeeds, it is a great joy. When it fails, you experience great frustration. But the most stimulating thing, of course, is obtaining new knowledge and passing it to other people through your publications.

Q: What activities as a young person eventually led you into engineering?

Wagih: In my earlier days, I had an interest in building toys. My brother and I used to have a lot of fun making our own kites. We didn't go to a store and buy a kite and fly it. We had to make them ourselves and then make them fly. Of course, sometimes we would meet with disappointment. An accident, like punching through the paper or cutting the tail off, might happen. The kites might have to be rebuilt. We would try to make them better, always improving on the previous design.

In my final year in school, I started building my own electromagnets. With one of these, I built a doorbell. Later on, my brother and I started to wind transformers and motors, and to repair radio receivers.

Aside from these things, I was interested in painting, drawing, and drafting. That interest gave me an edge later on when I went into engineering and studied engineering drawing.

Some Project Ideas

1. Build a device that operates with power from a solar cell.
2. Construct a windmill generator that converts the energy from the wind into electrical energy. Use a small windmill directly coupled to a small DC generator. Use the electrical power output from the generator to light up a few flashlight bulbs or to drive a small electric motor.

For Assistance

Advice and assistance can be obtained from electrical engineers, electronics technologists, and electronics teachers. Most of the materials needed can be purchased at an electronics supply store or through a scientific supply house. Your teacher can give you an address for one of these. Other materials may be found at home.

Science in Action

Science and Technology

Rocket Science

Rockets were used by Chinese armies in A.D. 1232. How effective they were as weapons we do not know for sure, but the use of rockets for military purposes spread throughout the world. In the War of 1812, for example, rockets carrying explosives were used by the British against the United States—which is what gave us the words "the rockets' red glare" in our national anthem. And certainly, we are familiar with the terrible destructive power of rocket-launched warheads in modern warfare.

The majority of rockets in the world today are still military. But the largest rockets, by far, are the ones used for space technology. It takes tremendous amounts of power to send loads into space.

The earliest rockets were fireworks.

The Saturn V launch vehicle was used to carry Apollo spacecraft to the moon.

Measuring Rocket Power

The power of rocket engines is described in terms of the amount of force, or thrust, they produce. For a rocket to get off the ground, its thrust must be greater than the weight of the rocket itself, including its fuel. Otherwise, it would just sit there.

The most powerful rocket of them all was the Saturn V, which carried astronauts to the moon. The cluster of five engines in its first stage had a combined thrust of over 33,000,000 newtons. By comparison, the engines or *boosters* that lift today's space shuttle missions off the ground produce a combined thrust of 25,800,000 newtons.

260 Force and Motion

The Fuel Dilemma

To get heavier loads up into space, you must use more fuel to produce more thrust. But when you add more fuel, you also add more weight to be carried—an enormous amount of it, as it turns out. A major part of the work that rocket fuel must do is to carry its own weight, as well as the weight of all the tanks and equipment that hold and transfer the fuel. In fact, the weight of the fuel in a vehicle like Saturn V is much greater than the weight of the *payload*—the object that is actually sent into space.

Getting up to Speed

For space exploration, it's not enough just to get a load off the ground. It must be brought up to a high velocity. In the case of a mission to the moon or the planets, a spacecraft must reach a velocity known as the escape velocity. For an object leaving the earth, this velocity is 40,000 km per hour. If escape velocity is not reached, a spacecraft will start to slow down the moment its engines are shut off and will eventually stop. Then it will start to fall back to Earth.

Mars lander

The Long Haul

In designing a rocket engine, you would also have to consider how long the trip will last. In the case of a mission with a human crew, a time of 10 years might be considered entirely too long! And remember—the longer the mission, the greater the amount of supplies you would need to take along. So, for a mission to be practical, you might have to achieve a velocity even greater than the escape velocity. But greater velocity requires more fuel, and this adds more weight to the rocket. Obviously, the engineering of a rocket for a space mission requires a lot of thought and planning.

Find Out for Yourself

Suppose you are a part of a team of scientists who are designing a mission to Mars. You want to land on Mars, explore the planet, and then return to the earth. What questions would you raise about the different forces and velocities involved? Do some research and try to find answers to your own questions.

Space technology is not cheap. The cost of building a space shuttle is over 2 billion dollars, and the average cost of a flight is over 200 million dollars! How much of the cost comes from taxpayers? Do you think we as a nation are sacrificing other things in order to pay for our space program? Do space efforts pay for themselves in benefits? Would you be in favor of sending missions, or a series of them, to the moon or Mars? Why?

Science in Technology

Unit 5
Structures and Design

THE SCIENCE OF STRUCTURE
1. Planning the Structure, p. 264
2. Response of Structures to Force, p. 272
3. Engineering Structures, p. 282

THE ART OF DESIGN
4. Beauty in Structures, p. 293
5. A Question of the Environment, p. 304

For Your Journal

Write a paragraph summarizing your thoughts on each of the following:
1. What is the difference between structure and design?
2. What features does this tower have that keep it standing?
3. How do beauty, the environment, and buildings relate to one another?
4. How does the size of a building affect its structure?

THE SCIENCE OF STRUCTURE

Planning the Structure

Meeting of the Council

The mayor of Zenith could not be heard over the noise in the Chamber of the City Council.

"Come to order! Come to order!" Mayor Jones pounded the gavel at his desk. The citizens in the gallery and the council members seated in the chamber continued arguing.

Seldom had there been an issue that stirred up so much debate among the citizens of Zenith. ABC Developers were seeking the approval of the Zenith City Council to redevelop the old neighborhood of Riverwood across the river from the city's business center.

"Fellow council members, fellow citizens," the mayor appealed. "We will not solve this problem by shouting at one another. Please, everyone will have a chance to speak. Council Member Kowalski, you have the floor."

Council Member Kowalski: I have received many letters and phone calls from residents and businesses in Riverwood. They support the proposal. With the money ABC has offered them, they will be able to relocate in other parts of the city. The proposal will result in many new jobs involving the construction of new buildings. The city of Zenith will benefit greatly from this proposal.

Council Member Casaro: Mr. Mayor, not all the residents and businesses in Riverwood support ABC's proposal. I present to the Council a signed petition from over 2000 residents and businesses opposing this proposal. I also present a letter from the Zenith Heritage Foundation that opposes the proposal as it now stands. They write that "the proposal would mean the destruction of many historic homes and buildings. The new buildings would obstruct the view of the beautiful mountains that surround our city."

264 Structures and Design

The present Riverwood

ABC Developers' proposed redevelopment of Riverwood. The original homes and businesses are gone.

The Science of Structure

Council Member Tufts: Mr. Mayor, Riverwood is an eyesore. The homes and businesses have fallen into disrepair. The woods and marsh are littered with garbage and debris. And the woods are becoming a hazard, especially at night. Something must be done with this area.

Mayor Jones: Many civic organizations have asked for the opportunity to speak to the Council about ABC's proposal. I call upon the President of the Environmentalists' Association, Mónica Rodriguez.

Mrs. Rodriguez: Mr. Mayor and Council Members, our organization is concerned about the proposed destruction of the existing marsh and woods. This area is a home for birds and other wildlife and a place where children from all the surrounding neighborhoods come to play. We should preserve this area for our children and for the future.

After much debate, the Mayor declared, "The future of this city belongs to the youth. Let's find out what they would like to do. I suggest we ask all the seventh, eighth, and ninth grade science classes in the city to send their proposals to the Council. Then we will invite the people whose homes and businesses might be affected to comment. The Council will make the final decision." It was agreed by the Council to proceed in this way.

Your Opinion

Suppose your class was asked to find the best solution to this problem. With your classmates, decide what you think should be done with Riverwood. Have one person record the decisions of your class. Later you will refer back to these decisions.

1. Would you have all the existing structures torn down?
2. Would you alter the environment to make more space for the new buildings? Explain why.
3. What new structures would you like to see built in this area?
4. Will the town benefit the most from a new business center or a historic neighborhood? Why do you think so?

An At-Home Assignment

As you learn about structures and design, complete your own plan for Riverwood. This plan should be in the form of a detailed sketch like the one from ABC Developers. Later your class will decide whose plan is best, and together with your classmates, you can construct a model based on this plan.

266 Structures and Design

Structures That Failed

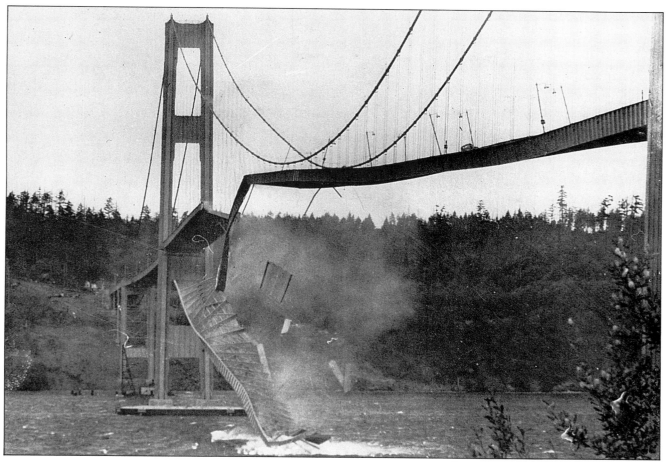

The Tacoma Narrows Bridge

Before you design new structures for Riverwood, you must ensure that your structures will not fall down. To build sturdy structures that are not likely to fall, engineers carefully study those structures that have failed. When the Tacoma Narrows Bridge collapsed in 1940, engineers around the world took notice. The engineers who investigated the collapse asked themselves:

1. Why did the bridge fail?
2. Was it poorly constructed?
3. Were the materials used inferior?
4. What were the forces that caused the bridge to collapse?
5. What changes should be made in bridge design and construction to prevent catastrophes of this kind in the future?

Answer these questions. Get together with two or three of your classmates and record your best answers in your Journal. Be sure to consider all of the forces affecting the bridge.

Picture Puzzles

What can you learn from structures that have failed? Study each of the pictures carefully. For each situation, try to answer the following questions:

1. What was the failure?
2. What might have caused the failure? Could it have been the wind, the weight of snow, the weight of the structure itself, an earthquake, or something else?
3. Was the structure being pulled apart (stretched) or being pushed together (compressed) before the failure?
4. How could the failure have been prevented?

Emily drew a concept map to analyze the possible causes of the Tacoma Narrows Bridge collapse. Perhaps you could use a similar diagram to answer the questions above.

Stone masonry beams supported by tall columns are a common sight among the ruins of ancient Greek civilization. Notice the large cracks in the beams.

A badly-shattered house in Gonen, Turkey. What might have caused this destruction?

Twisted and broken bones as seen by an X ray.

Structures That Haven't Failed Yet

Much can be learned about how to build safe, strong, and beautiful structures by examining those that have not failed yet. Nature and humans show their creativity in the variety and strength of the structures that they build. Study the following 10 pictures of structures that have been successful over the years.

In small groups discuss for each structure:

(a) What is it designed to do?

(b) What holds it up?

(c) What holds it together?

(d) What could cause it to fail?

Keep a record of your discussion in a table such as the one shown below.

Structure	Designed for	Held up by	Held together by	Possible causes of failure
spider's web	catching and holding insects	threads to a tree	a network of strong threads	—a very heavy insect —strong wind —heavy rain —? —?
Leaning Tower of Pisa, Italy				

A freighter

A spider's web

▲Leaning Tower of Pisa, Italy

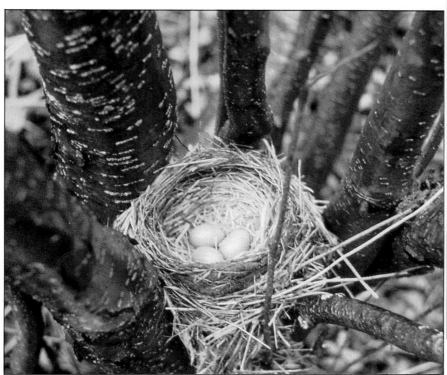

◄A bird's nest with eggs

The Science of Structure

A bat

Hot air balloons

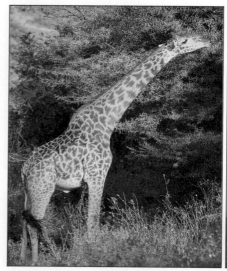
A giraffe

The "Gossamer Albatross," a human-powered aircraft

A sunflower

Hoover Dam, Colorado River

Constructing on Your Own

You have examined some structures that failed and others that did not. Can you design a strong, safe structure and build a model of it?

Suppose you decide to include a bridge in your plan for Riverwood. Perhaps you want the bridge to carry automobile traffic across the river between Riverwood and the business center of the city. Or maybe you want a footbridge to cross the inlet. In either case, your bridge must be strong enough and safe enough for those who will use it.

EXPLORATION 1

Constructing a Model

You cannot design and build the actual bridge, but you can build a model. Try to build the strongest model possible.

You Will Need

- 4 standard index cards, each approximately 7.6 cm × 12.5 cm
- a 20-cm length of narrow masking tape
- scissors
- a ruler

What to Do

1. As shown in the sketch below, your model must have the following dimensions. It must cross a gap of 11 cm. To do so, it will have to be at least 12 cm long. The width of the card should be 6 cm or more.

2. If supporting structures are used, they must leave an unobstructed gap to allow for boat traffic.

3. Your model may rest on books, but it cannot be taped to them.

4. You are allowed to bend, fold, and cut the index cards.

Now comes the fun part! Can you place a mass of 50 g on your bridge without making it collapse? Congratulations! Now continue to add masses to find out which design holds the greatest load. Who had the best design? Why was it the best?

Writing About Your Model

A toy company is requesting suggestions for the construction of cheap, easy-to-make model bridges. You decide to submit your design. Write a report to Trixie's Toy Company that includes:

(a) a design that works
(b) the steps in making the toy bridge
(c) the qualities that make it a good design
(d) the pitfalls to avoid in building such toy bridges

Response of Structures to Force

The designers of Riverwood will need to know whether the structures they design will stand up. Will the structures support their own weight? Can they support a load? How will the structures behave in the wind or during an earthquake? Also, the designers will have to convince the municipal authorities that the structures will be safe. To do this, they need to use the right words to communicate their ideas.

Consider your model bridge. What technical words did you use in describing your model bridge to Trixie's Toy Company?

Here is a list of words that you can use to describe the response of structures to force. Keep a list of the terms in your Journal. As you learn the meaning of each of these words, write down the definition. Remember to use what you discover in the Explorations to find the best definitions for the terms.

bending	load
compressing	relaxing
compression	response (respond)
compressive force	shear
contracting	shear force
contraction	stretching
deflection	tensile force
extension	tension

EXPLORATION 2

Three Kinds of Force: Tensile, Compressive, and Shear

You Will Need
- dull scissors
- heavy paper
- modeling clay

Part 1

1. Study the illustrations of the three structures in Diagram A: a beam, a rope, and a column. Do you observe a force on each? How does the structure respond to the force? How does each structure change in shape? For each case, describe both the force and the response of the structure to the force.

2. In Diagram B, the three structures have been redrawn. The agents of force have been replaced by arrows. The arrow indicates the direction of the force. Draw Diagram B in your Journal and correctly label the arrows with one of the following terms:

   tensile force—a force that causes stretching

   compressive force—a force that causes compression

   shear force—a force that causes bending or twisting

The Science of Structure 273

EXPLORATION 2—CONTINUED

Part 2

1. To get a better idea of shear force, take a pair of dull scissors and slowly begin to cut a piece of heavy paper. Notice how the paper bends before breaking. When a force bends or twists an object, you know that the object is experiencing a shear force. In the drawing on the right, the arrows represent the shear force of the scissors' blades.

2. Take modeling clay and form it into a small rectangular block. Grasp the top of the block between two fingers. With the other hand, use two fingers to grab the bottom of the block. Now, push the top half and the bottom half in opposite directions. You are exerting a shear force on the modeling clay.

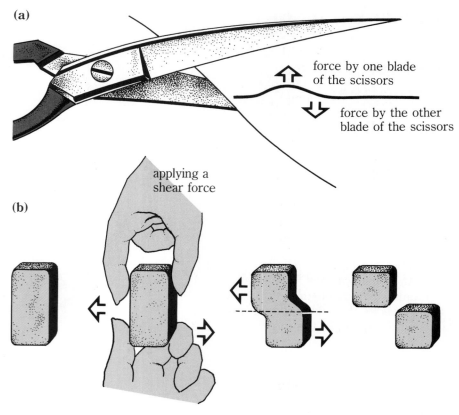

Part 3

The causes of failure in structures are forces that cause tension, compression, and shear. Look again at pages 267 and 268, and identify which of these forces were responsible for the failure of each structure. For example, the Tacoma Narrows Bridge collapsed from shear force caused by wind. Do you see where the bridge twisted and where shear force affected the roadway?

Part 4

Look at the structures in Diagram A on page 273. Read the statements that follow and answer the questions:

1. The beam is under a load, producing both tension and compression. It responds by bending. If the load is removed, how does the beam respond?

2. The rope is also under tension. It responds by stretching. If the force is removed, how does the rope respond?

3. A load is placed on the column. It responds by compressing. If the material of the column behaves like a rubber ball, how will it respond when the load is removed?

You may find that it was easy to answer these questions. However, the answer to the following question may not be so obvious.

4. If there is no weight placed on the column, does it experience any load? Imagine slicing the column in two. What happens if you remove the top half of the column? How does the bottom half respond?

Structures do not only react to external forces; they also react to their own weight. They must be strong enough not only to withstand external forces, but to support their own weight.

274 **Structures and Design**

EXPLORATION 3

Elastic Responses

In the unit "Force and Motion," you made a force meter using a rubber band. As the force increased, so did the length of the rubber band. The stretching was the response of the rubber band to the tensile force on it. This stretching is said to be an *elastic response*. Why? Because when the force was removed, the rubber band relaxed and returned to its original length. Most materials used in construction respond in the same way.

Activity 1 of this Exploration is a case study, and Activity 2 is an experiment you can perform yourself.

Activity 1
Stretching Steel— A Case Study

Does steel really stretch when a strong force is applied to it? Naomi knew she could increase the tension on the strings of her guitar by turning the pegs. But did the steel strings actually get longer? Naomi went to a mechanical engineering lab where they had an apparatus to measure the change in length of steel wire. Here are Naomi's results for stretching a 1-m long steel wire with a thickness of 1.0 mm.

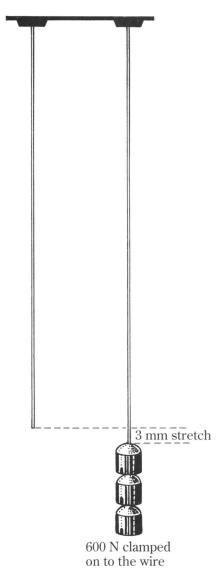

Simulation of the apparatus

Tensile (Stretching) Force (N)	Stretch (mm)
0	0
200	1
400	2
600	3
800	4
1000	5
1200	With this force, the steel string broke.

1. How much would the steel wire stretch when the force pulling it is:
   (a) 100 N?   (c) 300 N?
   (b) 500 N?   (d) 900 N?
2. Suppose a 10-m steel wire of the same thickness as the guitar string is used to tug a heavily loaded canoe behind a motor boat.
   (a) If the boat pulls with a steady force of 200 N,
      (i) how much will 1 m of the wire stretch? (See the results of Naomi's experiment. She used a wire that was 1 m long.)
      (ii) how much would you expect the 10-m wire to stretch?
   (b) If the motor boat suddenly yanked the wire with a force of 1500 N, what would happen?

EXPLORATION 3—CONTINUED

Activity 2
Bending Wood

A common example of an elastic response is the motion of a diving board. The diver uses the elastic response of the board to spring into the air.

Your purpose in this Activity is to compare the amount of bending, or **deflection**, with the different forces exerted on the end of a model diving board.

You Will Need

- a piece of wood (like a meter stick)
- a meter stick for measuring
- a clamp or other means to fasten the wood to the end of a bench or desk
- masses (or use your force meter for applying known forces)
- string
- scissors

What to Do

1. Set up the apparatus as shown in the diagram.
2. Hang a mass on the free end of the stick. Measure the amount that the end deflects. Repeat for several different masses.
3. Record your results. See how Doug, an eighth grade science student, recorded his results.

I used a meter stick made of spruce wood. It was 2.5 cm wide and 5 mm thick. I fastened the meter stick to the edge of a table, leaving 90 cm free to bend. Before putting weights on the end of the meter stick, I first measured the height of the meter stick above the floor. I found the height to be 78.3 cm. Then I measured the height with a weight fastened onto the end of the meter stick. The difference in the heights showed me how much the meter stick deflected. I recorded this in millimeters so it would be easier to graph.

For weights, I suspended metal blocks that had a mass of 200 g each. I remembered from studying forces and motion earlier that a mass of 100 g weighed about 1 N.

Weight (N)	Height of free end from floor before adding weight (cm)	Height after adding weight (cm)	Deflection (mm)
0	78.3	78.3	0
2	78.3	76.9	14
4	78.3	75.6	27
6	78.3	74.1	42
8	78.3	72.8	55
10	78.3	71.4	69

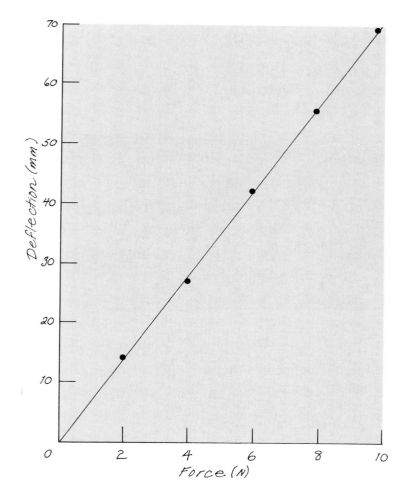

Using the Language of Tension, Compression, and Shear

Did you find definitions for most of the words at the beginning of this lesson? Use as many of those words as you can by rewriting your report to Trixie's Toy Company.

Is it easier to describe your model bridge? Using the language of engineers helps you describe structure and force precisely. By mastering scientific language, you can begin to write and think like an engineer!

4. Now graph your results. Doug let the vertical axis represent deflection and the horizontal axis represent force (weight). After plotting his points, he drew a line through them. Plot your data and draw your line. Is it a straight line or a curved one?

5. What do you conclude about the relationship between force and deflection?

6. If the length of the meter stick over the edge of the table was shorter, what difference in results would you expect? Test your prediction by repeating the experiment, but this time clamp the stick at a different position. Plot the new data on the same graph. What do you conclude?

The Science of Structure 277

Strong Materials

The designers of Riverwood need to know which materials are best for building the structures they design. The materials they select must be strong, but not too expensive. This is a problem you will face many times in your life. Perhaps you will need to select the right fabric to make your own clothes or to upholster furniture. You may want to repair a skateboard or build bookshelves.

Consider the spider's problem.

Which material do you think is the strongest: copper wire, nylon fishing line, spider's web, or wood?

What is meant by strength? Tensile strength is defined as the force needed to pull materials apart. A cord or string made out of any of these materials has its own tensile strength. If you pull hard enough on the ends of each cord, the cords will break. Remember our example of the steel wire. It took a 1200-N force to pull the wire apart. Which of the materials mentioned above will break with the least amount of force? the most?

Suppose you have long, thread-like pieces of each material. Suppose also that each "thread" has the same thickness of 1 mm. Does this change your answer? Which is the easiest to pull apart? Which is the most difficult? What is your prediction? Choose the list below that you think places the materials in order from strongest to weakest.

(a) copper, wood, nylon, spider's web

(b) wood, nylon, copper, spider's web

(c) nylon, spider's web, copper, wood

(d) copper, nylon, spider's web, wood

Here are the facts! After many tests it has been found that it takes 820 N to break the nylon line; 190 N to break the spider's web; 110 N to break the copper wire; and 81 N to break the wood. Aren't these results surprising? How do they compare with your prediction? If you thought differently, remember that the thickness of these materials plays a role in judging their strength. For example, most spider's webs that we see have threads less than 1 mm thick.

Strong Shapes

Some materials are stronger than others. But the strength of any material can be increased by changing its shape. Recall the bridges you built using index cards. What shapes had the most resistance to bending? Did you discover that a thin, flat card has comparatively little strength to support loads that cause it to bend?

The following Exploration should give you many ideas for building models of structures for Riverwood.

EXPLORATION 4

Testing for Strong Shapes

In this Exploration you will test the strength of five different shapes made out of different materials. Work with a small group, and then share your discoveries with the rest of your class.

You Will Need

- a force meter from the unit "Force and Motion," or a set of standard masses (10-g or 20-g intervals)
- a ruler
- index cards for Investigations 1 through 4
- masking tape for Investigations 1, 4, and 5
- paste or glue for Investigations 2 and 3
- heavy paper, a small pail filled with sand, and a hard board (approximately 10 cm × 10 cm) for Investigation 5

Investigation 1

Shapes Unlimited

1. Use only a single index card to make each of the five shapes shown to the right. You may use a small amount of tape to fasten the edges of the index card for shapes d and e.
2. Predict which of the five shapes you think would have the most strength to resist bending.

3. With a force meter or masses, apply force to the middle of each shape until it no longer returns to its original position when the force is removed. This happens when a permanent crease or fold is caused by the force.

(a)

(b)

(c)

(d)

(e)

4. Make and test at least one shape of your own design.
5. Record your results in a table in your Journal. In the left column of your table, sketch each of the shapes. In the right column, give the bending force that was required to cause the shape to buckle.
6. Which of the five shapes was the strongest? Was your prediction correct? How did the shape you designed compare with these five?

The Science of Structure 279

EXPLORATION 4—CONTINUED

Investigation 2
Lamination

1. Make a table like the following to record your results.

Number of cards laminated together	Amount of sag (mm)

2. Arrange the books and index card as shown in the illustration above.
3. Place a standard mass on a single card so that it sags about 2 or 3 cm. Measure the sag.
4. Laminate (glue together) two cards so they have doubled in thickness. Using the same standard mass, measure the sag.
5. Repeat laminations for triple thickness and quadruple thickness. Measure the sag.
6. Make a graph similar to the one shown above and plot the sag against the number of cards laminated.
7. Extend your graph so that you can predict the amount of sag when six cards are laminated.
8. Test your prediction.

Investigation 3
Corrugation

1. Laminate three index cards together.

2. Find the force the laminated structure will sustain before sagging 1 cm.
3. Take another three cards and fold the middle card as shown in the diagram. Glue the cards together.

4. Find the force the corrugated structure will sustain before sagging 1 cm.
5. What do you conclude?

Investigation 4
Tubular Girders

1. Roll an index card into a tubular girder. Hold it together with small pieces of masking tape.
2. Measure the diameter of the tube.
3. Find the force required to cause the girder to buckle.
4. Using the same amount of material and keeping the tube the same length, vary the diameter. Repeat for a total of five different diameters. Record your results in a table like the following:

Diameter	Maximum force

5. Make a graph of your findings (maximum force versus diameter).
6. What do you conclude?

Investigation 5
Tubular Columns

1. Using heavy paper, make a tube to be used as a column.
2. Measure the diameter.
3. Construct a platform as shown. Place the pail over the center of the column as shown in the diagram.
4. Find the force required to buckle the column. You can do this by filling the pail with sand until the column buckles. Weigh the platform, pail, and sand together. This gives you the force required to buckle the column.
5. Repeat two more times, using the same amount of material for the column and keeping it the same height, but varying its diameter. Record your results in a table like the following:

Diameter	Maximum force

6. What do you conclude?
7. Instead of placing the pail over the center of the column, try placing the pail at different locations on the column to find out the effect of off-center loading.

The Science of Structure

3 Engineering Structures

Many human-made structures are quite complex. However, they are made up of simpler components that give them their strength and shape. You can include these components in the structures you plan for Riverwood.

Beams and Trusses

Where have you seen *beams* before? What are they made of? Beams can be made of a variety of natural and artificial materials. Have you ever seen a *truss*? A truss is a framework of connected planks or steel bars that add strength and support to a structure. Trusses are frequently found in the roofs of houses and other buildings where long distances are spanned without columns.

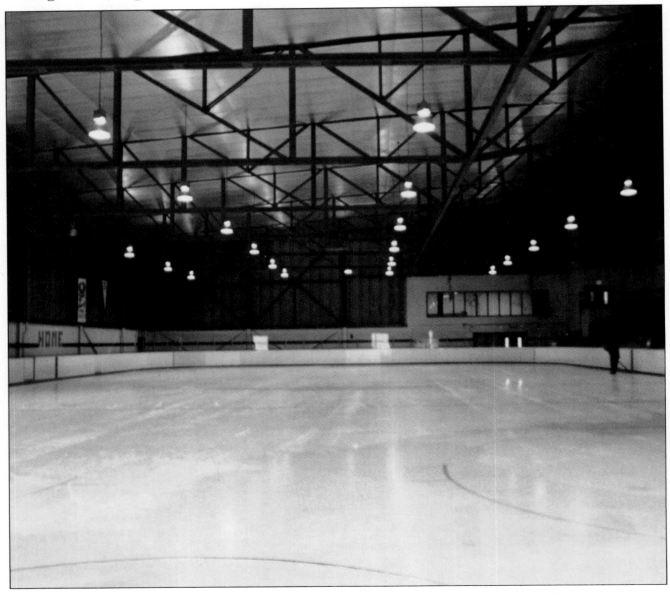

EXPLORATION 5

Activity 1
Comparing a Simple Beam with a Truss

You Will Need
- a board with one hole drilled at each end (the board can be a thin plywood strip, about 10 to 12 cm wide by 30 cm long or a similar size strip of corrugated cardboard)
- a short wooden post or piece of cardboard (10 to 15 cm long)
- standard masses
- supports (blocks or books)
- string
- 2 eye-screws

What to Do
1. Set up a beam as shown in the diagram below. Put a load of standard masses on the middle of the strip so that it sags noticeably.

2. Convert the beam into a truss, as shown in the diagram. The screws fit through the holes in the beam, and the string should be fastened to the screws. Allow the short post to sit on top of the string.

   Place the same load on it as you did before. What do you conclude?

Truss

Activity 2
The I-Beam

You Will Need
- 6 boards, all the same size (boards may be strips of plywood or corrugated cardboard)
- nails
- a hammer
- tacks or glue
- standard masses
- supports (blocks or books)

What to Do
1. Nail or fasten three boards together as shown below in diagram (a).
2. Nail or fasten three boards together in the shape of an I-beam as shown in diagram (b).
3. Put a load on the middle of the beam shown in diagram (a) so that the beam sags measurably.
4. Now place the same load on the I-beam.
5. What do you observe? Which is stronger, the I-beam or a simple beam made from the same amount of material?

Why do you think one beam is stronger than the other?

The Science of Structure

Cantilever

A *cantilever* is a simple beam that is fixed at one end. Here are some examples of cantilevers. Explain why each of them fits the definition of a cantilever.

(a)

(b)

(c)

Show the difference between a cantilever and a simple beam by drawing each one supporting a person.

A Cantilever Bridge

After single-beam bridges, cantilever bridges were probably the next kind to be built. It is thought that the first cantilever bridge was built in China. Diagram (a) illustrates how they were built. Diagram (b) shows the basic principle of how they work.

Identify in diagram (b):

1. the two cantilevers
2. how the cantilevers are supported
3. the load that the cantilevers are supporting

Explain in your own words how cantilevers are able to support heavy loads without collapsing. How does the cantilever bridge in diagram (a) support its load?

(a)

(b)

Arches and Domes

By now you know there are many ways to support a load. You can support loads by using columns and beams. You can shape your beams in various ways, forming them into I-beams, for example. You can also transform a beam into a truss. And you can make a beam into a cantilever by attaching it at one end for support.

284 Structures and Design

Another way to support a load is by making an *arch*. Its widespread use represented one of the most important advances made in the earliest civilizations. By the time of the Roman Empire, arch bridges and arch aqueducts (for carrying water) had become the major means of spanning distances such as valleys and rivers.

The *dome* is closely associated with the arch. A dome is basically a three-dimensional arch.

The domed cathedral Hagia Sophia in Constantinople (now Istanbul, Turkey) was the cultural center of the Byzantine Empire.

An igloo is another type of dome. Can you suggest what supports an igloo? Exploration 6 will help you answer this question.

The Science of Structure 285

EXPLORATION 6

Comparing Beams and Arches

In this Exploration you will learn about the advantages of using an arch for spanning distances.

You Will Need

- index cards
- standard masses (or paper cups to be filled with sand)
- modeling clay
- a pencil and paper
- a protractor
- a ruler
- a can, with top and bottom removed, cut in half vertically
- supports (blocks or books)

What to Do

Activity 1

Find the maximum load that a simple beam can support before it sags 1 cm in the middle. Compare this to the maximum load an arch can support. Compare high arches with flat arches. You can hold the ends of the arch with blocks or books. These supports are called *buttresses*.

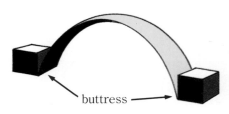

Now think about these questions:
(a) How does the strength of an arch compare to that of a simple beam?
(b) How does the shape of the arch itself affect its strength?
(c) What role do the buttresses play in an arched bridge?
(d) In what way do the blocks support the arched card differently than they do the beam?

Activity 2

Examine the photographs on page 285. Consider these things:
(a) the materials from which each arch is made
(b) the weight of the arch itself
(c) what supports each arch

Arches usually sit on the tops of walls. Each wall must not only support the weight of the arch and any load on the arch, but must also act as a buttress. In what two directions must a wall exert forces on the arch?

Study the two diagrams of structures below. One shows walls supporting an arch. The other shows walls supporting beams and trusses. Why do the walls in the arched structure need to be much thicker than those in the beam structure?

Thick walls are required to withstand force.

Thin walls are possible since the force is straight down.

Activity 3

This part of the Exploration is designed to help you understand how an arch is able to support a load. The basis of creating arches is using a wedge for support. An arch can be considered as a series of blocks, each wedged into place.

The bottom dimensions of each wedge are smaller than the top dimensions. Imagine trying to push one of the wedges downward. What would happen? You can see that an arch should have great resistance against collapsing from a downward push.

Now, make a bridge using wedges formed from modeling clay, as shown. You will need half of a can or a bent card to act as a mold for shaping the arch. With paper, pencil, and protractor, you can make a template for shaping the wedges. Make the template the same diameter as the can plus the desired thickness of the wedges.

Find the load that the clay bridge can support without collapsing. What factors would cause it to collapse?

Reshape the clay into a beam bridge. Use all of the clay. Compare the strength of the beam bridge to that of the arched bridge. For a fair comparison, what factors (variables) should be the same for both bridges?

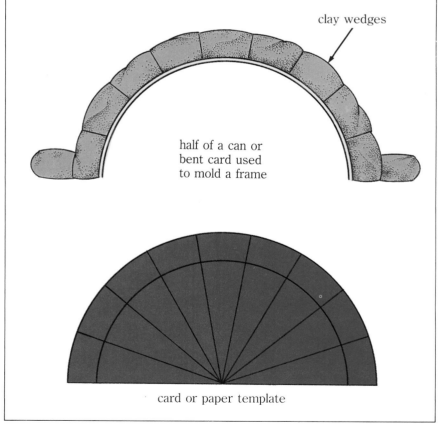

clay wedges

half of a can or bent card used to mold a frame

card or paper template

The Science of Structure 287

Bridge Classification

Think of the bridges you have seen. Most bridges can be grouped as one of four types: arched, cantilever, suspension, or girder. Examples of these four types are illustrated below. What is the main characteristic of each type?

1. Arched Bridges

(a) Simple arch

(b) Elongated arch

(c) Arch with suspended roadway

(d) Bowstring girder beam or tied arch

(e) Arch with supported deck

2. Cantilever Bridge

3. Suspension Bridge

288   Structures and Design

4. Girder Beam Bridges

(a) Warren girder

(b) Pratt truss

(c) Howe truss

(d) Lattice truss

(e) Stayed girder

In the next section you will try your hand at classifying some bridges from around the world into these four basic types.

Famous Bridges

On page 290 there are photographs of well-known bridges from different parts of the world. Some are very old and yet are still standing. Others are quite modern. Examine each bridge and answer the following questions.

(a) Which common type does each bridge most closely resemble? (See bridge classification for help.)

(b) How is each bridge designed to resist the different forces that it might receive?

A helpful way to record this analysis is to use a table similar to this one. Copy the table into your Journal and fill in the information. Do not write in your book.

Bridge Identification

Name	Bridge Type	Bridge Design to Resist Forces
Bridge of Sighs	Simple arch	Arch supplies strength to support the passageway and people.

The Science of Structure 289

Bridge of Sighs, Italy (top left); Thousand Islands Bridge linking Ontario, Canada with New York State (top right); Ebert Bridge over the Rhine, West Germany (center left); Astoria Bridge, Oregon (center right); Golden Gate Bridge, San Francisco (bottom)

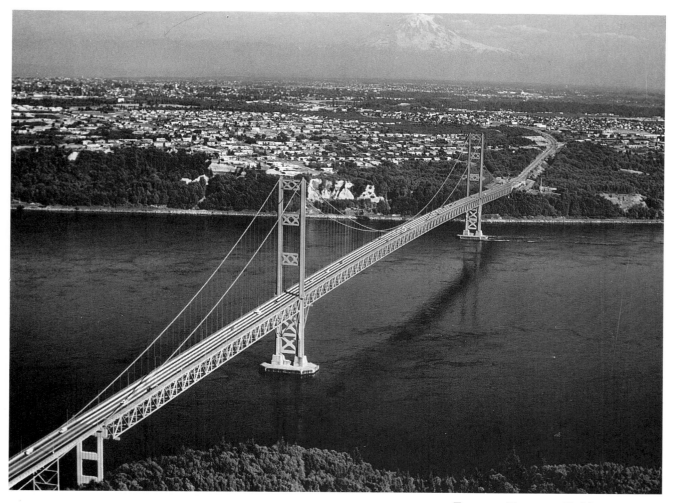

Compare this photo of the Tacoma Narrows Bridge with the photo on page 267. The old bridge pitched and rolled in a light wind until it collapsed. What changes were made in the structure of the bridge? How do these changes prevent the bridge from failing again?

Tacoma Narrows Bridge, Washington (rebuilt in 1950)

Further Analysis

1. Choose two of the bridges on page 290 and consider the following questions for each.
   (a) Which parts of the bridge are under compression? What are the compressive forces?
   (b) Which parts are under tension and what are the forces causing this tension?
   (c) Which parts might experience bending or twisting? What would be the possible cause of these shear forces?
2. Visit a bridge and sketch it, showing all the parts that give it strength and support. Label your sketch to show where you think tension, compression, and shear may occur. What type of bridge is it?

The Science of Structure

Challenge Your Thinking

From the photographs below, identify examples of the following structures:

    (a) simple beams   (d) arches
    (b) cantilevers
    (c) trusses

THE ART OF DESIGN

④ Beauty in Structures

Our Human-Made Environment— Yesterday and Today

So far you have studied some of the scientific and technical features of structures. But that is not all you must consider when you plan structures for Riverwood. What will your structures look like? Will they blend in with their environment? How will they relate to each other? Engineers, architects, designers, and urban planners look to the past to help them make these decisions. In Exploration 7 you will take a journey through history. This will help you make your own decisions about the building style you would like to see in Riverwood.

EXPLORATION 7

A Journey Through History

On the following pages are photographs and illustrations of some important buildings that were built at different times and places. You are going to write and illustrate your own story about the people and buildings from one of these times in history.

What to Do

1. Team up with one or two classmates, and look through the next seven pages. Choose one period in history that interests you.
2. Research the period that you have chosen so that you can answer the questions below.
   (a) What was the culture of the society like? What kind of government did the society have? Who owned the land? Who owned the buildings? Who built the buildings?
   (b) What purposes did the buildings serve?
   (c) What structural features are evident in the buildings of this period (columns, beams, trusses, arches, domes, cantilevers)?
   (d) What materials were used?
   (e) What construction methods were used?

The Art of Design   293

EXPLORATION 7—CONT.

3. Using your knowledge of the period, write your own story. Include at least one illustration. You may choose one of the buildings shown on these pages or choose one from another source. The length of your story should not be more than two to three pages.

Sources of Information

- Encyclopedias—look up information on architecture.
- Libraries—ask the librarian to help you find books on architecture and on the specific period of history that you are studying.
- Your social studies teacher—ask your social studies teacher to help you find textbooks that describe the period of history that you have selected.

Nomadic Peoples

Nomadic people who formed many of the earliest human communities did not build permanent dwellings. Their lives involved constant movement in search of food and other materials.

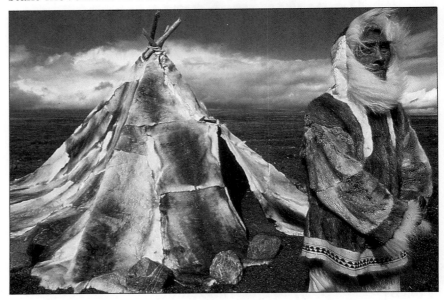

The teepee shown above is similar in design to those used by the nomadic Plains Indians of North America. These people had a highly developed society but no permanent architecture. What is the shape of the framework that gives support to the teepee? Where else can you see a structure shaped like this? Do you know of any nomadic cultures in modern times?

A nomad family in the South Gobi, Mongolia—outside their *ger*, the felt-covered, collapsible tent of Mongolia.

294   Structures and Design

Early Civilization (9000–1000 B.C.)

When people learned to raise crops, they were able to settle in one location. Staying in the same location gave them the chance to build permanent buildings. These first settlements represent the earliest civilizations.

These ruins of the Kingdom of Cush are located in the Sudan.

The Great Pyramid of Cheops

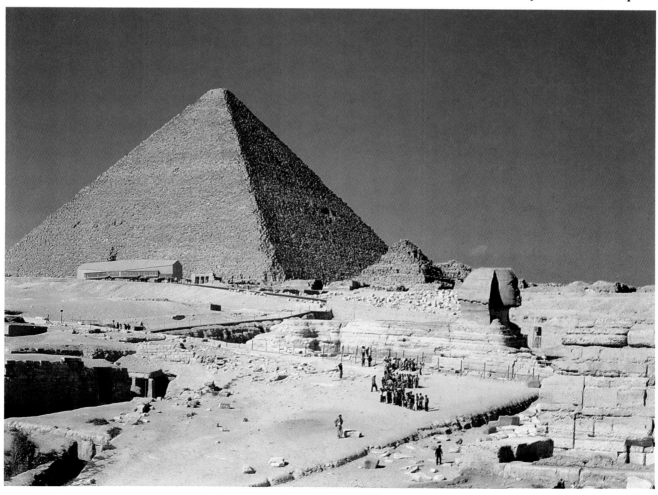

The Art of Design

Ancient Greece (800–338 B.C.)

The architecture of ancient Greece has had an enormous influence on designers for many centuries. This drawing is from a model of the Acropolis at Athens, Greece, as it looked about 450 B.C. Name some buildings you have seen that appear to be patterned after the buildings of ancient Greece.

Drawing of a model reconstruction of the Acropolis

Plan of the Acropolis, Athens

Ancient Rome (27 B.C.–A.D. 476)

Ancient Rome is characterized by its classic structures. In the center of the reconstructed model, you can see the Roman Circus Maximus. On the upper right side of the model, you can see the famous Colosseum. At the bottom of the model is the aqueduct that carried water to the citizens of ancient Rome. Many of the classic structures are still standing in modern Rome.

The Art of Design

Medieval Europe (1100s–1500s)

Medieval Europe is characterized by the famous cathedrals that were built during this period. Is there a church or other building in your state that appears to be patterned after the cathedrals of medieval Europe? In your Journal, make a list of examples you can think of.

Notre Dame Cathedral, Paris, France

Renaissance (1300s–1600s)

After the Middle Ages, a widespread cultural revolution occurred in Europe. It was called the Renaissance, and it affected science, the arts, and education. Architects and artists, like Donato Bramante and Michelangelo, worked together to create beautiful buildings. Do you see any similarities between Roman and Renaissance architecture?

◄ Chateau de Fontainebleau, France

St. Peter's Church, Rome ►

The Cathedral of Florence, Italy ▼

The Art of Design

Modern Civilization

Advances in technology have changed how we build structures. Structural steel and elevators have made building skyscrapers possible. Pre-stressed concrete is used to make beautifully curved roofs for stadiums, theaters, and other buildings. Even huge, enclosed stadiums can now be built through the application of modern knowledge of materials and aerodynamics.

▲ Sydney Opera House, Australia

◀ Superdome, New Orleans

▼ New York skyline

EXPLORATION 8

Conversations with Walter Gropius and Frank Lloyd Wright

Daniel and Karen are two junior-high students who became very interested in modern architecture. They did research and created an imaginary interview with two pioneers of modern design. As you read their interviews, make a list in your Journal of the most important ideas. Later you can apply these ideas to your plans for Riverwood.

Daniel and Karen met Walter Gropius (1883–1969) first. Imagine it is March 15, 1952. Gropius is the head of the Department of Architecture at Harvard University. Listen in as Karen and Daniel conduct their interview.

Karen: We are pleased that you could take time to meet with us. We have heard so much about you. Many architects in the United States give you the credit for developing the modern ideas of architecture and design.

Walter: Thank you. The new International Style of architecture and design is the result of ideas from many artists, designers, and engineers from around the world. Much of the pioneering work was done at the Bauhaus School of Design in Germany when I was its director.

Daniel: What was it like working there?

Walter: It was very exciting. Many young artists and designers worked there. Several of them became famous for their work.

The Bauhaus School of Design, Dessau, Germany, headed by Walter Gropius and later, Ludwig Mies van der Rohe. Although this style of building is common today, it was very unusual when it was built. How does it compare to school buildings constructed before 1935 in your community?

We came up with many new ideas about the way things should be designed. We considered buildings, furniture, books, and much more. Unfortunately, most of us had to leave the country when the Nazis came to power. I first went to England in 1934 and then came to Harvard University in 1937. Here, I have had the good fortune to work with many talented architects.

Karen: What did you and the other designers achieve?

The Art of Design **301**

EXPLORATION 8—CONTINUED

◄ A model for a building that was never constructed. Can you imagine how it might have looked? Have you seen any buildings like this sculpture?

▲ Naum Gabo—Translucent Variation on Spheric Theme. This is a 1951 version of the 1937 original. The sculptor used plastic, glass, and wire to create a sense of space and weightlessness in his modern sculpture.

Walter: Those of us who worked at the Bauhaus School linked art and design with modern technology. The sculptors were particularly important in leading the way. They were using modern materials. From these artists we have learned to think of structures as forms of sculpture.

Daniel: What are your goals as a teacher of design and architecture?

Walter: I want people to recognize that everything built by humans has a design. Some people use a lot of decoration in their designs. However, I prefer simplicity. I believe designers should be able to adapt their designs to mass production.

Daniel and Karen's next stop was in Taliesin, Wisconsin. There they met Frank Lloyd Wright (1869–1959), the most famous American architect. The date is July 15, 1955.

Daniel: When did you begin your career in design, and with whom did you work?

Frank: In 1887 I lived in Chicago and began to work with Louis Henri Sullivan. He was a pioneer in the design of large commercial buildings. In those days, we were learning to build all those skyscrapers. We tried to make them look friendly by including many of the decorative elements of earlier architecture.

After six years with Louis, I set up my own studio where I could devote myself to the designing of homes. I wanted to tackle the problem of the house as a shelter. That has been a life-long interest for me. Of course, I continued to design many public buildings, including an art gallery and a university campus.

302   Structures and Design

Armchair designed for Bauhaus by Marcel Breuer, 1924.

Karen: It is known that you have opposed the International Style of architecture. How do your views differ from those of Walter Gropius and his followers?

Frank: Modern architecture has been influenced by Gropius' views and my own. I believe homes and buildings should blend in with their natural environment. Their construction should include materials from that environment. I call this organic architecture because it links structures with nature.

It seems to me that many architects and designers forget about nature when they design. So there is a definite contrast between my views and those of Gropius. I think a building's relationship to nature is one of its most important characteristics.

Fallingwater, Bear Run, Pennsylvania (1936). Designed by Frank Lloyd Wright. How has Wright used cantilevers to create a home that blends in with its environment?

Modern furniture design, inspired by the work of the Bauhaus School.

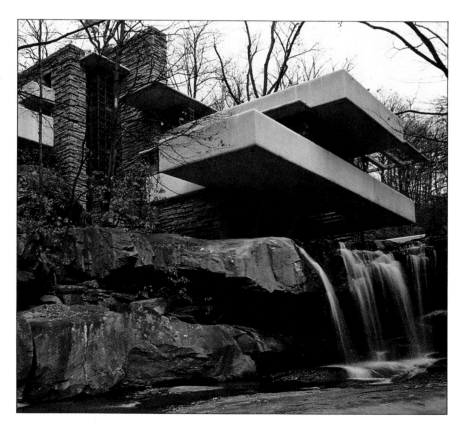

The Art of Design

5 A Question of the Environment

The photos on these two pages illustrate what architects and planners can do to make the urban environment more pleasant for people. Describe which features of these structures make them pleasant to view or to visit.

The designers of Riverwood have to consider what kind of city they want to live in. They have to know what regulations apply that may restrict the kinds of buildings they can build. An example of such a regulation would be a *zoning law* that does not allow commercial buildings in a residential area.

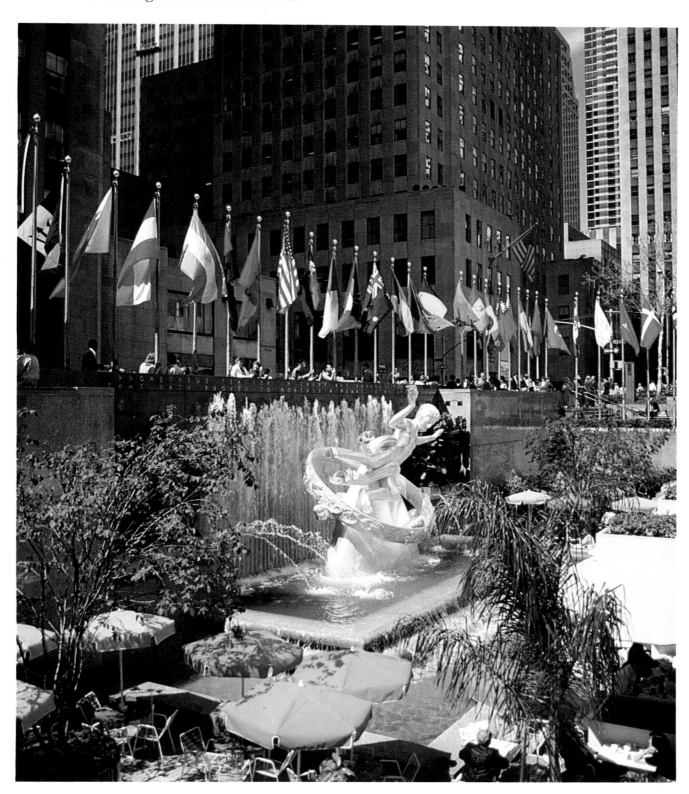

The Art of Design 305

Consider the Views of Phyllis Lambert

Canadian architect Phyllis Lambert has had an influence on the appearance of several North American cities. One of her greatest achievements was her work as the Director of Planning for the Seagram Building in New York. She worked for the renowned architects Mies van der Rohe and Philip Johnson. The building they created helped to establish the International Style as a major trend in North American architecture.

She is now setting a different trend as a conservationist, working to restore and preserve buildings and entire communities.

Think about and discuss the following comments that she has made:

It was not the architects who first publicly complained of urban disaster... Those who first cried out and pointed to the real problem and created the basic structures of reform have been citizens' groups. These groups are made up of people who are concerned about not having their neighborhoods disrupted... Such citizen groups not only make better urban developers than people in government and business, but also become more competent themselves in directing their own lives... I consider that my major task ... is seeing to it that there is an effective means for action and participation by citizens' groups in the decisions that affect our daily lives.

Phyllis Lambert

The Seagram Building, New York City

Challenging Research Activities

Here are some activities that may interest you and some of your friends.

1. Ask your parents or other adults how the appearance of your city has changed in the past years. What areas of the city have had new buildings constructed? Find out if any buildings or neighborhoods were torn down.
2. From your council representative, find out the laws your city has about the kinds of buildings that can be built in your neighborhood. These are called zoning laws. Look for answers to these questions:
   (a) What kinds of buildings are allowed in your neighborhood (commercial buildings, apartment buildings, single family homes, factories, office buildings)?
   (b) Is there a limit on the height of a building?
   (c) How close together can builders construct homes and other buildings?
3. Find out about the work of historical preservation groups in your city. What buildings and neighborhoods have been preserved because of their work?
4. Consider the design problems of urban centers. What is meant by the term *inner city*? Find out if your city council has plans to combat the problems of the inner city by changing the environment. If you do not live in a large city, research the history of design and structures found in Harlem in Manhattan, New York.

Sources of Information

- The library
- Architectural firms
- The planning department of the city government
- Parents and other adults

A Project for Everyone

In a sketch, redesign a familiar environment (your room, the school gym, the classroom, or the school). Decide how the changes you propose would affect other people who share that environment.

What could residents and city officials do to improve the environment here?

A Design for Riverwood

You have now considered many problems in designing and building structures. You know about some of the scientific and engineering ideas involved.

If you were making decisions about a home or other structure, you might be influenced by your knowledge of the history of architecture and design. In making your decisions, you would be more likely to consider the effects on other people. You would want any structure you built to be safe and to meet the standards of zoning laws. In these respects, you now think like a professional designer. More importantly, you can now play an active role in future decisions about your environment.

It is time to apply some of your new knowledge to the design of Riverwood. As soon as you have a plan for the community, proceed to Exploration 9.

EXPLORATION 9

Constructing a Model of Riverwood

1. Submit your own plan for Riverwood to your class. Explain how it meets the following criteria:
   (a) usefulness
   (b) beauty
   (c) safety
   (d) the interests of people who live in Riverwood
   (e) the concerns of those who live in the surrounding neighborhoods

2. As a class, decide which plan best meets the needs of the Riverwood community.

3. Build models of the structures that will appear in Riverwood. As a class, decide:
   (a) what materials are to be used in building the structures (spaghetti, string, balsa wood, index cards, cardboard, or other materials)
   (b) what materials are to be used for structures that need enclosure, such as buildings and homes (paper, cloth, plastic wrap, or other materials)
   (c) the size of the models
   (d) how the models will be tested for strength
   (e) the architectural style or styles that will be used

4. On your own, or with a partner:
   (a) Build a model of the framework for one of the new structures for Riverwood. Test it for strength. Your model must be safe under a load that is one and one-half times greater than the maximum load that might be expected. Remember to consider the effects of natural disturbances as well as the weight placed on the structure.
   (b) If your model is not safe, repeat (a) using more material or a different design.
   (c) Add walls, windows, and roofs as needed.
   (d) Add decorations to your structures in harmony with the architectural style or styles agreed upon by your class.

Challenge Your Thinking

1. Take a look at the buildings around your school or home. Do you see the use of triangles, circles (or parts of circles), rectangles, and squares in the buildings? Do you think that these shapes have anything to do with support or strength? Record your findings in your Journal.

2. Try your hand at architecture in the style of either Frank Lloyd Wright or Walter Gropius. Sketch a home for a family of four. Be sure to include bathrooms, a kitchen, a garage, a place for a washer and dryer, and anything else that the family might need.

3. Below is a photograph of the White House. Analyze the structure, and determine which period of architecture it represents or imitates. Is it attractive to look at? Why or why not? How does it fit into its natural environment?

Unit 5 Making Connections

The Big Ideas

In your Journal, write a summary of this unit, using the following questions as a guide:

- What makes a structure strong?
- What causes a structure to collapse?
- How do you distinguish between tensile, compressive, and shear force?
- How do the three different forces affect a structure?
- How can shape, lamination, corrugation, and tube girders increase strength?
- What is the difference between a beam and a cantilever? Where would you find each?
- How are bridges given strength and support?
- Why are arches and domes stronger than simple beams?
- What are some characteristics of the different architectural periods?
- How can structure and design make buildings beautiful?
- What must engineers and builders consider before constructing a building?

Checking Your Understanding

1. Study the skeleton of the horse shown below. Then answer the questions.

   (a) Which parts of this structure function like columns?
   (b) Which parts function like beams?
   (c) Which part functions like a cantilever?

310 Structures and Design

(d) If the horse had a rider, which parts of the horse would be under compression?

(e) Which parts would be under tension?

2. Design a solution for this structural problem: Support a glass of water by using three stainless steel knives and three soft drink bottles. Each knife can touch only one bottle. Using a paper cup, try out your design. Draw the support arrangement you created.

3. Draw a concept map using the following words: *nomadic, stone, skyscraper, teepees, modern, architecture, Parthenon, animal skin, Ancient Greek,* and *reinforced concrete*.

Reading *Plus*

Knowing about structure and design helps you understand how buildings, bridges, highways, and skyscrapers are constructed. Now read pages S77–S88 of the *SourceBook* to explore the variety of architecture found in other cultures.

Updating Your Journal

Reread the paragraphs you wrote for this unit's "For Your Journal" questions. Then rewrite them to reflect what you have learned in studying this unit.

Science in Action

Spotlight on Architecture

Sydney Dumerasq, like his father, grandfather, and great grandfather before him, is an architect. He and a team of architects he works with design homes, buildings, and other structures. We met with Sydney at his office. He showed us how he uses a computer to help him design buildings.

Q: What do you find most appealing about your job?

Sydney: The creative aspect. You take a site with nothing on it, put your imagination to work, and actually design a building that wasn't there before. That building may stand for generations.

Q: What are some of the considerations you have when designing a building?

Sydney: First of all, I always have to bear in mind that I have two clients. I have the people who are paying the bills, but I also have the general public as a client. If the building is going to stay for generations, people are going to have to look at it for that time. People will live in it and work in it. It must be safe in all respects.

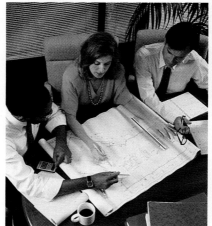

Q: What school subjects turned out to be the most important for your career?

Sydney: Everything. You should study absolutely anything and everything. Never turn down an extra course. Always keep your eyes and ears open.

Q: How do these relate to architecture?

Sydney: History helps a lot. For example, when you study Roman civilization or Greek civilization you will probably become familiar with their architecture. Geography is also very important because buildings are related to their environment. And mathematical concepts and tools are essential.

Q: Which aspects of science are important in architecture?

Sydney: Physics is especially important, particularly the physics of how structures work and under what conditions they stand or fall. Chemistry is also important. How do some materials react with others? How much does a given material expand or contract when the temperature changes? You don't need to know anything in great detail but you need to know the basic things.

Q: What qualities are most essential for architects?

Sydney: Optimism and perseverance.

312 Structures and Design

Some Project Ideas

1. Build a scale model of the home you designed. Use simple materials such as cardboard, craft sticks, and balsa wood.
2. Design a solution to a real construction problem. To find such a problem, visit homes and buildings in your area. Perhaps an additional room or a new set of stairs is needed. Design a solution and construct a model. If the solution is a good one, you might even convince someone to apply it. If it is your home, you might show your parents your suggestion and see if they agree.
3. Consider the special needs of the handicapped. Find out what architects include in their designs in order to accommodate handicapped people. Redesign your airport or local grocery store so the needs of the handicapped are met.

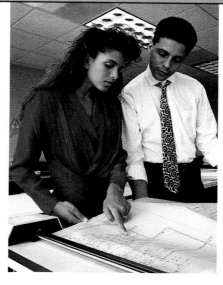

4. Investigate the factors that can make a building energy efficient and design improvements to an existing building.
5. Talk to your teachers about desirable characteristics of a classroom, a laboratory, or other school facility. If you have already identified a problem that needs a solution, design an improvement to the existing school or classroom structure. Check with teachers, other students, and the principal. Making an improvement to the school could be a class or school project.

For Assistance

Look under the heading *Architecture* in the yellow pages of your telephone directory. Also, there may be a Department of Architecture at a university in your region. Once you have identified a project and have given it some thought, arrange to visit an architect who is willing to advise and encourage students doing architectural projects.

Useful advice and assistance can also be obtained from builders and from businesses that sell building materials.

Industrial arts teachers in your school can be especially helpful. They can advise you on how to go about designing a solution, including the choice of materials. Later they can give you advice about constructing your solution (either a model or the real thing). You may be able to arrange to do your work in an industrial arts workshop.

Science in Action

Science and the Arts

An Architectural Marvel

In the early 1400s the people of Florence, Italy, had a real problem on their hands. Their plan for enlarging and rebuilding their cathedral had called for a magnificent dome, 42 meters in diameter, to be built at one end of the cathedral. They had begun this enormous project over a hundred years before, and now they didn't know how to finish it.

The new cathedral was to be the largest in the world. It would establish the power and prestige of Florence over all the other cities in central Italy. It was a matter of great civic pride. But if they couldn't figure out how to build the dome, they would be the laughing-stock of the neighboring cities.

The dome of the Florence cathedral presented a major engineering problem.

The dome was built using a double shell.

The Problem

The usual method of building a dome, according to the techniques known at the time, was to build a supporting structure of wood, lay brick over it, and then remove the wooden structure. But for such a huge dome, there was no way to get timber long enough or strong enough to build the supporting structure.

An Artist's Solution

In 1418 the cathedral officials announced they would give a cash prize of 200 florins to anyone who could figure out how to construct the dome. One person who entered the competition was a famous artist and goldsmith who had turned his attention to architecture. His name was Filippo Brunelleschi (broon uh LESS kee). He demonstrated, by means of models, how the dome could be built without a supporting structure. His models were carefully studied, along with his inventions for lifting construction materials into place. In 1420 the decision was made. Brunelleschi was made the chief architect of the project.

Brunelleschi's solution to the dome problem consisted of using a double shell with ribs formed by pointed Gothic arches. The double shell construction saved weight, and the pointed arches kept the side-thrusts to a minimum. The spaces between the ribs were filled in with bricks, one layer at a time, in a series of circles of decreasing diameter.

Completed in 1436, after sixteen years of work, the dome assured Brunelleschi's fame. Scholars and architects still study the Florence cathedral for insights into Brunelleschi's methods.

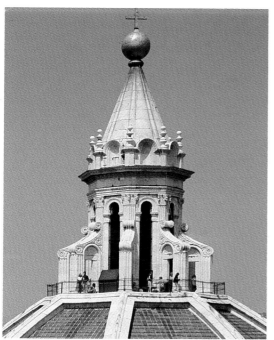

Brunelleschi's design for the "lantern" at the top of the cathedral shows his artistic side.

The Crowning Touch

The *lantern* on top of the dome was designed by Brunelleschi, but it was not completed until after his death. A lantern, in architectural terms, is an open structure on the top of a building to let in light and air. The graceful design reflects Brunelleschi's past experience as an artist and goldsmith.

Find Out for Yourself

Brunelleschi was an artist with a scientific mind. His mathematical interests led him to the rediscovery of *linear perspective,* a method artists use to show depth on a flat surface. This method had been known to the Greeks and Romans but was lost during the Middle Ages. Do some research to find out about this method of drawing.

Brunelleschi took an interest in the works of the ancient Romans. He filled his notebooks with sketches and data of their buildings. He was particularly impressed by a beautifully preserved domed building known as the Pantheon. This ancient building influenced him in his thinking about the cathedral project.

Brunelleschi wasn't the only person interested in ancient architecture. Ancient Greek and Roman ideas were beginning to have a major impact on European art and architecture in many ways. What can you find out about this historical development, which was known as the *Renaissance?* What was its influence on architecture?

This ancient Roman building, the Pantheon, convinced Brunelleschi that the cathedral dome could be built.

Science and the Arts

Unit 6: The Restless Earth

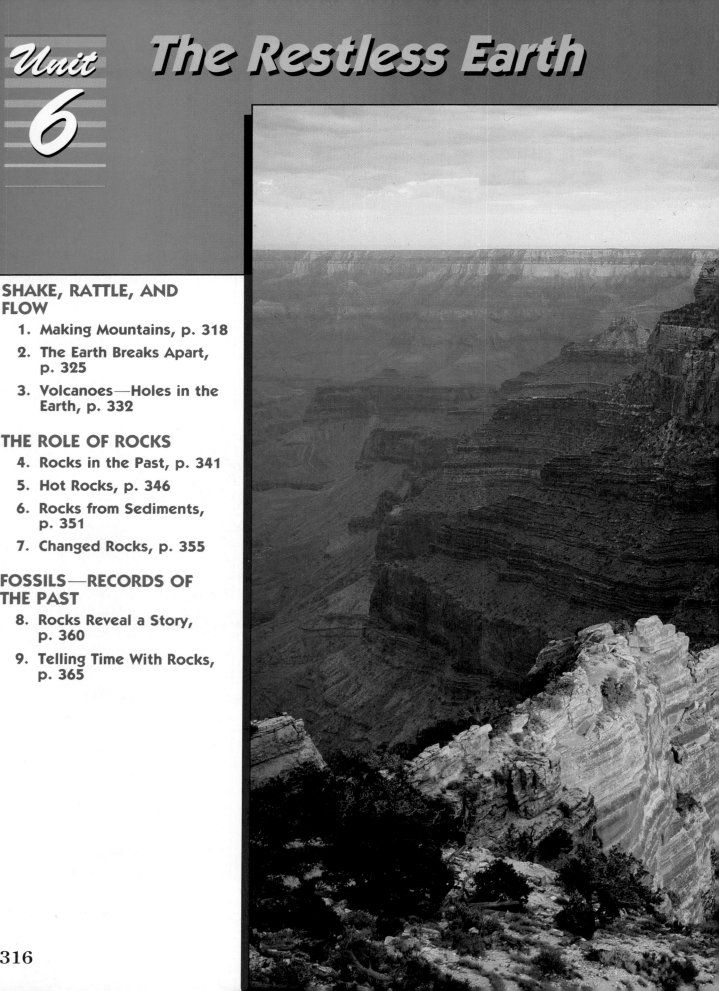

SHAKE, RATTLE, AND FLOW
1. Making Mountains, p. 318
2. The Earth Breaks Apart, p. 325
3. Volcanoes—Holes in the Earth, p. 332

THE ROLE OF ROCKS
4. Rocks in the Past, p. 341
5. Hot Rocks, p. 346
6. Rocks from Sediments, p. 351
7. Changed Rocks, p. 355

FOSSILS—RECORDS OF THE PAST
8. Rocks Reveal a Story, p. 360
9. Telling Time With Rocks, p. 365

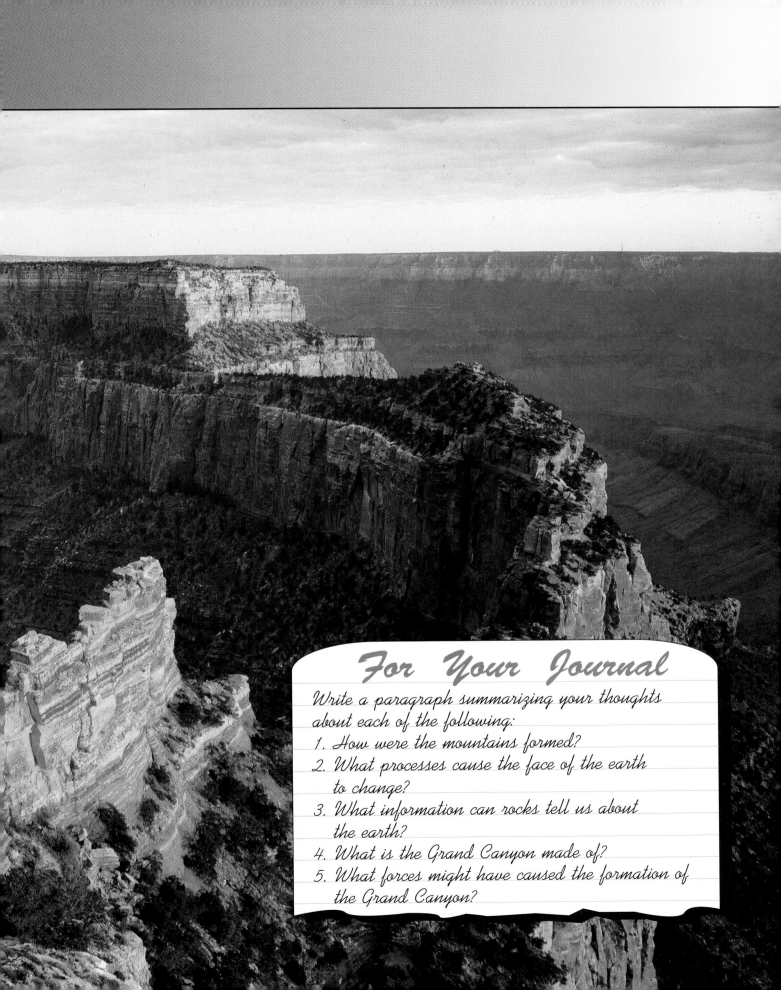

For Your Journal

Write a paragraph summarizing your thoughts about each of the following:
1. How were the mountains formed?
2. What processes cause the face of the earth to change?
3. What information can rocks tell us about the earth?
4. What is the Grand Canyon made of?
5. What forces might have caused the formation of the Grand Canyon?

SHAKE, RATTLE, AND FLOW

1 Making Mountains

Among the earth's most spectacular features are the mountains. At 8848 m, Mount Everest is the tallest of all, but it has many rivals of impressive size scattered over the face of the earth. Throughout the ages, people have marvelled at and wondered about mountains.

The Rise and Fall of Mountains

We know from early writings that people have been fascinated by mountains for thousands of years. Early civilizations tried to answer the same questions about mountains that we raise today. Examine each of the ideas described on pages 319 and 320, and discuss them with someone.

Sarah's Idea

Just think, someday there will be no mountains to climb.

Why is that?

Last year, when we studied erosion and weathering in science class, the teacher told us that mountains are worn down about 30 cm every 1000 years. Since Mount Everest is the tallest mountain in the world, at 8848 m, by the time it is worn down to sea level, all the other mountains will be gone too.

Do you agree with Sarah's reasoning? If not, how would you convince her that she is wrong? According to her figures, how long would it take Mount Everest to wear down?

Back to the Greeks

The ancient Greeks and Romans believed that the mountains were formed by giants who wanted to reach the heavens. The giants then piled one peak on top of another to form a stairway to the sky. Zeus, the Greek king of the gods, struck down the peaks with his thunderbolts, and scattered the remains into rugged mountain chains. Then Zeus imprisoned the giants beneath the mountains. During their struggle to escape, the giants broke the earth's crust and hurled liquid rock against their enemy, Zeus.

Could mountains have been formed in this way? What explanation does the myth give for earthquakes? for volcanoes?

Zeus, with thunderbolts in his right hand

Agricola's Proposal

In 1546, Agricola, a German doctor, suggested that mountains came into being in five different ways:

1. through the eroding action of water
2. through the piling up of sands along seacoasts and in desert lands
3. through underground winds
4. through earthquakes
5. through volcanic fires

Agricola believed, however, that most mountains were formed by the eroding action of water. Read how he described it. Do you think any of Agricola's ideas make sense? Are there ideas you disagree with? Why?

It is evident that many and indeed most mountains owe their origin to the action of water. The little brooks first wash away the surface soil and then cut into the solid rock. By carrying it away grain by grain, the water can finally cut even a mountain range in two, removing great blocks of rock in the process.

In a few years, a small stream can dig a deep depression or river bed across a level or gently sloping plain. Over the course of years these stream beds reach an astonishing depth, while their banks rise majestically on either side. From those banks small fragments of rock are continually detached by rain or frost. Large rocks that have cracks and fissures fall because of their grand weight and land in the stream below. In this way the steep cliffs gradually recede and convert into gentle slopes, and so the original plain converts into a series of elevations and depressions. The elevations are called mountains and the depressions valleys.

(Adapted from De Ortu et Cavsis Subterraneorum, *Book*, 1546)

Could water do this?

Plate Tectonics and Mountain-Building

The theory of continental drift was proposed in 1911 by the German scientist Alfred Wegener. It helps explain the gradual motion of the continents as they drift in the ocean. The Canadian geologist J. Tuzo Wilson took this idea one step further.

Wilson suggested that the outer crust of the earth (the *lithosphere*) is broken up into many large rigid plates and several smaller ones. The continents ride on top of these plates. It is the motion of the plates that causes continental drift. Wilson's idea became the theory of **plate tectonics**. The map shows the boundaries of these plates.

J. Tuzo Wilson

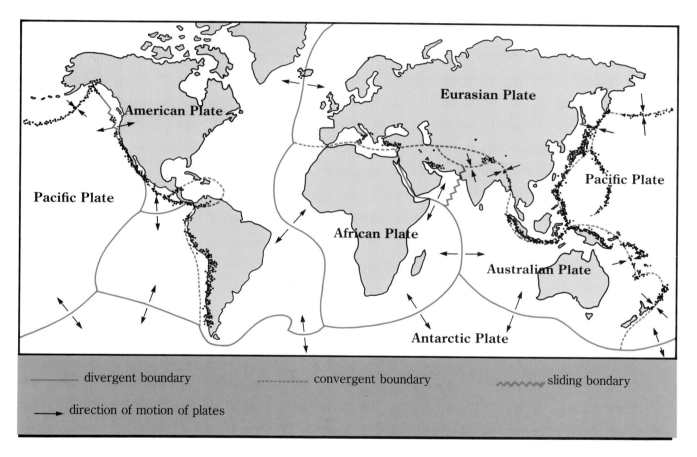

What is happening at the boundaries of the plates, according to the theory of plate tectonics? Check the dictionary for definitions of *divergent* and *convergent*. Then do the Thought Experiment on page 322 to find an answer to this question.

An Earth-Shaking Thought Experiment

Think of the earth as a jigsaw puzzle with six huge pieces and several smaller ones.

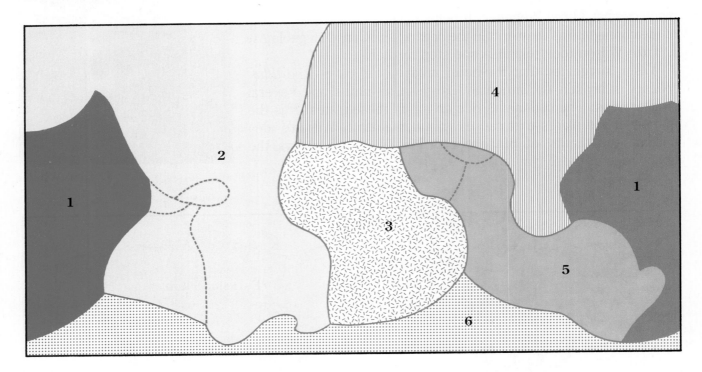

- Compare the jigsaw pieces shown here to the map of the earth on page 321. What does each puzzle piece represent? Why are there two plates with the number 1?
- Find the common border between piece 2 and pieces 3 and 4. Is this border on land or in the ocean?
- Scientists know that molten rock that comes from deep in the earth forms new crust at the ocean bottom along this border. In other words, the border is spreading outward. What effect will this movement have on pieces 2, 3, and 4?
- What will happen to the other pieces in the jigsaw puzzle? Check the key on the map of the plates to see what is happening at the boundaries of the jigsaw pieces. Work with three of your classmates to draw a map predicting the change in the plates that will occur millions of years from now. Where will each plate be? How will the puzzle change?

Applying the Theory of Plate Tectonics

Review these observations and explanations made by geologists. Then apply the theory of plate tectonics in order to answer the questions.

- As the new crust made of younger rock is formed, the ocean floor spreads. This causes the plates to move. They move from 5 cm to 20 cm per year. Estimate how long it has taken the continent of Africa to drift away from South America. Check an atlas or a world map to determine the current distance between these two continents.

- When an oceanic plate drifts up against a continental plate, the oceanic plate slides under the continental plate. Look at the map on the opposite page to see where this may be happening. Which puzzle pieces are involved? This type of motion may cause various disruptions. What might they be?

- When two continental plates collide, the crust of the earth buckles and moves upward. Locate a place where this may be happening. Which puzzle pieces are involved?

Rocks Under Stress

It is difficult to comprehend that there are forces strong enough to push the earth's crust into mountains. Yet when geologists study existing mountains, they find rocks which have been bent, folded, crumpled, and broken, as in the pictures below. The following Exploration illustrates how the theory of plate tectonics explains the folding of the rocks in mountains.

EXPLORATION 1

Folds and Faults

Models can help explain how folds and faults are formed in the crust.

Part 1

You Will Need

- 2 colors of modeling clay
- a knife

What to Do

1. Cut or flatten the clay into strips approximately 1 cm × 3 cm × 10 cm. Make two strips of each color.
2. Stack the strips, alternating the colors.
3. Place one narrow end of the block against a wall. With your hand, apply pressure to the other narrow end.

4. Compare our model to the picture on the previous page. What process does this model illustrate? What do the layers of clay represent? Where do the forces that deform rocks originate?

Part 2

What to Do

1. Look at the photograph above. Why do you think the boundary in the rock formed as it did? Use another clay model to demonstrate your answer.
2. Compare your model to the photograph. The boundary between two rock sections that move relative to each other is called a **fault**. Explain how the fault in the rock might have formed.
3. If 1 cm in the photograph corresponds to 75 cm in the actual rock, find out how far the rocks have moved along the fault.

Drawing Conclusions

Rocks deep within the earth are under extreme pressure and temperature. As a result, they behave like clay, bending when forces are applied. On or near the surface, however, rocks tend to break rather than bend. This is because the pressure and temperature are not as great on the surface of the earth.

In this Exploration, you have found out that the formation of folds and faults can result in mountains. Mountains can be formed in other ways as well. For example, Mount Shasta in northern California was formed by the action of volcanoes.

It appears that mountain-building is an ongoing process. Where do you think the next mountain will be formed?

324   The Restless Earth

② The Earth Breaks Apart

A Picture Study

Carefully examine the picture above. What observations can you make about the fence in the foreground? Make several inferences that might explain the appearance of the fence. Do your observations support your inferences? What added information would you need to verify your inferences?

EXPLORATION 2

Another Kind of Fault

In Exploration 1, you made models that helped explain how mountains are formed. In this activity you will make a model of a different type of fault that will help explain how earthquakes occur.

You Will Need

- the clay block from Exploration 1
- a knife

What to Do

1. Separate the layers of clay carefully, and reshape each layer into rectangles measuring 1 cm × 3 cm × 10 cm.
2. Restack the layers as before.
3. Cut the block lengthwise into two equal pieces.
4. Holding the left portion of the block in your left hand, push the other block forward with your right hand until the blocks slide apart, as in the photograph below.

Questions

1. What does each block of clay represent?
2. What does each layer represent?
3. What is simulated by the boundary between the two clay blocks?
4. What would happen to a fence if it crossed the fault?
5. Compare your model to the picture of the fault in the photograph on page 324.

326 The Restless Earth

Another Picture Study

Starting at the bottom of the photo, trace the course of the stream as it crosses the fault. How does this picture support your inference about the fence in the first Picture Study on page 325?

What Causes Earthquakes?—An Elastic Theory

What do earthquakes and elastic bands have in common? H.F. Reid was one of the investigators of the earthquakes that devastated San Francisco in 1906. He devised a theory to explain the cause of earthquakes. That theory is called the **elastic rebound theory.**

Look at each drawing and the related facts on the next page, and you too will discover the connection between earthquakes and rubber bands!

After the San Francisco earthquake

Shake, Rattle, and Flow   **327**

The diagrams represent aerial views of the boundary between two plates, P and Q.

Diagram A: The Facts

- P and Q are blocks of crust slipping in opposite directions along a fault.
- The broken line is a street lying across the fault.
- The rocks forming blocks P and Q are locked by friction created by the pressure of surrounding rock layers. The friction keeps the blocks from moving.
- The rocks store the energy of the frictional force just like a stretched rubber band stores energy.

Diagram B: The Facts

- As stress builds up, the rocks begin to strain.
- The street bends with the rock.
- The rocks continue to build up energy from the pressure.

Diagram C: The Facts

- Finally, the stress breaks the friction lock at the **focus** or **hypocenter**.
- On each side of the focus, the rocks spring back into their original shape. This is similar to how a rubber band returns to its original shape after you've stretched it.
- The stored energy is released in the form of intense vibrations known as **shock waves** or **seismic waves**. This release is what we call an earthquake.

Diagram D: The Facts

- The street is displaced at the fault.
- The blocks have moved in opposite directions.

You Be the Teacher

To determine whether her students understood the elastic rebound theory, Ms. Morris asked them to use the theory to explain the photograph of the stream on page 327. Manuel was absent from school when Ms. Morris explained the theory, and now he needs some help.

Imagine that you are a member of Ms. Morris's class and that Manuel has come to you for help with the assignment. Write down what you might say or do to explain the theory to him.

Diagram A

Diagram B

Diagram C

Diagram D

Measuring the Strength of Earthquakes

Have you ever felt the earth vibrate? Perhaps you have felt the ground shake as a train passed by. Earthquakes also cause the earth to vibrate. As you can imagine, huge amounts of energy are released when the earth moves like this.

Every earthquake is given a number describing its **magnitude,** or strength. But what does this number really mean? Measuring the energy released in an earthquake is a long, complicated process. To permit rapid calculation of the strength of an earthquake, **seismologists** (people who study earthquakes) have adopted the **Richter Scale of Magnitude.**

The number expressing the magnitude of an earthquake is calculated by means of a formula. The information contained in the formula is provided by instruments known as **seismographs.** These instruments record the vibrations of the earth during a quake. An increase by just one number in the measurements corresponds to a tenfold increase in earthquake strength. For instance, an earthquake that is 7.4 on the Richter Scale is about ten times stronger than an earthquake with a magnitude of 6.4.

A vertical seismograph and the recording drums

Shake, Rattle, and Flow

The Modified Richter Scale of Magnitude

Magnitude	Estimated number of earthquakes recorded each year	Estimated damage
10.0	Possible but never recorded	Would be felt all over the earth
9.0	Possible but never recorded	Would be felt in most parts of the globe
8.0 to 8.6	Occur infrequently	Very great damage
7.4 to 7.9	4	Great damage
7.0 to 7.3	15	Serious damage; railway tracks and bridges bent
6.2 to 6.9	100	Widespread damage to most structures
5.5 to 6.1	500	Moderate to slight damage
4.9 to 5.4	1400	Felt by everyone within the affected area
4.3 to 4.8	4800	Felt by most
3.5 to 4.2	30,000	Felt by a few
2.0 to 3.4	over 150,000	Not felt, but recorded

Seismologists in the field

Analysis, Please!

1. What is the magnitude of the strongest earthquake ever recorded by seismographs?
2. Which magnitude of earthquakes occurs most often?
3. An earthquake that measures ___?___ on the Richter Scale will bend bridges and railroad tracks.
4. What does the prefix *seismo,* as in *seismograph,* mean? Find other words that use this prefix and explain them.
5. Look at the diagram, then state what you think the **epicenter** of a quake is.
6. About 100 J of energy are released from each cubic meter of rock at the time of an earthquake—the equivalent of a firecracker per cubic meter. This may not seem like very much energy, but consider the following situation:

    Suppose the fault is 1000 km long, extends 100 km downward, and bends streets as far as 50 km on either side of the fault. How many cubic meters of rock would be under strain? How much energy would be released when the rocks break the friction lock?

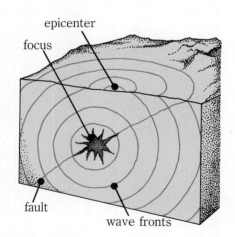

330   The Restless Earth

EXPLORATION 3

Where are the Earthquakes?

Here is a research project. Consult the school library or the local public library for information recorded on earthquakes during the past fifteen years. In your Journal, plot their locations on a simple sketch map of the earth.

Next compare the pattern, if one exists, to the map on page 321 that shows the earth's plates. What is happening at the edges of the plates? Review the elastic rebound theory on page 328. Why are the plates moving in opposite directions? What might P and Q represent?

The map below shows the epicenter locations for earthquakes between 1969 and 1979. Did earthquakes after 1979 occur in similar locations? Are you surprised by the number of earthquakes recorded?

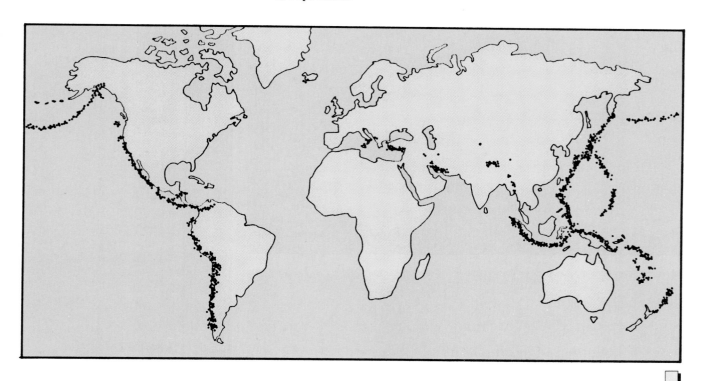

Shake, Rattle, and Flow   331

3 Volcanoes—Holes in the Earth

Mount St. Helens erupting

332 The Restless Earth

A Survivor Speaks—Mount St. Helens, May 18, 1980

At the moment of the eruption, David Crockett was there.

He shouted into his recorder, "The road exploded in front of me . . . it's hard to breathe . . . my eyes are full of ash."

Grabbing his equipment, he left the car and raced up the near ridge which served as a wall between him and the blast. Since it had been clear-cut of timber, he was not threatened with falling timber but a blizzard of ash turned day into night. "Right now I think I'm dead," he recorded, coughing and fighting for his breath. It was then he realized that the deadly blast had gone over him, leaving him uninjured. He dropped to his knees and thanked God for his miraculous escape.

(from *Mount St. Helens: A Sleeping Volcano Awakes,* Marian T. Place, 1981)

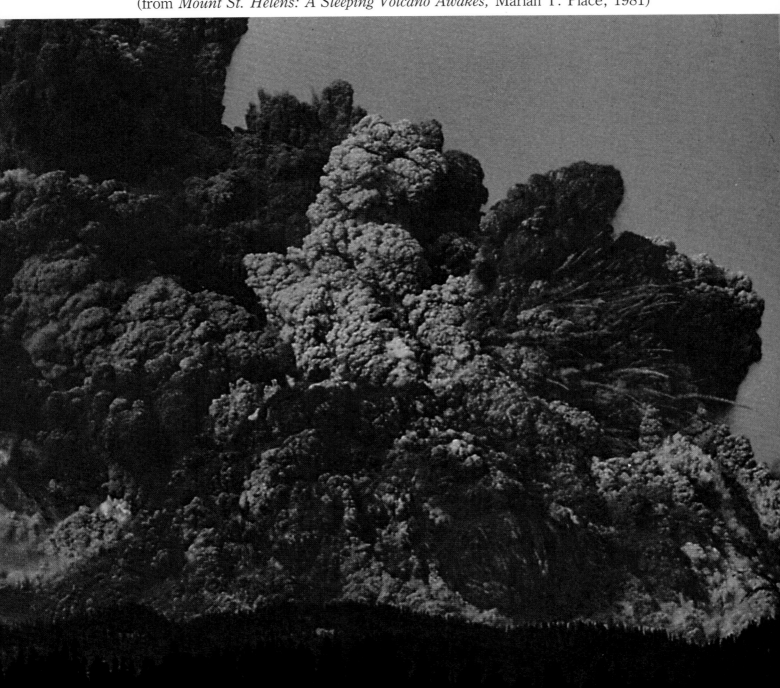

Mount St. Helens after the eruption

Billions of volcanic ash particles, like the one above, covered the area.

Earth tremors and escaping gases had warned of a possible eruption for several weeks. However, Mount St. Helens had been dormant for over 120 years. Few of the people living in the shadow of the volcano expected it to erupt this time. Certainly no one anticipated such an eruption!

When Mount St. Helens did erupt, the explosion was tremendous. It released 500 times the energy of the atomic bomb dropped on Hiroshima in 1945. Much of the mountain was blown away. One cubic kilometer of debris—rock, ash, and steam—shot up in a column 15 km to 18 km high. The eruption left a crater 1 km wide, 2 km long, and 0.3 km deep.

We not only have first-person accounts from survivors like David Crockett, but we also have pictures of the eruption taken before, during, and after its occurrence. Television brought the destructive power of the explosion to people around the world.

But many scientists had been confident that Mount St. Helens would erupt this time. As early as 1975, in fact, some geologists were predicting an eruption before the end of the century. How could they foresee this?

334   The Restless Earth

A truck destroyed by the blast

Clouds of volcanic ash

Re-seeding the mountain

In June of 1991, a major volcanic eruption occurred at Mount Pinatubo in the Philippines. Just 90 km northwest of Manila, the eruption left over one hundred thousand people homeless. Ash shot 30 km into the air and lava poured down the mountain at 100 km per hour.

A Volcanic Word Search

No doubt you have studied or read about volcanoes before. You may also have seen films of eruptions or visited a volcano. Make a list of all the words you can remember that are associated with volcanoes. As you read the following pages, add new words to your list. Then make a word search puzzle in your Journal like the one started here. The words can run in straight lines in any direction—up, down, across, and diagonally. Fill in all the unused boxes with other letters, and then give the puzzle to a friend to solve.

V					
	O				
		L		C	
			C	R	
				A	
				T	N
				E	O
				R	

Famous Volcanic Eruptions

Vesuvius, A.D. 79

This is history's most famous volcanic eruption. For two days, dense clouds of black ash and pumice poured from the vent of Vesuvius. The Italian city of Pompeii was buried under 5 m to 10 m of mud, ashes, and pumice. Several thousand people were trapped and killed.

Vesuvius, like Mount St. Helens, is a *composite cone volcano,* or *stratovolcano.* Such a volcano is made up of alternate layers of lava, ash, and cinders, rising about 2 km above the base. Vesuvius is a highly explosive volcano even today.

Krakatoa, 1883

Volcanic eruptions can be described as "quiet" or "explosive." On August 27, 1883, an explosive volcano erupted. A volcanic island in Indonesia literally blew apart. The explosion was so loud it could be heard 5000 km away, in Australia. The eruption left a hole in the ocean floor over 350 m deep and 10 km wide. Ash and steam rose to a height of 80 km and circled the globe, producing red sunsets for several years.

Mont Pelée, 1902

On May 8 at 7:50 A.M., Mont Pelée, a volcano on the island of Martinique, exploded. A huge cloud of glowing ash and steam, believed to have had a temperature of at least 1000 °C, rose and hung in the air for a moment; then it roared down the slope to engulf the town of St. Pierre. In less than two minutes the city was in ruins with 30,000 people dead. There were only two survivors.

Paricutín, 1943

On February 20, 1943, an astonished farmer observed a column of white fumes rising from a crack in a cornfield in Mexico. By the next day a cone of cinders and lava had grown 8 m high. At the end of the first week the cinder cone was 128 m in height; immense quantities of lava were jetting from it into the sky. This volcano continued to be active until 1952, when the cinder cone reached a height of 375 m.

Surtsey, 1963

The island of Surtsey, off the coast of Iceland, began as a crack in the ocean floor in November, 1963. Although spectacular, the formation of Surtsey can be described as a "quiet" eruption. In this type of eruption, lava flows through cracks or fissures in the earth's crust. As Surtsey was born, the lava flow contacted the seawater and caused numerous explosions.

Why is the Surtsey volcano described as "quiet"?

EXPLORATION 4

Where Are the Volcanoes?

Can you find evidence of volcanic activity in this photograph?

Where are the active volcanoes in the United States? Does it surprise you that Mount St. Helens lies on the west coast?

Is there a reason why volcanoes occur only in certain places? Scientists believe they have an answer to this question. Here is how you can share in their discovery.

In an atlas, locate as many famous volcanoes as you can. Add at least five more names to the list given below. Then sketch another map of the world in your Journal, and mark the locations.

Fujiyama (Japan)
Mayon (Philippines)
Hekla (Iceland)
Etna (Italy)
Katmai (Alaska)

The more volcanoes—active, dormant, or extinct—you can locate, the more easily you will be able to understand the scientists' explanation.

When you have completed your map, describe in general terms where volcanoes occur. Compare this map to the earthquake map you made in Exploration 3, and to the map on page 321 showing the earth's plates. Is there any similarity among the maps? Why do you think volcanoes occur only in certain places?

338 The Restless Earth

The Cause of Volcanoes—Those Plates Again!

When the earth's crust shifts restlessly or a volcano erupts somewhere, the explanation usually lies in the movement of the earth's plates.

As plates move apart, volcanic rock called **basalt** either moves quietly into the gap or erupts from below to fill the gap. This is one way in which volcanoes are formed.

Another way that volcanoes are formed is by the collision of plates. When plates collide, one of them is usually driven under the other. When the underlying plate reaches the hotter temperatures of the interior of the earth, it begins to melt. This melted rock, known as **magma,** is lighter than the rock around it. Thus, it slowly rises, triggering earthquakes and creating volcanoes. The magma that emerges on the surface—in either a quiet or a violent outburst—is called **lava.** The violence of the volcano is determined by the gases in the magma. If the gases can escape gradually, as they do in the volcanoes of Hawaii, there is usually no explosion. Instead a constant flow of lava allows the volcano to erupt quietly. In the case of a stratovolcano like Mount St. Helens, however, the magma gases are bottled up and they burst forth in a tremendous explosion.

As the gas bubbles from the magma (now called lava, since it is on the surface), it creates **pumice,** a type of lightweight rock filled with tiny holes. Quite often, the pumice gets broken up into volcanic ash, cinders, and dust. These small fragments make up the clouds that billow high above a volcano. Eventually they rise into the atmosphere and spread around the earth, often causing spectacular sunsets.

Lava forms polygonal columns (top). A lava flow (bottom)

A Field Trip

Imagine that you are taking a field trip with your class to a volcano. You climb to the top of the volcano and look over the edge. How might you tell if the volcano is active, dormant, or extinct?

Suddenly you can hear rumblings under your feet, and the ground begins to move. In your Journal, record the sequence of events inside the volcano as you think they would occur.

Challenge Your Thinking

1. Compare the photographs of the mountains shown. Which mountains are older? What evidence supports your inference? What factors differ for each mountain?

2. Sharpen your logic skills! Use the basic facts presented to deduce the answers to the questions that follow.

Facts

(a) The Atlantic Ocean is getting wider. New ocean crust is being formed.

(b) Plates collide.

(c) When plates collide, the continental crust thickens and buckles as the plates are compressed.

(d) Plates slide by each other.

Predictions

What is happening to the plates at the mid-Atlantic boundary? What evidence of crustal movement would you expect to find in Iceland?

Find three examples of convergent plate boundaries. What kinds of activities will happen here?

Locate the Himalayas and the Rocky Mountains. Could they have been formed this way?

Find two places where plates are sliding by each other. What do you predict will happen there?

340 The Restless Earth

THE ROLE OF ROCKS

4 Rocks in the Past

The Uses of Rocks

Have you seen pictures of the impressive arrangement of huge rocks at Stonehenge in England? If so, you have probably wondered how people in those early days constructed such a sophisticated observatory. Study the photograph and explanation to discover what those ancient people could do with this vast arrangement of rocks.

Stonehenge
Stonehenge was built in southern England around 1650 B.C. The people who built this stone observatory apparently used it to keep track of the motions of the sun, moon, and stars. By standing at the center of the circle, an observer sees the sun, moon, or stars line up in between specific gaps in the stone. The sighting stones outside of the circle increased the accuracy of the measurement.

The Role of Rocks 341

Study this collage of pictures of rocks. How and why have these rocks been used? Think of more uses for rocks. Create a drawing of at least one other use for rocks.

Great Wall of China

Egyptian pyramid

Sculpture of Ramses at Abu Simbel

Tombstones

EXPLORATION 5

Getting Acquainted With Rocks

Imagine picking up a rock just outside your home. What does it look like? Where did it come from? How did it form? Studying the rock can give you clues about its age, its formation, and the changes in the earth it has experienced.

The nature of rocks can give you information about many of the things you wish to know about the earth. In this activity, you will begin to explore the nature of rocks.

Part 1
Examining Rocks

You Will Need

- a magnifying glass
- 2 different types of rocks, each about 4–6 cm in diameter

What to Do

1. Divide into groups of four. Each group will have eight rocks to study.
2. Select two rocks from those given to your group. Study and record your observations about each on a sheet of paper. Your description should be clear enough so that your classmates can identify the two rocks.
3. When everyone in your group has finished writing the descriptions, mix up the rocks on a table. Try to pick out the two that you wrote about. Can you find them?
4. After finding your own rocks, put them back on the table and mix them up again. This time, exchange descriptions with another member of your group. Find the rocks described on your classmate's sheets. Were you successful? If not, perhaps the descriptions need to be improved. If you were successful, could you suggest anything to add to or delete from the descriptions? Repeat this exchange with each member of your group.
5. As a group, compare the descriptions written by each member. Which properties of the rocks did each describe? Which observations were the most helpful? least helpful?
6. Using all the observation sheets of your group, make up a new sheet listing those properties of the rocks that were most helpful in identifying them. Below is an example.

Part 2
Classifying Rocks

You Will Need

- rocks from Part 1
- index cards

What to Do

1. Join with another group, so you now have sixteen rocks to work with. Choose one property that half of the rocks share, and divide the rocks into two groups according to this property. Using an index card, make a label naming the property for each group.
2. Determine how each of the two large groups can be divided into smaller and smaller groups that have more similar characteristics. Make labels each time you subdivide the groups.
3. Continue until you have separated all the rocks into either groups having one rock or groups whose properties are all the same.
4. Describe your classification scheme in your Journal. Then display your rocks and labels in the classroom.

Property (Characteristic)	Observations Related to Property
Color	Ranged from plain gray to black, white, and red
Composition	The gray rock was smooth. The mixed rock was made up of small rocks stuck together. The red rock had shiny crystals and was rather rough.
etc.	etc.

The Role of Rocks   343

The History of Rock

Each rock you examined has a history to tell. The crucial points of the history include:

- how the rock was formed
- where it was formed
- how it ended up in your hands

The observations you made about the rocks you gathered are clues to discovering the history of your rocks. In this section, you will learn to interpret the data and reconstruct the events that led to the rocks' formation. The history of each rock will include one of the following three descriptions.

Deep within the earth, *molten* materials slowly cooled and hardened, forming crystals that were easily visible; an **igneous** rock was born.

Particles of sand and soil were carried by wind and water and eventually deposited on an ocean bottom. Increasing pressure from the accumulated weight of the particles, combined with chemicals from the sea, cemented the particles together to form a **sedimentary** rock.

Heat and extreme pressure reorganized the crystals of an already existing rock to change it into a new type of rock, a **metamorphic** rock.

A sedimentary rock

An igneous rock

Metamorphic rock

Rock Groups

Most of you would have little trouble classifying rock bands into groups according to the type of music they play. But can you classify rocks into groups? There are so many different kinds of rocks that scientists once searched for ways to organize them. Read the three descriptions above again. On what basis have geologists classified rocks? Perhaps a clue will help: *ignis* is the Latin word for "fire."

EXPLORATION 6

Rock Group Identification

It can be difficult to decide whether a rock is sedimentary, igneous, or metamorphic, but the identification key shown will help you to classify the rocks.

You Will Need

- a magnifying glass
- rock samples

What to Do

In your Journal, make a table with the following headings for recording your results: *Sample Number, Observations, Type of Rock*. As you study each rock, record your observations and identification in a table like the one below.

Sample Number	Observations	Type of Rock
1	flat sheets	?

Choose one rock to begin your study. Starting with 1(a) and 1(b), pick the sentence that best describes your rock. Then look at the number at the end of that line and go to the step number indicated. Continue until you reach a line that identifies the rock as either sedimentary (S), metamorphic (M), or igneous (I). Repeat this procedure for eight different rocks.

You may find that some rocks do not seem to fit readily into any category. Expert help may be required. Remember, identifying rock types is not easy!

Identification Key

1. (a)	The rock is grainy or is made of more than one material.	2
1. (b)	The rock is made of only one material.	5
2. (a)	The particles or grains are closely fitted together (interlocking).	3
2. (b)	The particles or grains are held in place by natural cement (non-interlocking).	S
3. (a)	All the grains are of the same type.	M
3. (b)	The grains are of two or more different types.	4
4. (a)	The grains are arranged randomly.	I
4. (b)	The grains are arranged in bands.	M
5. (a)	The rock is glassy or porous.	I
5. (b)	The rock is made of strong, flat sheets.	M

2(b)

3(b)

4(a)

4(b)

5(a)

5(b)

The Role of Rocks

5 Hot Rocks

Exploring Igneous Rocks

Before doing Exploration 7, look at the rocks on page 348. All of the rocks shown there are igneous. At one time each rock was molten. Many contain interlocking crystals. Why do they look so different? Performing the Exploration that follows will help you answer this question.

EXPLORATION 7

A Simulation: Forming Igneous Rocks

This activity will help you understand how igneous rocks are formed. Work in groups of three or four.

Be Careful. *Since you will be using hot materials, you must use caution. Safety goggles must be worn! Use care when handling stearic acid.*

You Will Need

- a metal ring
- a ring stand
- wire gauze
- sand
- 3 beakers (400 mL)
- test-tube tongs
- a small paper cup
- a piece of paper towel
- a piece of glass or aluminum foil
- 10 g stearic acid
- an alcohol lamp or bunsen burner
- ice
- water
- a knife
- a scoop
- 2 test tubes

What to Do

1. Fill a beaker three-quarters full of sand. Scoop out a hole in the sand large enough for the paper cup, and place the cup in the hole.

2. Prop up a piece of glass or foil on a folded paper towel. It should form a slight slope.

3. Fill a beaker halfway with ice and water.

4. Set up the ring stand. Place the wire gauze on top of the metal ring and the burner below it. Fill the third beaker halfway with water and place it on the wire gauze.

346   The Restless Earth

5. Place 5 g of stearic acid into each test tube and stand the test tubes in the beaker of water. Heat the water until all of the stearic acid melts.

6. Using the test tube holder, carefully pour a small portion of the melted stearic acid onto the sloping glass or foil, allowing it to run down the slope.

7. Quickly pour the rest of the stearic acid from the first test tube into the paper cup sitting in the sand. Observe both samples of stearic acid as they cool. Record your observations in your Journal.

8. Pour the melted stearic acid from the second test tube into the beaker of ice water. Record your observations.

9. When the stearic acid in the ice water is completely cool, cut it in half with a sharp knife.

Interpreting Your Observations

1. Compare the appearance of the stearic acid in the paper cup, the stearic acid poured down the slope, and the stearic acid poured into the ice water. In your Journal record the differences you observe. What do you think caused these differences?

2. Take a close look at the stearic acid that you cut in half. Is there a difference between the appearance of the material in the center and the material on the edge? If so, what do you think caused the difference? Was there a difference in the time it took for the edges and the center to cool? Why?

3. This experiment is meant to simulate the formation of igneous rocks. What is simulated by each of the following?
   (a) the melted stearic acid in the cup
   (b) the sand
   (c) the melted stearic acid that was poured down the sloping glass or foil
   (d) the melted stearic acid that was poured into the ice water

Pedro's Report

Here is what Pedro wrote in his report. What would you add to make the report more complete?

"This was one of the most interesting experiments we've done. When the melted stearic acid cooled quickly by being poured into cold water, a rubbery mass was made. Our teacher said that when lava flows over the earth's surface, it cools down fast and the rock formed has extremely small crystals. When the stearic acid cools more slowly, larger crystals are formed. This must be what happens when magma cools slowly beneath the earth's crust."

Now examine the six igneous rocks pictured here. On the basis of Pedro's report, classify each rock according to the rate at which it cooled. Use the terms *quick* or *slow* to describe the rate that each rock cooled.

On the basis of what other properties could you group these igneous rocks?

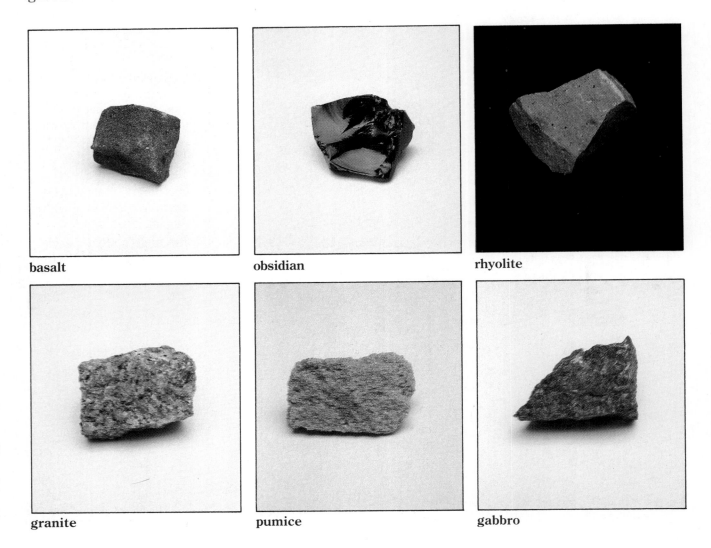

basalt   obsidian   rhyolite

granite   pumice   gabbro

The Restless Earth

Fiery Language

There are many terms used in connection with igneous rocks. You have seen some already: lava, magma, molten. Beginning with these words, compile a dictionary of rock terms in your Journal.

The Formation of Igneous Rocks

When igneous rocks form from the molten lava that flows onto the surface of the earth, they are called **extrusive** or **volcanic** rocks. Basalt is the most abundant rock of this type on earth. Other volcanic rocks include rhyolite, obsidian, and pumice. These extrusive rocks are made up of tiny crystals.

When lava flows into the ocean, it forms pillow-like shapes about one meter across. Would this lava have small or large crystals? Why do you think so?

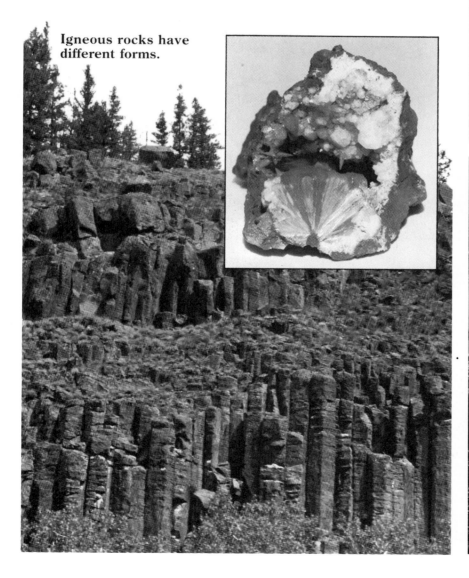

Igneous rocks have different forms.

The Role of Rocks

Liquid rock, or magma, tends to cool slowly deep within the earth's crust. Recall the stearic acid experiment from Exploration 7. Based on your experimentation, what size do you think the crystals of these rocks would be? Rocks formed in this way are known as **intrusive** or **plutonic** rocks. Granite is the most abundant example. Gabbro is another common plutonic rock. This illustration shows the formation of igneous rock by the cooling of magma. Copy the drawing into your Journal, and place the following labels in the correct positions:

- extrusive (volcanic) igneous rocks
- rocks formed from the slow cooling of magma
- rocks formed from the fast cooling of lava
- rocks with large crystals
- rocks with small crystals
- intrusive (plutonic) igneous rocks

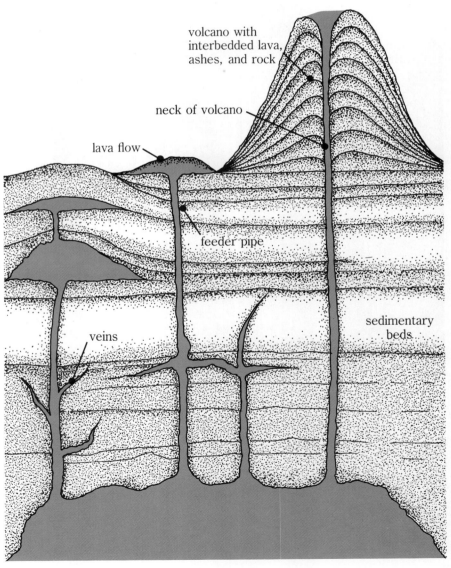

6 Rocks from Sediments

Hanging Out With Granite-Man

"Yeah, yeah, yeah . . . so I was on my way to the concert when I ran into my old friend Sam Sandstone.

'Where ya been?' I asked him.

'Up the river,' he says.

So I say, 'Naw, really? Tell me what's happen' up there lately.'

So he starts tellin' me about these rocks, man, that fall into the river. Now, the river, being one swift dude, carries these rocks, like FAST, down toward the ocean. On the way, the heavy rock dudes drop to the bottom first, the pebbles settle out just a little later, and the sand goes swirling and twirling on top. Then the river dumps these cruisers into the ocean where the heavy salt water presses them together.

'Solid?' I ask him.

'Totally solid,' Sam says. 'These dudes have their own group, man. They are their own kind of rock, if ya know what I mean.'

'Man,' I say, 'That's too cool!'

'My sediments exactly,' says Sam."

Granite-Man certainly tells a good story, doesn't he? Think about the process that Sam described. What do you think sediments really are? Why do they separate out? How does the speed of a river contribute to the formation of sedimentary rocks? What does the ocean have to do with sediment? Reread what Granite-Man has to say with these questions in mind.

Where Are Sediments Deposited?

Examine the illustration to the right. Think about why sediments get deposited in layers. Sketch the drawing in your Journal, and label the locations where sediments might be deposited.

Where would you expect to find larger materials, such as stones, pebbles, and gravel? Where might sand be deposited? Where would you expect to find deposits of fine materials such as silt and mud?

Getting to Know the Sedimentary Rock Group

The formation of sedimentary rocks is closely associated with water. One type forms when water carries soil, pebbles, and other particles to the ocean floor where these sediments become rock. The second method involves chemicals dissolved in water. By evaporation and precipitation of substances like calcium carbonate, sedimentary rocks can form.

Several sedimentary rocks are pictured here. However, try to work with actual rocks if samples are available. Then consider the following questions.

- Which of the rocks are obviously made up of sediments?
- What kinds of sediments make up these rocks? (shells of sea animals, salt, clay, sand, pebbles, etc.)
- Which rocks do not appear to be composed of sediments? (These are the kind formed by changes in the water's solutes.)

If you need help, check the library for a geologist's field guide or a geology textbook.

shale

sandstone

coquina

conglomerate

gypsum

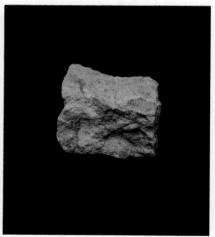
limestone

EXPLORATION 8

Learning About Layered Rocks

Earth scientists tell us that certain sedimentary rocks are formed from layers of sediment. In other cases, evaporation plays an important role in the formation of sedimentary rocks. These activities will help you find out more about both processes.

Part 1

You Will Need

- about 6 mL of soil that contains a variety of particle sizes
- 10 mL of water
- a test tube with a stopper or a glass jar with a tight lid

What to Do

1. Place the water into the test tube or jar, and pour the soil into the water.

2. Before shaking the mixture, predict what will happen. Which particles will settle out first—the largest or the smallest? Sketch what you think the results will be.
3. Now shake the mixture vigorously for one minute. Observe the results. Does it resemble your sketch; was your prediction correct? If not, draw a corrected sketch.

Part 2

Marilyn read that if 100 g of seawater evaporates completely, then 3.5 g of dissolved solids remain behind. She was curious to know whether the solid material left behind looked and tasted like table salt. Marilyn collected some seawater and left it in the sunlight until all the water had disappeared. She discovered that the material that remained was not as white as table salt, although it tasted the same. Should she have tasted the salt to identify it?

When Marilyn told her teacher, Mr. Delaney, about her experiment, he told her that what she had collected was mostly ordinary table salt. However, other compounds like magnesium, sulfur, calcium, potassium, bromine, and iodine were also present.

Marilyn asked Mr. Delaney to explain why huge salt beds are found under certain areas of the ocean. What is your theory about how such large salt beds are formed?

The Role of Rocks

A Rocky Problem

Patrick and Nancy discussed the last Exploration they did in class.

Nancy said, "I'm confused. I can see how the layers formed in the test tube, but trying to understand how pieces of dirt end up as rock on the side of a cliff is beyond me."

"Maybe it would help if we tried to make some rocks," Patrick suggested. "Let's ask Mr. Delaney to help." Their teacher then suggested that they read the information below for some hints about what to do.

Layers of Sediment Become Rock

When soft sediments become hard rock, **lithification** has occurred. One form of lithification is called *cementation*. In this process, the spaces between particles of loose sediments are filled with a cementing agent that is carried in solution by the water. As the cementing agent fills the gaps between the particles, the loose materials are cemented together to form sedimentary rocks.

During *compaction,* the pressure of overlying sediments compresses the space between the particles. As the particles are pressed closer together, they stick to each other forming sedimentary rocks.

There are still other lithification methods. *Drying* and *crystallization* are two of them. How do you think these processes of lithification work?

Patrick and Nancy asked their classmates to help them come up with some ideas to simulate lithification. Loni remarked that popcorn can be made to stick together when mixed with honey. Juan had watched his neighbor make concrete the weekend before, and he thought this was a good simulation. Randy suggested pressing small marbles of modeling clay together.

What was each student thinking about in proposing these ideas? Can you come up with your own lithification model?

7 Changed Rocks

The Meaning of Metamorphic

Metamorphosis means change. You might have used the word to describe the life cycle of frogs or insects. For example, when a caterpillar spins a cocoon and comes out as a butterfly, metamorphosis has occurred. What do you think a **metamorphic** rock is like? Look at the examples of change at the right.

For rocks, metamorphism occurs when the environment changes, such as when there are movements in the earth's crust. How could igneous and sedimentary rocks change into a new kind of rock? Read "Historical Flashback!" to find out.

(a) caterpillar   butterfly

(b) ingredients   cake

Historical Flashback!

Charles Lyell became a geologist in a roundabout way. He was born in 1797 in Great Britain. Lyell attended Oxford University to study law, but was very interested in geology as a hobby. This scientific interest led him to the latest geological theories. These theories had been proposed by James Hutton just before Lyell was born. One of Hutton's theories concerned **intrusions.** Hutton thought that magma sometimes pushed upward into the cracks in existing rocks. Then the magma cooled and formed igneous rocks, including granite. Such rocks are called intrusions because the magma *intrudes* (is forced into) the rock.

Two types of intrusions exist. **Sills** are intrusions that form in horizontal cracks in an existing rock structure. **Dikes** are intrusions that form vertically across existing rocks.

In the following two photographs, locate what appear to be sills and dikes. Where did they originate?

The Role of Rocks

Hutton observed that the rocks surrounding an intrusion had been through a process of change. The reason for this seemed simple: molten rock is very hot, and the heat affected the sedimentary or igneous rock surrounding the intrusion.

The rocks surrounding the intrusion do not become hot enough to melt. However, certain chemical changes take place as gases and fluids from the cooling magma move into cracks, joints, or small pores within the rocks. These surrounding rocks then change in appearance and characteristics.

Lyell was convinced that Hutton's theory was correct. He placed these "changed" rocks into a third group, which he named metamorphic. Here are some pictures showing original rocks and their metamorphosed forms:

- shale and slate
- limestone and marble
- sandstone and quartzite

What differences do you see between the metamorphic rock and the original rock?

Charles Lyell

shale · limestone · sandstone

slate · marble · quartzite

356   The Restless Earth

EXPLORATION 9

Rocks: Before and After

You know that metamorphic rocks are changed igneous or sedimentary rocks. In this activity, you will compare original rocks with their metamorphosed forms.

You Will Need

- 4 pairs of rocks, each pair consisting of an igneous or sedimentary rock and its metamorphosed form

What to Do

1. Examine each pair and compare the properties of each. Record your observations in a table in your Journal.
2. When you have finished your examination, check the distinguishing property or properties that you identified against the rock identification key on page 345. Do your observations include all these characteristics?
3. How closely do the metamorphic rocks resemble the original rocks that they formed from?

Can Rocks Be Recycled?

Do you suppose that metamorphic rocks can be changed back into their original forms? Can igneous rocks be changed into sedimentary rocks? In other words, can rocks be recycled? Discuss the following questions with a classmate, and decide whether you think rocks can be recycled.

1. How are igneous rocks formed?
2. What conditions change igneous rocks into metamorphic rocks?
3. How are sedimentary rocks formed?
4. What is the source of sediments?
5. What conditions change sedimentary rocks into igneous rocks?

What pattern, or cycle, have you discovered? Using the words and phrases listed below, write out a description of how recycling occurs.

plate tectonics
molten rock (magma or lava)
metamorphic rock
heat
cooling
uplift
igneous rock

erosion
pressure
sedimentary rock
melting
cementing
sediments
intrusions

Those Plates Again!!

The theory of plate tectonics supports the idea that rocks can be recycled. The collision of plates creates conditions that cause the formation of both igneous and metamorphic rocks. Look at the diagram below. It shows the collision of a continental plate with an oceanic plate. With a partner, find the locations of each of the following occurrences:

- The oceanic plate is moving underneath the continental plate.
- Parts of the continental plate are under great pressure because of the collision. This leads to the formation of metamorphic rocks.
- The rocks of the oceanic plate are being heated to a high temperature and are melting.
- Magma is rising in cracks in the continental plate, forming new igneous rock.
- Igneous and metamorphic rocks erode and then are carried to the bottom of the ocean. This is the first step in forming sedimentary rock.

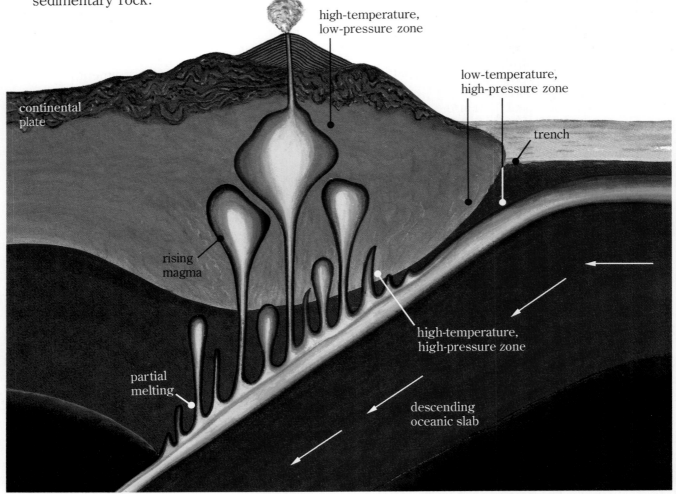

358   The Restless Earth

Challenge Your Thinking

1. Copy this diagram into your Journal. Do not write your answers in the book. For each arrow, fill in the process that occurs as one type of rock is recycled. The first answer has been completed for you.

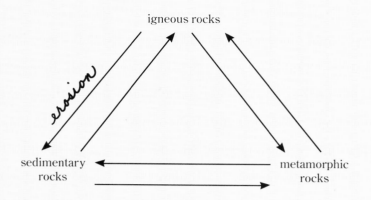

2. Nigel is having trouble classifying some rocks. Using the following information, help Nigel determine whether the rocks are igneous, sedimentary, or metamorphic.

   Rock 1: This rock is grainy and has a lot of different particles. I think I see half a pebble stuck in there!

   Rock 2: This rock is grainy too, but the grains fit together tightly. There are at least four different grains here—two of them are gray, one is white, and one is black. You can see that each color has its own band.

   Rock 3: This rock isn't grainy at all, but it is made of several flat layers.

   Rock 4: I like this rock because it's smooth and glossy. It's only made of one material and it isn't grainy either.

3. Imagine that you are a talk show host. You have invited the following people to appear on your program:

   (a) a geologist who lives in the Rocky Mountains
   (b) Charles Lyell and James Hutton
   (c) a rock collector who is organizing a field trip for junior high school students

   Make a list of questions you would ask each of your guests about geology.

The Role of Rocks

FOSSILS—RECORDS OF THE PAST

8 Rocks Reveal a Story

Imagine the scene: dense clumps of trees and huge ferns crowd the many swamps and river banks. Amphibians large and small crawl on the land and swim in the warm waters. The air is alive with the sounds of giant insects, but no birds can be seen. A many-legged arthropod resembling a giant cockroach makes its way across a mudflat.

Can you imagine that such a scene took place in your state? Fossil evidence suggests that plants and animals such as these lived throughout North America. Three hundred million years later, the scene has changed dramatically. You have learned about the forces that brought about some of these tremendous changes. Now you are going to learn how scientists use fossils to reconstruct the events of the past. Fossils are abundant across the United States, and provide a rich source of information about the plants and animals that lived long ago.

What does fossil evidence suggest about your state's history? Can you discover more about it? Consult an encyclopedia or your librarian for information about the prehistoric environment of the United States.

A Classroom Collection

Do you live in an area where fossils are found? Perhaps you have seen fossils embedded in the rock. The fossils could be as small as a leaf or as large as a dinosaur. Look carefully at the rocks around your neighborhood. Also, look at the stones used in constructing large buildings—sometimes fossilized leaves or insects can be seen. If possible, make a display of fossils brought in by your class.

How Fossils Are Formed

Fossils are the remains of prehistoric animals and plants preserved in rock. The rocks containing fossils lie undisturbed beneath the surface of the earth for millions of years. Finally, through weathering and erosion, the fossil-bearing rocks are exposed on the surface.

Types of Fossils

All the fossils you have seen were formed in one of four different ways. Therefore, fossils fall into four groups:

(a) the actual bones, shells, or other remains

(b) an imprint or mold of the remains

(c) a cast, made when a mold was filled with sediment that later hardened into stone

(d) some form of track or trail left by an animal

At least one example of each of these groups is pictured here. Examine each picture and identify the group it belongs to. Record your conclusions in your Journal.

Fossils—Records of the Past

Who Gets Fossilized?

Examples of life are all around us, yet the chance of a single organism being preserved in the geologic record is very unlikely. Why is this so?

There are many ways in which the remains of organisms can be destroyed. If they are left lying on the surface, they can be attacked by scavengers or bacteria. Bones or shells can be crushed by overlying sediments, or they can be eroded along a seashore or in a stream bed. Why, then, do some survive?

If the organism has hard body parts, or if it is covered by some protective material like soft mud or sand shortly after its death, the remains have a better chance of being preserved.

Using the list below, sort the organisms into these three categories: *Good chance for fossilization*, *Fair chance for fossilization*, and *Poor chance for fossilization*. Record your work in your Journal.

(a) a tree growing in a swamp
(b) deer tracks in the sand
(c) a seashell on a beach
(d) a horse drowned while crossing a river's flood plain
(e) a jellyfish in the ocean
(f) a freshwater clam in a stream
(g) dinosaur tracks in mud
(h) a small animal caught in a tar pit
(i) a reindeer killed by disease on an island
(j) an evergreen growing on a rocky ledge

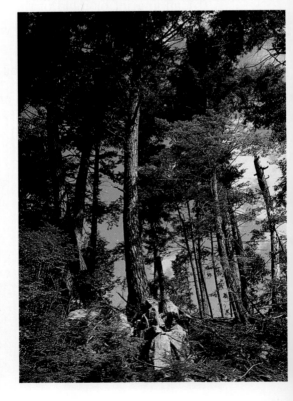

Something to Think About

In which rock group would you expect to find the greatest number of fossils? the fewest fossils? Give an explanation for your answers.

Something to Talk About

Imagine that you have been transported several million years into the future. Discuss with your classmates what you might find in the rocks that exist in your area. What are some human-made objects that might survive? Which ones might not? What could a scientist of the future learn about our civilization from these fossils?

Interpreting the Fossil Record

Perhaps the most fascinating fossil remains are those of the dinosaurs. You know that the dinosaurs disappeared from the earth. This occurrence marked the beginning of our own geological era—the Cenozoic Era.

What Killed the Dinosaurs?

There are many hypotheses that try to explain the sudden death of the dinosaurs 65 million years ago. Can you think of any yourself? Science relies on hypotheses that can be tested. A hypothesis will begin to be accepted only after a great deal of data has been collected and examined and the hypothesis still seems correct. Many hypotheses have been suggested to explain the extinction of the dinosaurs. Some of them are listed below.

(a) Radiation from an exploding star burned them.
(b) Flowering plants developed lethal poisons that the dinosaurs ate.
(c) The climate became too hot or too cold for reproduction to be successful.
(d) A huge asteroid collided with the earth, creating a vast dust cloud that blocked the sun. This lessened photosynthesis and drastically lowered the temperature.
(e) An epidemic of some sort killed them.

How many of these hypothesis could be tested? How likely do you think they are?

You Be the Scientist

Paleontology is the study of fossils, ancient life forms, and their evolution. Suppose you are a paleontologist whose task is to interpret the fossilized footprints in the drawings shown. What can you discover about the creatures that made these tracks?

Use these questions to get you started.

1. How many animals were involved?
2. In which direction did each animal move?
3. What can you tell about the size of each animal?
4. How many legs did each animal have?

Now, piece together these drawings so that they follow a logical sequence of events. What story do these footprints tell?

Telling Time With Rocks

The Rock Record

You know that geologists are interested in events that took place long before humans were around to record them. You also know that the rocks of the earth's crust contain evidence of these events. In this lesson, you will see how this evidence has been organized into a geological history of the earth. The first step is to learn what is meant by *relative time*.

A Relative Time Scale

In your Journal, place the events in this list in the order that they occurred. Put the event that occurred the longest time ago at the top of the list.

(a) You started kindergarten.
(b) Columbus discovered North America.
(c) Elvis hit the top of the charts.
(d) The first humans landed on the moon.
(e) Marie Curie won the Nobel Prize for her discovery of the elements radium and polonium.
(f) The Berlin Wall was torn down.
(g) Magellan sailed around the world.
(h) Rap music became popular.
(i) You were born.
(j) Today.

Did you have trouble deciding the order of events? Is the time span between all the events the same? Did most of these events occur recently or before you were born?

When you list events in the order in which they occurred, you make a relative time scale. You do not need the exact dates; you simply need to know what happened first. This gives you a scale based only on the sequence of events. Such a scale does not tell you how much time is involved or exactly when an event occurred. However, it does show whether one given event happened before or after another.

Fossils—Records of the Past 365

The Geologic Time Scale

Just as you ordered the historical events on page 365, geologists and paleontologists have interpreted the sequence of events of the story told in sedimentary rock. In doing this, they developed the **geologic time scale.** On this scale, rocks are placed in order from oldest to youngest. The geological and biological events recorded in these rocks are also placed in order.

Subdividing the Geologic Time Scale

Like many ideas in geology, the geologic time scale started out simply. Then, as more data became available, the scale became more complex. First, rock formations and layers were placed in order. Then geologists faced the task of subdividing this time scale. To see how they approached this task, look back at the relative time scale you made earlier. Classify the events into various groups, calling each group an "era." For example, you might consider Columbus' discovery of North America and Magellan's voyage as "the era of exploration." Do some eras contain only one event? If so, think of at least two more events to add to each era. Now, subdivide the eras into smaller groups, giving each its own name.

On what basis did you make your subdivisions? Geologists subdivide their time scales based on the period of formation of certain rocks. They call each subdivision an **era.** There are four eras. Here are their names, starting with the oldest era.

Precambrian: the era of only primitive life forms on earth

Paleozoic: the era of the earliest complex life forms on earth

Mesozoic: the era of the "middle" life forms on earth

Cenozoic: the era of the most recent life forms on earth

The four eras are then subdivided into **periods.** For instance, the Mesozoic Era consists of three periods: the Triassic, the Jurassic, and the Cretaceous. Periods themselves are further divided into **epochs.**

The dividing line between any two time units—between one era and the next, or one period and the next, for example—is based on a recognizable change. The change could be anything from a change in life forms to a period of mountain-building. For instance, the extinction of the dinosaurs separates the Mesozoic Era from the Cenozoic Era. The Paleozoic Era is considered to have ended with the uplift of the Appalachian Mountains. It is important to keep in mind that the dividing line is never sharp or sudden. Rather, it spans the time that it took for a major change to occur.

Making Sense of What You Read

1. How is the geologic time scale subdivided?
2. What is the purpose of subdividing the large units of time?
3. How do geologists decide where one division ends and a new one begins?
4. How sharp is the dividing line between subdivisions?
5. What is still missing from the geologic time scale? (Hint: Check the next paragraph.)

Geologic Time Scale

Era	Period	Epoch
Cenozoic	Quarternary	Holocene (recent)
		Pleistocene
	Tertiary	Pliocene
		Miocene
		Oligocene
		Eocene
		Paleocene
Mesozoic	Cretaceous	
	Jurassic	
	Triassic	
Paleozoic	Permian	
	Pennsylvanian	
	Mississippian	
	Devonian	
	Silurian	
	Ordovician	
	Cambrian	
Precambrian		

Clocks in Rocks

The geologic time scale merely puts the events of the earth's history in sequence. It does not reveal anything about the dates the events occurred. The age of the earth itself was an unanswered question for the early geologists. For hundreds of years, scientists debated the age of the earth, but it was not until about 1850 that serious study of these matters began.

Various techniques for dating the earth were proposed. One of the simplest was proposed at the the end of the nineteenth century. It related to the accumulation of salt in the ocean. A scientist named John Joly measured how much salt was being added to the ocean each year. By estimating the total amount of salt in the ocean, he determined that the earth was about 100 million years old. Unfortunately, this number was not accurate.

The discovery of radiation finally helped answer the question. In 1906, the famous physicist Ernest Rutherford first discovered the age of uranium in his laboratory at McGill University in Montreal. Geologists now had a method for putting dates on the geologic time scale, and even for discovering the age of the earth. The result is pictured on the next page.

Ernest Rutherford's laboratory

Investigating the Geologic Time Scale

Relative and absolute time and history of the earth

Use this diagram that shows the eras, periods, and epochs, and the times when they occurred, to answer the following questions:

1. Which was the longest era? the shortest era?
2. Which era provides the least fossil evidence? the most fossil evidence?
3. When did plants and animals first appear on land?
4. (a) How long were the dinosaurs on earth?
   (b) When did they become extinct?
5. In which period did the first forests grow? How long ago was it?
6. Suggest some changes your grandchildren might see when they study the geologic time scale.

368   The Restless Earth

Challenge Your Thinking

1. Test your skills as a paleontologist! Give one or more possible explanations for each of the following statements.

   (a) Some sedimentary rocks contain a large number of marine fossils.

   (b) Rocks of the Precambrian Era have few, if any, fossils.

   (c) Fossils of plants are less common than animal fossils.

   (d) The fossils of similar species are found in Africa, China, Paraguay, and the United States.

   (e) Fossils are seldom found in igneous rocks.

2. Fossil evidence leads scientists to believe that, one hundred million years ago, the western plains were covered by a warm, shallow sea. Check the geologic time table to discover what animals existed then. Imagine that you were one of these animals and could express yourself. Write a description of what life and your surroundings were like.

3. Kristi and Julianne went hiking in the Rocky Mountains. While gathering rocks for their classroom's geology display, they came across a fossil of a small fish. How can you explain that?

Fossils—Records of the Past

Unit 6 Making Connections

The Big Ideas

In your Journal, write a summary of this unit, using the following questions as a guide.

- How are mountains formed?
- What causes earthquakes?
- What is the elastic rebound theory, and how does it relate to earthquakes?
- What causes a volcano to erupt?
- What does the plate tectonics theory tell us about the location of earthquakes and volcanoes?
- How do you distinguish among igneous, sedimentary, and metamorphic rock? How is each formed?
- How are rocks recycled?
- How are fossils formed? Where are fossils the easiest to find? Why?
- What is a relative time scale?
- How is the geologic time scale divided?

Checking Your Understanding

1. Fill in the blanks for each of the following sentences.

   (a) When an earthquake occurs, a ___?___ measures the magnitude of force. Strong quakes often happen at ___?___ ___?___ where plates meet.

   (b) The change in location of continents can be explained by the theory of ___?___ ___?___. In this theory ___?___ boundaries are points of collision and ___?___ boundaries are points of separation.

   (c) In a volcano, ___?___ is released either through small cracks on the crust or a large vent. When it emerges, it is called ___?___. A ___?___ volcano no longer erupts with this material.

2. How good are you at making connections? How are the words in each pair related to each other?

   (a) plate tectonics and earthquakes

   (b) the Cenozoic era and the extinction of dinosaurs

   (c) the formation of igneous rock and volcanoes

   (d) shale and slate

3. Write a story that traces the formation of one of the following items.
   - a sedimentary rock
   - a fossil of a prehistoric turtle
   - the Himalayan mountains in India
   - a volcanic island

4. Draw a concept map that relates the following words:
   basalt, pumice, metamorphic, shale, rocks, igneous, slate, limestone, sedimentary, and *marble.*

Reading *Plus*

Now that you've been introduced to the movements of the earth's crust and the rocks of which that crust is composed, let's take a closer look at the make up of rocks. On pages S89–S110 of the *SourceBook,* you will find new and interesting information about rocks and the stories they reveal. Read to discover more about our restless earth.

Updating Your Journal

Reread the paragraphs you wrote for this unit's "For Your Journal" questions. Then rewrite them to reflect what you have learned in studying this unit.

Science in Action

Spotlight on Geology

Rebecca Jamieson has a Bachelor's degree, a master's degree and a Ph.D. in geology. She enjoys her work as a professor and research scientist at a university.

Q: How did you become interested in geology?

Rebecca: My interest began with rock collecting trips with my parents and family friends. I also became quite interested in the outdoors, particularly camping. But even though I always had an interest in rocks and fossils, I certainly didn't think then that I would become a geologist. That came later after I enrolled in a geology course at the university.

Q: What do you find most exciting about your work?

Rebecca: Aside from teaching, I find that field work is by far the most stimulating work that I do as a geologist. You go out in the countryside, look at rocks that you haven't seen before, and try to understand them. It's a real challenge. You can't ignore anything.

Q: What are field trips like?

Rebecca: Well, for example, I was recently on a field expedition to the Bay of Islands area in Canada. One of the things that geologists always want to do is to find out how old rocks are. But the kinds of rocks found there don't usually have fossils in them, so we had to use dating techniques that can only be done in the laboratory. We collected big buckets of old fragments of oceanic crust to take back with us to the lab for dating.

Next we took a float plane to a nearby lake. Then we had to climb a mountain, where we got more samples and carried them back down with us and out through the woods. Finally we took a boat to look at more rocks at the Bay of Islands. But we got caught in a terrible thunderstorm. It was so dark you couldn't even see the rocks. In the hour going back by boat we were soaked. It was terrible, but that doesn't happen too often.

You can tell that most geologists travel a lot. The rocks won't come to you. You have to go to them. So I get to visit many different places.

Q: What subjects should high school students take if they're interested in studying geology at a university?

Rebecca: Geology is in many ways a science that brings together the other sciences. For this reason, it's really important to have a good background in math, chemistry, physics, and biology. History and English are also important. I encourage students to take all these courses.

Some Project Ideas

Rock and mineral collections can be the basis of many worthwhile research projects. Here are some of the possibilities:

1. Collect samples from rock outcroppings near your home. Look along the sides of highways, roads, and railroad tracks. Consult a geological map, a reference book on the geology of your area, or a geologist to find out (a) whether the rocks you collect are sedimentary, igneous, or metamorphic; (b) what minerals are present in them, and (c) how old they are. Compare this information to the geological history of your area. Is there more than one major rock type near your home? Which type is oldest? Which is youngest?

2. Collect stones, pebbles, and fragments of boulders near your home. With the help of a reference guide to rock types, identify the different types. How did these rocks get to the location where you found them? Where did they probably originate? You may need the help of a geologist to answer such questions.

3. Geology collections make excellent displays. In a display you can summarize what you learned by making use of maps, drawings, and photos, as well as samples.

For Assistance

Reference books about geology, paleontology, and geophysics are good resources for information. In addition, geology teachers at high schools and universities can provide information and advice.

Science in Action

Science and the Arts

Jewelry

The land masses of the earth are constantly on the move, producing tremendous heat and pressure. Under these conditions certain precious stones may be formed—like diamonds, emeralds, and rubies. Volcanoes, for example, may contain a rock known as *kimberlite,* which often contains diamonds.

Precious stones as they are found in nature bear little resemblance to the beautiful finished products of the gem-cutter's art. This art requires patience, skill, and—at times—nerves of steel!

The Cullinan Diamond

The largest diamond ever discovered was found in Pretoria, South Africa, by a workman in a mine. He was attracted by something reflecting the sun, which he first thought was a large chunk of glass. It turned out to be a diamond weighing 3106 carats—about half a kilogram—and measuring 5 × 6 × 10 cm. The diamond was named the Cullinan Diamond after the founder of the mine. It was bought by the Transvaal government for $750,000 and presented to King Edward VII of England in 1907 for his sixty-sixth birthday.

Cutting the World's Largest Diamond

The Cullinan Diamond was sent to Joseph Asscher, a famous diamond-cutter in Amsterdam. The task of cutting it was truly nerve-wracking. With larger diamonds, the first cut is often made by cleaving the diamond along its grain. The difficulty with this is that the grain is invisible. Asscher studied the diamond for months and, finally, decided on his line cleavage. He drew a V-shaped groove along that line.

The tension was almost unbearable as he lifted his mallet to strike the metal rule he held along the groove. If his line of cleavage was not correct, the world's largest diamond could shatter completely. He brought the hammer down and, strangely, the metal rule broke in two! The diamond was still intact, but Asscher's nerves were shattered. He went to a hospital to recover.

When he finally had the strength and courage to try again, Asscher brought his personal physician with him. This time, his mallet struck cleanly, and the diamond cleaved exactly along the line he had chosen. Asscher himself did not learn of his success until some time afterward, for he had fainted at the moment he struck the blow!

The Cullinan Diamond was eventually cut into nine major gems and ninety-six smaller ones, plus a number of fragments. The largest gem, the Cullinan I or Star of Africa, is mounted in the imperial scepter of the British government. It is on display in the Tower of London.

The world's largest diamond

DIAMOND–BRILLANT CUT (round)

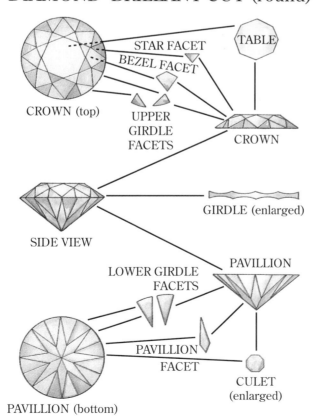

The Diamond Cutter's Art

A diamond is judged by the four C's: color, cut, clarity, and carat weight. Perfect cutting can bring out the full beauty and value of a diamond. The traditional cuts are designed to maximize the brilliance and "fire" of the diamond. One of the most popular cuts is the *brilliant cut* used in many rings. In this cut there are 58 facets, or flat surfaces, including the tiny one at the tip called the *culet*.

Find Out for Yourself

Why do you think people like to wear gems? What can you find out about the history of jewelry? Do some research on different gemstones and precious metals. Where are they found? What are their chemical compositions? How are they formed? What are different methods and styles of cutting and polishing them?

Science and the Arts

Unit 7: Toward the Stars

THE OBSERVERS
1. What Is Astronomy? p. 378
2. An Ancient Science, p. 382

THE EARTH MOVES
3. A Scientific Revolution, p. 392
4. Motions and Their Effects, p. 398

EXPLORING THE SOLAR SYSTEM
5. Visitors from Space, p. 406
6. The Space Probes, p. 411
7. Colonizing the Solar System, p. 416

OUR UNIVERSE
8. Messages from the Stars, p. 422
9. Earth's Place in the Universe, p. 429
10. A Likely Beginning, p. 436

For Your Journal

Write a paragraph summarizing your thoughts about each of the following:

1. If you could travel from Earth into outer space, what might you pass first? next?
2. How do you account for day and night? summer and winter?
3. Why are kilometers often inadequate for measuring distances in space? What do we use in their place?
4. What motions of the Moon make the Sun look like it does in this photograph?

THE OBSERVERS

What Is Astronomy?

Imagine!

You are an astronomer atop the Mount Kea volcano in Hawaii, using a telescope to take photographs of the heavens. The next morning while studying the photographs you took the night before, you discover a most amazing sight.

Quickly, you call other members of the team for consultation. Together, you decide that you have photographed two galaxies colliding in outer space. Even more exciting, it looks as though there's a dark, mysterious "black hole" within the galaxies. A breakthrough? You and your fellow astronomers write an announcement about your discovery for the media.

The Honolulu Daily Tribune

75 cents • Since 1952 • January 1, 1992

Enormous Dark Object Discovered by Astronomers

by Alan Moore
HRW News Service

MAUNA KEA, HAWAII — Using the 88-inch telescope atop Mauna Kea volcano in Hawaii, astronomers have discovered a huge, mysterious object within two colliding galaxies. If the object is a black hole, it is 10 to 100 times as massive as any black hole previously known or even believed possible.

The object was discovered while astronomers were studying a bright galaxy named NGC 6240, which lies some 300 million light-years from Earth. The Mauna Kea astronomers estimate that the object must be some 100 billion times as massive as the Sun. This means it would have a mass equal to that of some entire galaxies. Despite such a large mass, the object occupies a space less than one millionth the diameter of Earth's galaxy, the Milky Way.

The discoverers say that if the object isn't a black hole, it might be a dormant quasar of the kind from which galaxies were born more than 10 billion years ago, when the universe was still young. The Mauna Kea astronomers acknowledge that many more observations must be made to reconcile established theories with the new discovery.

NGC 6240 as photographed by the Mauna Kea Telescope.

Supreme Court Splits on Key

Students Scoring Higher in Science

monitors student achievement was quoted as saying that test scores have risen significantly in

How important is their discovery? What is a galaxy? a black hole? a light-year? a quasar? A simple answer is that they are all part of the science called **astronomy**.

What do you think astronomy is? In your Journal, write as many phrases as you can to complete the statement "Astronomy is ___?___." The photos on this page and the next will help you add to your list.

Artist's impression of the Milky Way galaxy. The red circle shows the location of our Solar System.

Astronaut on our nearest neighbor—the Moon

Volcanoes on Mars

Saturn's rings

Star cluster

The Discovery shuttle, with the Hubble space telescope aboard

380   Toward the Stars

Artist's impression of colliding galaxies

Solar eclipse

Ask an Astronomer

What do astronomers do? If you could ask an astronomer some questions about the universe and Earth's place in it, what would you ask? In small groups, make a list of questions that really interest you. The drawing shows some questions to get you started.

Share your group's questions with the other groups. As you study this unit, you'll discover answers to many of these questions.

The Observers

2 An Ancient Science

Observing the Heavens

Did you know that astronomy is the oldest science? Since the earliest times, humans have looked toward the stars. In fact, ancient observers were excellent viewers of the skies. By careful observations, early Babylonians, Egyptians, Aztecs, and Mayans could accurately predict the locations of heavenly bodies. They relied on the skies for answers to such questions as when to plant crops and when the seasons would change.

You are like the ancient astronomers when you look up at the sky on a clear night. What you see is what they saw. Make a list of all the heavenly bodies these ancient gazers might have seen.

Did you include in your list NJYLCRQ, KMML, QRYPQ, QFMMRGLE QRYPQ, and the QSL (or SUN, if you crack the simple code)? Ancient peoples saw all of these. In the following Exploration, you will think about and investigate some of the motions that occur in these heavenly objects. Part 1 has a list of observations that all sound like they might be true; some may even be observations you have made yourself. The trick is, some are false. Your mission is to discover which statements are not true. In Part 2, you will have the chance to check out some of these statements experimentally.

EXPLORATION 1

Heavenly Motions

Part 1
A Celestial Quiz

Work in groups of three. Decide if each statement is true or false. Is there complete agreement on each statement among members of your group?

1. Each day the Sun rises due east and sets due west.
2. Every 7 days, the length of daylight changes by several minutes.
3. In your area, there are times of the year when the Sun is directly overhead.

4. Your winter shadow at noon is shorter than your summer shadow at noon.
5. Shadows shorten as the morning progresses.

6. Like the Sun, many stars appear to rise and set at various times of the night.
7. If you look at the Big Dipper at two different hours of the same night, its orientation (position) in the sky will change.
8. If you see the Moon in the early evening in the east, it must be a full Moon.
9. On two consecutive evenings at the same hour, the Moon is farther east and bigger on the second evening.
10. You always see the same side of the Moon from your position on Earth.
11. All stars that you see appear to be the same color.
12. The Moon appears larger when it is seen on the horizon than when it is seen higher in the sky.
13. The only time you can see the Moon is at night.
14. The Moon is closer to Earth than the stars are.
15. Human behavior is affected by the phase or shape of the Moon.
16. Not all stars are equally bright.
17. The star patterns, or constellations, change shape from night to night.
18. Planets are indistinguishable from stars when seen with the naked eye.
19. Some constellations are seen only in certain seasons.
20. Planets wander in and out of the constellations.

Do we always see this face of the Moon from Earth?

Did everyone agree on these statements? Which ones created the greatest debate? In the next part of this Exploration, you will check out some of these statements as an observer. From your investigations, you may be able to suggest other observations that could be added to the Celestial Quiz.

EXPLORATION 1—CONTINUED

Part 2
Checking It Out

Choose at least two of the following observational activities to help you verify statements from the Celestial Quiz. When reporting your conclusions, indicate which of the statements on the Celestial Quiz you have confirmed or disproved. What other observations have you made?

1. The first sundial that was used to determine the passage of time was a simple upright stick. Using modeling clay, mount a paper towel tube on a piece of cardboard, as shown below. For several days during the next week, place it outdoors at the same time each day. Be sure to put the cardboard in exactly the same position each time. Draw the shadow's position and length at that time, and record the date for each day. Does the direction of the shadow change? The length of the shadow? What does the shadow indicate about the Sun's position in the sky?

2. Use the shadow stick from #1 to find out how shadow length and position change during the day. What direction does the shadow point in early morning? at noon? a little before the Sun sets? Will there ever be a time when the Sun is directly overhead and therefore not cast a shadow? How does the length of the shadow vary?

3. Observe the rising or setting Sun. **Caution:** *Don't look directly at the Sun.* Several times during the week, record the time when the Sun first appears above or disappears below the horizon (or the top of a building). Make sure you view the Sun from the same position each time. Use some fixed object, such as a tree, to determine the exact point at which the Sun rises or sets. Does this point change? If it changes, in what direction does it change? Would this direction be the same in spring as in fall? Does the time of rising or setting change? By how much?

384   Toward the Stars

4. Over the period of a week, chart the position and shape of the Moon. Do this at the same time each night. As the month progresses, in what direction does the Moon move? How does its shape change from night to night?

5. On a clear night, spend some time observing the stars. Are all stars equally bright? Are they all the same color? Using a star chart, try to identify the brightest stars.

6. Sight a star near the western horizon. Follow its position at various times on the same night. Does it set as the Sun does? Do stars rise in the east?

7. Locate the constellation of the Big Dipper in the night sky. Draw what you see on a piece of paper. An hour or two later, observe the Big Dipper again and draw what you see, using the same piece of paper. Did its shape change? Did its orientation change? Is there a point in the sky about which it and other stars seem to be revolving?

Lining up two fixed points with a star to see if the star moves

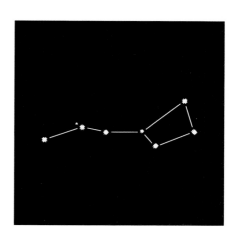

8. Line up two fixed points, such as a house and a tree, with a star. Each night make another sighting at the same time from the same spot. What do you observe? Does the star's position appear to be different from night to night? If so, in what direction does it move?

9. With or without binoculars, carefully observe the features of the Moon. Make a diagram of some of the features you see. Do the features change from night to night, or do they remain the same?

Think About It

Look over the statements in the Celestial Quiz again. Are there some statements not yet verified? What observational activities could you plan to check them out? Are there other statements you could add to the Celestial Quiz? How would you verify these new statements? Are there still other observational projects you would like to carry out? What are they?

A Celestial Picture Study

The following illustrations and their accompanying stories will add to the large number of observations you have already made. Team up with another person to study each illustration, read its story, and consider the discovery questions. How many new ideas about the motion of the heavenly bodies can you discover?

Picture Study 1

Native Americans in Wyoming's Bighorn Mountains used a calendar wheel to predict the day when the Sun would reach its highest point in the sky. According to legend, "the growing power of the world" is strongest on the day the Sun is highest. On this day, the rising Sun was in line with two piles of rocks, or cairns.

- Is the Sun higher in the sky at one time of the year than another?
- Does the position on the horizon where the Sun rises and sets change from day to day?

Picture Study 2

In 1610 a famous scientist named Galileo made the diagram of the Moon shown below by using a new invention called a telescope. (You will be hearing more about Galileo later in the unit.) Next to his diagram is a recent photograph of the Moon in a similar phase.

- Does it appear that the face of the Moon you see at night is the same as the one that Galileo saw almost 400 years ago?

Picture Study 3

If the lens of a camera is held open for a time while aimed at the night sky, each star becomes a star track.

- Do stars appear to revolve? If the arrow points to the beginning of a star track, in what direction are the stars revolving?
- What other observations regarding stars can you make from this photograph?

Picture Study 4

The early observers noted objects in the skies that they called "wanderers." These are the planets. At that time, only five were known. How many do we know about now? This time-lapse photograph shows the motions of several planets over a long period of time.

- What strange motions do planets seem to make that the Sun and Moon do not?
- Does each planet seem to follow more or less the same path that the other planets do?

Picture Study 5

Star charts show the positions of stars and constellations for a particular time of the year. Study the star charts for both a summer and a winter sky.

- What features are common to both charts? How do they differ?

Star chart of a summer sky   Star chart of a winter sky

The Observers 387

Making Sense of the Skies

If you lived 2000 years ago at a time when there were no telescopes or spaceships, and the only means of observing celestial objects was with the naked eye, how would you explain your observations so far this unit? With a friend, weave your explanations into a theory that explains as many of your observations as possible. Can your theory explain the reasons for night and day? the star tracks? why the calendar wheel works? Are there observations your theory cannot explain? Does this mean that your theory is not useful?

In Exploration 2, you will be able to compare your theory with those of a number of early thinkers and observers.

EXPLORATION 2

The Early Observers

Living in Greece and Egypt over 2000 years ago were many keen observers and thinkers. How did they explain their observations of the heavens? As you read about three of them, think of the observations they may have been basing their ideas on.

1. Aristotle (384–322 B.C.)

Aristotle (pronounced air us TAHT ul) was the most influential thinker of his time. He suggested that everything in the universe is composed of four elements: earth, air, fire, and water. He proposed that each element has a natural resting place, with earth at the center, surrounded by spheres of water, air, and fire.

His model of the universe contained a total of eight spheres revolving around Earth. The first seven spheres carried the Moon, the Sun, and the five known planets. Aristotle proposed that beyond these seven spheres, the stars were tiny lamps fixed to another revolving sphere.

Aristotle

He theorized that the sphere carrying the Sun revolved around Earth each day, causing day and night. The motion was much like following the threads on a corkscrew. With each revolution around Earth, the Sun was carried higher (or lower) in the sky. He proposed that this movement causes the seasons.

- How is this theory based on everyday observations?
- Try drawing a diagram based on Aristotle's ideas.

2. Aristarchus of Samos (310–230 B.C.)

Aristarchus (pronounced air us TAR kus) combined mathematics and good observation to come up with another theory that is surprisingly close to the picture of the Solar System known today. He suggested that the Sun is the center of the universe—not Earth. He proposed that Earth and the other planets revolve around the Sun, causing the year. Day and night, Aristarchus suggested, are caused by the rotation of Earth on its axis. He was the first to measure the distance to the Sun and Moon. He concluded that the Sun is much farther away and larger than previously thought. The stars, he reasoned, are unimaginably far away. *Although his theory explained observations as well as Aristotle's theory, it was not accepted by the people of that time and was forgotten for nearly 2000 years.*

- What do you think about Aristarchus' theory?
- Why do you think that good ideas and reasoning are sometimes disregarded in favor of other ideas?
- Try drawing a diagram that illustrates Aristarchus' ideas.

3. Ptolemy (about A.D. 127–145)

Ptolemy

Ptolemy (pronounced TAHL uh mee) supported the ideas of Aristotle. He reasoned that if Earth really moved, the tremendous winds caused by Earth's movement would blow birds off their perches and leaves off the trees. He supported Aristotle's idea that the Sun revolves around a stationary Earth by noting that every day, we see the Sun actually moving across the sky.

But Ptolemy recognized that Aristotle's model had two serious flaws. First, it could not explain why the "wanderers," or planets, varied in brightness from year to year. If they orbited Earth at fixed distances, their brightness should not vary greatly. Second, the paths that the planets traveled were more erratic than the paths suggested by Aristotle's theory. A planet such as Mars, for instance, appears to make a loop in the sky going backwards for a while before continuing its forward motion. This curious backward motion is called **retrograde motion.**

As seen from Earth, Mars appears to make a looping motion in the sky.

EXPLORATION 2—CONTINUED

Ptolemy's solution was to add another motion to a planet. He suggested that a planet not only revolves in a circular orbit around Earth but also makes a smaller revolution about points on its main orbit—like a small wheel turning on the circumference of a larger wheel. To get a good picture of Ptolemy's idea, imagine the rocking seat of a ferris wheel actually making a complete turn as the ferris wheel itself turns. The motion of the seat is like the motion of a planet in Ptolemy's model.

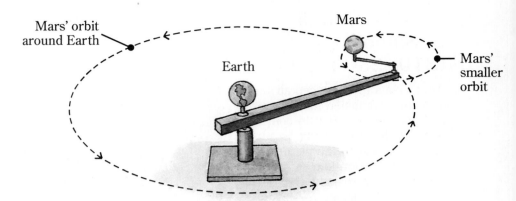

Ptolemy gave Mars two motions. He thought that Mars moves in small circles as it moves around Earth in a larger orbit.

- Study the model above that shows how Ptolemy thought Mars moved. How does Ptolemy's "solution" explain the changing brightness and occasional backward (retrograde) motion of a planet such as Mars, as viewed from Earth?
- Can you think of another explanation for retrograde motion?

Thinking Further

1. Is the theory you proposed on page 388 similar to any of the early Greek and Egyptian observers' theories? Which theory seems to account best for your observations of the heavens? Why?

2. Imagine that you lived at the time of Aristotle, but your ideas matched those of Aristarchus of Samos. What could you say to convince Aristotle that Earth revolves around the Sun? Remember, they had no telescopes and spaceships back then.

3. For 1500 years, people believed that Earth was the center of the universe. Why do you think this was such a popular view?

4. Aristarchus of Samos proposed a *heliocentric* model of the Solar System. Aristotle and Ptolemy both proposed *geocentric* models. Based on their theories, what do you think these terms mean?

390   Toward the Stars

Challenge Your Thinking

1. Many early astronomers suggested that Earth is flat. For example, one astronomer pictured Earth as a flat circular disk floating in a sea of water. Another early observer considered Earth to be a flat object freely suspended in space. What may have been some reasons for thinking Earth is flat? What are some reasons for thinking Earth is spherical?

2. Polaris (the North Star) will appear noticeably higher and higher above the horizon the farther north you travel, until at the North Pole, it is directly overhead. Does this observation support a flat-Earth or a spherical-Earth theory? Study the diagrams to the right and explain how they help support your answer.

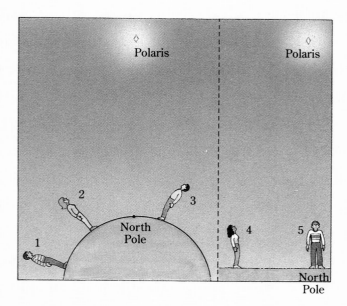

3. The scenes below show various positions of Polaris over a horizon. Copy the scenes into your Journal and label each one with the observer (from the spherical-Earth diagram) who could have seen such a view. Which scene below is closest to what Observer #4 would see? Would the view that Observer #5 sees be much different from that of Observer #4?

The Observers

THE EARTH MOVES

3 A Scientific Revolution

The Trial of Galileo

It is June 1632. The atmosphere in the courtroom is electric. The greatest scientist of the time, Galileo Galilei, is on trial for teaching the views of Copernicus: that the Sun is the center of the Solar System—not Earth. In this model, Earth is just one of a number of planets that revolve around the Sun. This is the second time Galileo has been on trial for his views. After the first trial, he was ordered by the Roman Catholic Church not "to hold or defend" Copernicus' theory. At the second trial, Galileo must answer to charges that he has willfully disobeyed this order.

The verdict—guilty on all counts. The sentence is house arrest for the remainder of his life. Legend says that as Galileo was led sadly from the courtroom, he was heard to mutter ". . . and still the Earth moves."

Imagine you are living at that time and have been asked to be part of a team of lawyers preparing for the trial of Galileo. The task is great, and much needs to be done. Here are some thoughts that your team has jotted down. Are there other questions that may be important in working toward an acquittal ("not guilty")?

- Who was Galileo? What was his background?
- Who was this man Copernicus? Where did he come from? What were his ideas? Why does Galileo consider himself a Copernican?
- Why would Church officials care whether the Sun went around Earth or vice versa?

EXPLORATION 3

Copernicus' Model: The Planets and the Ecliptic

To prepare for the trial, you and your team need to make a model of the Solar System based on Copernicus' ideas. You want to see whether the Copernican theory explains celestial motions that the theories of Aristotle and Ptolemy could not explain.

Part 1
The Model and Its Use

At the time of Galileo, only six planets were known. How many are known now? Galileo taught that the six planets orbited the Sun. Your model will show the orbits of the first four planets you would see if you started at the Sun and traveled outward. The distance of each planet from the Sun is expressed in terms of Earth's average distance from the Sun. This distance is referred to as one astronomical unit (AU), and it is equal to 150 million kilometers. (Try writing out this figure.) The time it takes for each planet to revolve once around the Sun is given in Earth days.

Your model will have a scale of 1 AU = 10 cm. Using this scale, Earth would be 10 cm from the Sun on your model. Why?

Observers from the earliest times noticed that, when seen from Earth, the planets moved against a background of 12 different groupings of stars, called **constellations**. These constellations make up what is popularly known as the **zodiac**. The early observers knew that the planets did not wander randomly through the heavens. They all appeared to move counterclockwise through the zodiac, which consists of the constellations Leo, Virgo, Libra, Scorpio, Sagittarius, Capricorn, Aquarius, Pisces, Aries, Taurus, Gemini, and Cancer. Your model would not be complete without the 12 constellations of the zodiac.

Planet	Average Distance (AU)	Time It Takes to Revolve Once Around the Sun (in Days)
Mercury	0.4	88
Venus	0.7	225
Earth	1.0	365
Mars	1.5	687

EXPLORATION 3—CONTINUED

You Will Need

- a pencil
- a compass
- a protractor
- a meter stick
- star charts
- glue or tape
- four sheets of unlined paper, fastened together to form one large rectangle
- several small plastic-foam balls or marbles to represent the planets
- modeling clay to hold the plastic-foam balls or marbles in place

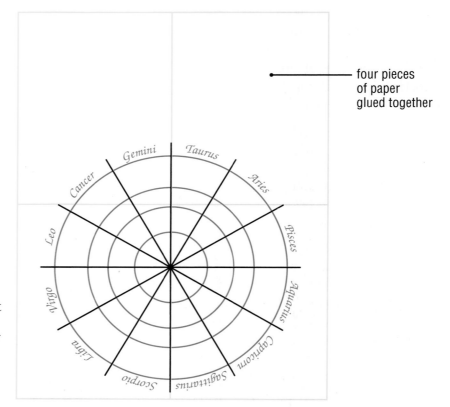

What to Do

1. Turn your glued paper so that its longer direction is up and down. Using a compass, draw a circle with a radius of 15 cm. The center of the circle should be in the lower half of the paper so that the circle comes within 4 cm of the bottom of the page. Which planet would have the orbit you just drew? Now draw other circles to represent the orbits of the remaining planets from the table on page 393.

2. Where is the Sun in this model?

3. Using a protractor and ruler, divide the circles into 12 equal segments. Each segment will form an angle of 30 degrees at the center of the circles. Extend each line 4 cm beyond the orbit of Mars.

4. In the segments outside the orbit of Mars, place the names of the constellations that form a backdrop to the moving planets. Refer to the diagram above to place the zodiac constellations in their proper order. Place Taurus at the top of your model, as shown in the diagram. Each constellation will occupy one 30-degree segment, although in reality they vary in size. Of course the stars making up these constellations are actually a tremendous distance beyond the orbit of Mars. Have you come across the names of these constellations before? Where?

5. Examine a star chart for this month. Which constellations of the zodiac are visible in your night skies? Where must Earth be placed on your model such that those constellations would be visible this month? Use a marble or plastic-foam ball to represent Earth, and place it in its proper location on the model. Use a small piece of modeling clay to hold it in place.

394   Toward the Stars

Interpreting the Model

1. From this model, infer what ideas Galileo taught that eventually caused him problems with the Church authorities.

2. Demonstrate on the model the motion of Earth over one year. To make the model fit with observation, the ball must be moved in a counter-clockwise motion around the Sun. What might you observe about the stars' positions because of Earth's motion?

3. Use another plastic-foam ball or marble to represent Mars. Place Earth and Mars in their respective orbits with the constellation of Taurus as their backdrop of stars. Move Earth through one year. In this time, how far has Mars moved? After the planets have moved this distance, will Mars be visible from Earth?

4. Any model must show why you experience night and day. How would you move the ball to demonstrate this? The motion you just showed is called **rotation**. What is the difference between rotation and **revolution**? Refer to the picture of the star tracks on page 387 and the observations you made of the Big Dipper. In what direction must Earth be rotating to cause these effects?

5. As the year progresses, new constellations of the zodiac appear in the eastern sky, while others disappear in the western sky. Use your model to explain this observation.

6. Copernicus (and later Galileo) inferred that the Sun and planets lie in the same plane. In other words, they inferred that the Solar System is essentially flat. This plane is called the **plane of the ecliptic**. Where is the plane of the ecliptic in your model?

Are the people on the ride rotating, or revolving? Is the ball rotating, or revolving?

7. In your opinion, is this a good model? Could it be used as an exhibit in the retrial of Galileo? Does it explain the retrograde motion of Mars in a satisfactory way? Is it a simpler explanation than the two motions that Ptolemy suggested? Part 2 of this Exploration will help you discover some answers.

EXPLORATION 3—CONTINUED

Part 2
The Retrograde Motion of Mars

What causes Mars to appear to go backward in its path for a while, before continuing in its original direction? A Copernican would say that it is simply the result of the faster Earth catching up to and passing Mars. This causes Mars to appear to loop backward in the sky when viewed against the more distant stars.

You can demonstrate this by following the progress of Earth and Mars over a period of 100 days on your model. Every 10 days on the model, note where these planets lie among the more distant stars.

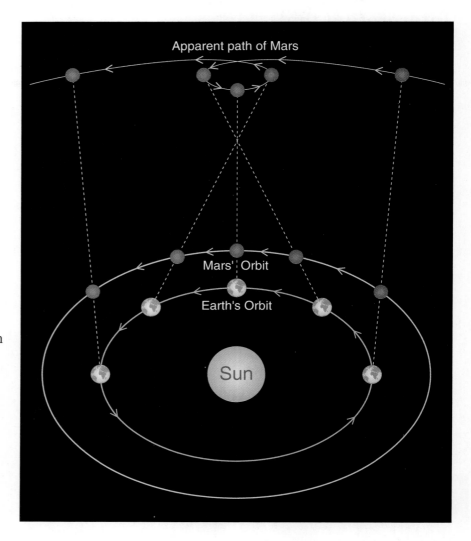

What to Do

1. The top of your page will represent where the stars appear to be when viewed from Earth. Call this the *star line*. In reality, this is a tremendous distance beyond the orbit of Mars. As Mars and other planets move, they will appear to move among the stars along this line.

2. Using a pencil, draw a dot to represent Earth where its orbit crosses the boundary between Taurus and Aries, as shown in the diagram below. This is the starting position of Earth. Label this position as Earth 1.

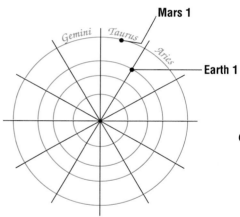

3. Draw Mars on its orbit so that it is halfway between the Taurus/Aries and the Taurus/Gemini boundaries. Label this position as Mars 1, as shown in the diagram above.

4. In 365 days, Earth will travel once around the Sun—a distance of about 63 cm on your model. How far will it travel in one day? Where will Earth be on the model in 10 days? Can you convince yourself that it has moved a distance of 1.7 cm from its original position? Using a compass, measure off this distance and label this position as Earth 2. Repeat this step to locate Earth's position in 20 days, 30 days, 40 days, and so on up to 100 days. Label these positions as Earth 3, Earth 4, and so on up to Earth 10. Be sure that Earth is revolving counterclockwise around the Sun.

5. Where will Mars be after 10 days, 20 days, and so on up to 100 days? Taking into account the larger orbit of Mars and the fact that Mars takes about twice as long as Earth to make the journey around the Sun, you can discover that Mars will move 1.4 cm along its orbit in 10 days. Mark off 1.4-cm segments from the original position of Mars, moving in a counterclockwise direction. Label the ends of these successive segments as Mars 2, Mars 3, and so on up to Mars 10.

6. Draw a line connecting Earth 1 with Mars 1. Extend this line to the star line at the top of the page. This is where Mars will appear to be for an observer on Earth.

7. Repeat this for each new position of Earth and Mars. Where does Mars appear on the star line for each position of Earth and Mars? Does Mars appear to change direction when viewed against the background of stars? In reality, does it go backward?

Defending Galileo

Are you ready to develop your defense of Galileo? How can what you just discovered be used to convince a jury that it is Earth that moves around the Sun—not the other way around? Practice your defense. Try explaining to an imaginary jury why retrograde motion is a natural consequence of Earth's motion. In the pages that follow, you will investigate in greater detail the effect of the motions of Earth as suggested by Copernicus and later by Galileo. This information may provide crucial evidence for the retrial of Galileo.

Motions and Their Effects

Many Journeys in a Lifetime

How many times have you traveled around the Sun? According to the Copernican theory, you make this journey once a year. The journey covers a distance of 942 million kilometers. If you were a tour guide for this journey, what sights would you point out? What major events would you experience? Besides the changing constellations of the zodiac, what other signposts would tell you where you are on your journey? For example, think about changes you experience throughout the day and year.

Referring to the model of the plane of the ecliptic you made in Exploration 3, write down three observations you would point out as tour guide for this trip. Then try Exploration 4, in which you will model such a journey.

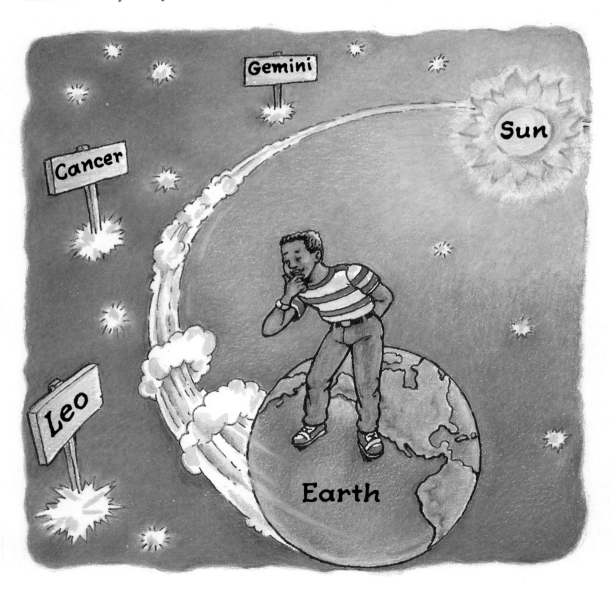

EXPLORATION 4

A Trip Around the Sun

You Will Need

- a large plastic-foam ball (20 cm in diameter or larger)
- an elastic band
- a plastic cup
- 2 toothpicks
- a light source
- your model of the ecliptic from Exploration 3
- a protractor
- a compass

What to Do

1. Form groups of four or five. Set up the model as shown in the diagram below. The elastic band around the large plastic-foam ball represents the *equator* of Earth. The toothpick represents the axis about which Earth rotates. Earth's **axis** is an imaginary line that runs from the North Pole to the South Pole, straight through the center of Earth. The axis is perpendicular to the plane of the equator. The light source represents the Sun.

2. You'll be setting up four situations that represent different positions on your journey at specific times of the year. At each position of Earth, you'll have several missions to accomplish and some thought-provoking questions to consider.

3. In setting up the models, keep two things in mind:
   - The motion of Earth around the Sun is counterclockwise when viewed from over the North Pole of Earth, as you can see from the diagram above.
   - Regardless of Earth's position around the Sun, Earth's axis always points in the same general direction. At any position along Earth's orbit, Earth's axis is parallel to where the axis was (or will be) at any other position in the orbit. Earth's axis is tilted at an angle of 23½ degrees from the perpendicular, toward the plane of the ecliptic. The axis always appears to point toward the star Polaris.

What season is represented in this set-up?

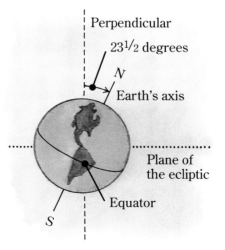

The Earth Moves 399

EXPLORATION 4—CONTINUED

Position 1

December 21 or 22—The Winter Solstice

In this position, Earth's axis is tilted at an angle of 23½ degrees from the perpendicular so that the Northern Hemisphere is pointing away from the Sun. Notice where on Earth the Sun is directly overhead. Do this by inserting a toothpick in a position where it will not cast a shadow. The toothpick should be positioned so that if extended, it would pass through the center of the plastic-foam ball. This latitude is 23½ degrees south of the equator—a latitude called the *Tropic of Capricorn*. At this time of the year, the constellation Capricorn cannot be seen from Earth because it is in back of the Sun.

1. In your Journal, use a compass to draw a diagram of Earth similar to the one shown here. Label the equator and the Tropic of Capricorn on your diagram.
2. Demonstrate a day by rotating Earth one full rotation. Which of Earth's poles is in darkness 24 hours a day? How does the length of the day change as you approach the equator? What two reasons can you give for the winter season in the Northern Hemisphere?
3. If Capricorn is on the other side of the Sun, away from Earth, what constellations must be in Earth's night skies? Set up your model of the plane of the ecliptic (Exploration 3) to represent this situation.

Position 2

March 20 or 21—The Vernal Equinox

Move Earth counterclockwise one quarter of the way around the Sun. Be careful to keep the axis pointing in the same direction at all times. Revolving Earth this far means that one-quarter of a year has passed. In this position, people are experiencing equal hours of day and night, all over Earth. Can you see why?

1. At what latitude is the Sun directly overhead at noon? Find out by inserting a toothpick so that it does not make a shadow. Draw a diagram like the one you drew representing Earth at the winter solstice. Label the latitude at which the Sun is directly overhead at noon.
2. Rotate the model Earth in this position so that the opposite side of the model is now experiencing sunlight. Is the Sun still directly overhead at the same latitude?
3. Go to your model from Exploration 3 and set it up to represent the vernal equinox. What constellations are visible in the night sky?

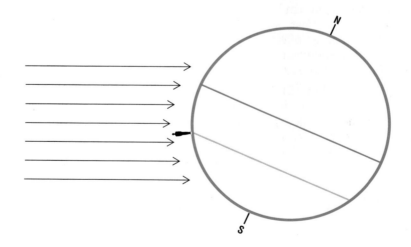

Position 3

June 20 or 21—The Summer Solstice

Earth has moved so that it is now directly opposite to where it was on December 21 or 22. Earth's axis should be parallel to what it was on that date. This is the first day of summer—the longest day of the year in the Northern Hemisphere.

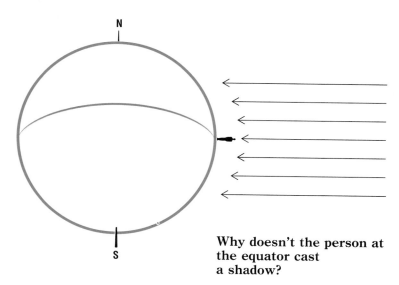

Why doesn't the person at the equator cast a shadow?

1. The Sun's hottest rays are hitting Earth 23½ degrees north of the equator at a latitude called the *Tropic of Cancer*. How could you prove that this is so? Draw a diagram like the one you drew representing Earth at the winter solstice, only this time, label the Tropic of Cancer. Will Earth be on the same side of the Sun as it was for the winter solstice?
2. Which pole is in darkness 24 hours a day? in sunlight 24 hours a day? What two reasons can you give for the warmer temperatures in the Northern Hemisphere?
3. Go to your model of the plane of the ecliptic and set it up to represent the summer solstice. What constellation is in back of the Sun? in Earth's night sky?

Position 4

September 22 or 23—The Autumnal Equinox

In this position on your journey around the Sun, the Sun's hottest rays are again hitting the equator at noon, as they were on March 20 or 21. This means that on this day, the Sun is directly overhead at noon. Again, everyone on Earth is experiencing equal hours of daylight and darkness.

1. During the complete day, is the Sun always directly overhead at the equator at noon? How could you find out?
2. If you were at the North Pole looking toward the Sun, would you see it rise above the horizon during the day, as it does where you live?
3. During a complete year, through what range of latitudes do the Sun's hottest rays move? Where would you most like to live?
4. Set up your model of the plane of the ecliptic to represent the autumnal equinox. What constellations are visible in the night sky at this point in your journey?

Now that you have made this trip, you know enough to be the tour guide on the next trip. What are the major sights and events you would point out on your next journey around the Sun? Record what you plan to highlight.

The Earth Moves

Kepler's Puzzle

The models you made in Explorations 3 and 4 lend support to Copernicus' theory. But during Galileo's time, those who thought that Copernicus' theory was valid still faced one serious problem: If the orbits of the planets were perfect circles, why does the observed brightness of Mars differ so much each time Earth comes between it and the Sun? The variation is much greater than it would be if the orbits were perfect circles.

Kepler musing over possible shapes for orbital paths

A German astronomer named Johannes Kepler happened on an answer to the problem in 1605 as he was trying to discover the shape of planets' orbits. For three years, Kepler attempted to fit the information known about Mars' orbit to circles of all sizes. At that time, everyone assumed that the planets traveled in perfectly circular paths because to them, the circle represented perfection. But no matter how hard he tried to make it work, no circle would fit the data. Kepler had fallen into the trap of anticipating what he believed to be the correct answer, thus blinding himself to possible alternatives. Has this ever happened to you?

In frustration Kepler finally resorted to an idea that he had previously disregarded—a shape called an *ellipse,* which looks like a stretched-out circle. When the data known about Mars' orbit was applied to this shape, it fit perfectly. Mars' orbit did not describe a circle but an ellipse, instead!

With this discovery, Kepler was able to describe the paths of all revolving objects bound by gravitational forces. His discovery applies to comets, planets, moons, human-made satellites, and even distant stars that have been caught up in each other's gravitational forces.

In the next Exploration, explore the shape for which Kepler at first had such low regard. Then answer the question that had long baffled astronomers: why does the brightness of Mars vary?

EXPLORATION 5

Investigating an Ellipse

You Will Need

- a sheet of corrugated cardboard or a stack of several sheets of cardboard
- unlined paper
- two thumbtacks
- a 40-cm piece of string
- a centimeter ruler

What to Do

1. Press the thumbtacks into a sheet of unlined paper that has been placed on top of the cardboard. Loosely tie a piece of string around the thumbtacks, as shown in the photograph below. Place a pencil through the string and trace out an ellipse. What you'll see is the shape that planets travel as they orbit the Sun. Using the same piece of string, can you make ellipses with different shapes? How?

2. Measure the long axis of several ellipses. What is the relationship between the positions of the thumbtacks and the shape of the ellipse? What must be the position of the thumbtacks to make a circle?

3. At the position of one of the thumbtacks, draw a diagram representing the Sun. This location is a focus of the ellipse. (The plural of focus is *foci*. An ellipse has two foci.) When Earth is closest to the Sun, it is 147 million km from the Sun. At this distance, it is winter in the Northern Hemisphere. At its farthest distance, it is 152 million km from the Sun. Earth reaches this position during the summer in the Northern Hemisphere.
   (a) Place this information on your diagram. Label the summer and winter positions of Earth.
   (b) How do you explain that when Earth is closest to the Sun, it is winter in the Northern Hemisphere? (Think of what you found out at Position 1 in Exploration 4.)

4. How does Kepler's discovery explain why Mars changes in brightness when seen from Earth?

5. Through his observations, Kepler made another discovery about planets moving in their elliptical orbits: The planets do not move with constant speeds. A planet increases in speed as it gets closer to the Sun. Compare the speed of Earth in its orbit during spring, summer, fall, and winter. If Earth's orbit were circular, what might you infer about its speed?

Drawing an ellipse

A Mock Trial

You have all the physical evidence needed for the retrial of Galileo. Which were the more persuasive arguments in support of the Copernican theory? What questions would you ask of the following witnesses who have been asked to testify at the retrial? Are there other witnesses you would call? What other witnesses could the prosecutor call?

Set up a mock retrial of Galileo with different class members playing the different roles. Form at least six small teams. Each team will be responsible for filling out the details of one or more roles and for choosing which team member will play that role. Those not playing a role will be the jury. Besides Galileo, his lawyer, and the prosecutor, include at least the following three roles. You may, of course, add other roles to the list.

- Church official: "I believe absolutely in the teaching of Aristotle. Earth, and therefore humans, are at the center of all things."
- Merchant: "What do I know of moving planets or suns? I believe what I *see*. I see the Sun moving, so I believe it moves."
- Astronomer/Scholar: "Galileo is very persuasive in his arguments. But I would like to examine both theories more carefully."

Challenge Your Thinking

1. You've made many observations of what happens in the Northern Hemisphere as a result of the motions of heavenly bodies. What happens in the Southern Hemisphere? Which way will the stars appear to revolve there—clockwise or counterclockwise? On days when you wear a winter coat to school, what would Australian students be most comfortable wearing?

2. The observations below are all true. Using plastic-foam balls, toothpicks, and any other materials you think you'll need, show why each observation is true. Work with others in groups so that you can share ideas as scientists do when they work on a problem. Then, with the aid of a diagram, explain in writing why each is true.

   The Observations:
   (a) The Sun rises in the East and sets in the West.
   (b) The Sun is higher in the sky in the Northern Hemisphere in the summer than in the winter.
   (c) In the summer in the Northern Hemisphere, the North Pole has 24 hours of daylight.
   (d) The Moon goes through phases from new Moon (when it is not visible in the nighttime sky) to full Moon (when the full face of the Moon can be seen).

Crescent moon **Quarter moon** **Gibbous moon** **Full moon**

   (e) Each night the Moon rises later (by approximately 50 minutes) than it did on the previous night.
   (f) The same face of the Moon is always visible from all parts of Earth.

3. In July 1991, a complete eclipse of the Sun was visible in Hawaii and Mexico. With a diagram or your model, show how a total eclipse of the Sun occurs and why one is not seen every year.

The Earth Moves

EXPLORING THE SOLAR SYSTEM

5 Visitors from Space

Meteors and Meteorites

The photo above shows evidence of visitors from space to Earth. Do you know what these visitors are?

Try viewing the Moon through binoculars. What is its most distinguishing characteristic? Is there any evidence of water? of wind and dust storms? What do you suppose caused the craters on the Moon? Much can be learned about your nearest neighbor by simple observation.

The Moon's features are similar to those on Earth. Compare the picture of the crater in Arizona, shown above, with a Moon crater, shown on the right. How are they similar? Do you see any differences? Could the craters found on both surfaces have been formed in the same way? You can simulate one possible cause for the craters in the following way. From a height of 1 m, drop several milliliters of water onto a bed of flour. What do you observe? What could have been "dropped" on Earth and on the Moon to create the craters there?

Lumps of rock and/or metals that sometimes enter Earth's atmosphere are called **meteors.** If large enough, they crash to Earth instead of burning up completely in the atmosphere. Meteors that reach the ground are called **meteorites.** The craters are visible evidence of their impact, even though the meteorite itself may not be found. That's because the heat generated on impact can vaporize a meteorite. How does your flour-and-water activity simulate this part of the process of crater formation?

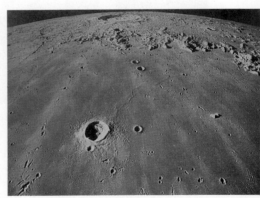

Moon crater

Much of the evidence of meteorite impacts on Earth has been destroyed through weathering and erosion by air and water. How could weathering and erosion be simulated in the flour-and-water activity? Why do there seem to be so many more craters on the Moon?

Have you ever seen a "shooting star"? If you have, what you saw was actually not a star at all, but something within Earth's atmosphere. (Recall that stars are located great distances away from Earth.) What observation might support the statement that "shooting stars" are a nearby phenomenon? What you saw were rock particles or "dust" that had entered the atmosphere. The heat created as the particles streaked through the atmosphere caused the meteor to glow white-hot, burning itself up before it could reach Earth's surface.

Meteor shower

During certain times of the year, meteor showers are quite common. The photo above shows a meteor shower. Many "shooting stars" can be observed in a short time-period. When this happens, Earth is passing through debris left behind by another visitor from the far outreaches of the Solar System. What might this visitor be? Read on.

Exploring the Solar System

The Hairy Star

The ancient observers were familiar with "hairy stars"—stars with tails. From the photo at the right, can you see why it might be called a hairy star? Of course, what they saw were not stars at all, but another visitor from the outer reaches of the Solar System. If you knew Greek, you would know that the ancient observers were talking about **comets.** *Comet* comes from the Greek word for "hair." The material in the tail of a comet gets left in space and can be responsible for meteor showers.

Throughout history, comets were thought to be a symbol of bad luck and disaster. To look at one was to invite something terrible to happen. For instance, the comet of 1066 appeared right before King Harold of England was overthrown by William the Conqueror. The comet was seen as a bad omen for the losing side. The arrival of this comet is recorded in the Bayeaux tapestry, shown below. Notice the comet at the top of the tapestry and the reactions of the people.

A comet

Today, comets are regarded as objects of interest. Both amateur and professional astronomers spend a lot of time trying to be the first to locate a new one. As a result of this interest, scientists now know the answers to questions such as the ones below. As you read further, see if you can discover answers to these questions too.

- Do comets revolve around the Sun, as do the other members of the Solar System?
- Why do comets have "tails"? Rocks and boulders called **asteroids** have been observed in the Solar System, but they have no tails.
- Are comets and asteroids made of the same material?
- What eventually happens to a comet?

408 Toward the Stars

Comets—What Are They?

The tail of the comet is a clue to its makeup. Think for a moment about what substances, when vaporized by the Sun's heat, can be made to glow. Did you name ice? If so, you've discovered one part of a comet's makeup.

In the far outreaches of the Solar System, trillions of frozen ice masses are held in a slow orbit around the Sun by their gravitational forces. Besides frozen water and other frozen material, these ice masses consist of dust and frozen gases. The orbits of some of the ice masses send them speeding toward the Sun, at which time they may be seen by observers on Earth. As the comet nears the Sun, the rise in temperature causes moisture and dust particles to be swept from the head of the comet. These particles form the comet's tail. The tail becomes visible when the particles reflect and scatter light.

The Comet of a Lifetime

Halley's comet

In 1682 Edmund Halley (HAL ee) observed the comet that now carries his name. After calculating its orbit around the Sun, he predicted that the comet would return in about 76 years. Few believed him, and he did not live to see if his prediction would come true, but it did. The return of the comet in 1759 proved that comets are members of the Solar System and that they revolve around the Sun.

Halley's comet made its last return in 1986. Examine the diagram below, which shows Halley's orbit. You'll learn a great deal about comets by using the diagram to answer the following questions.

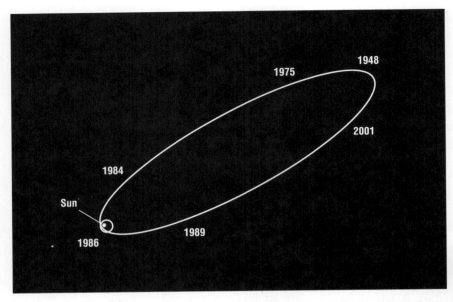

The orbit of Halley's comet. Earth's orbit is the small ellipse at the left.

1. Does a comet appear to obey Kepler's discovery regarding the paths of revolving bodies?
2. Where is the comet traveling at its greatest speed? at its slowest speed?
3. When will Halley's comet return to be seen again by observers on Earth?
4. Why can't Halley's comet be seen with the naked eye at the present time?

Halley's comet

6 The Space Probes

The *Voyager* Discoveries

Early voyagers sailed the seas and discovered new lands. They returned to report their findings. The *Voyager* spacecrafts are also discovering new lands, but they report by sending back photographs and information that will keep scientists busy for years to come.

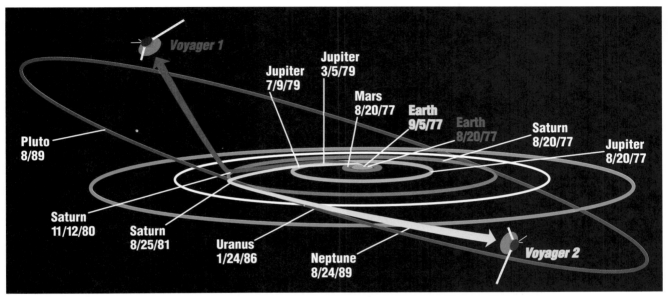

Your flight plans—the paths of Voyagers 1 and 2. The white labels show the positions of the planets on the dates indicated.

Imagine being on a journey that takes you from Kennedy Space Center in Florida, past Mars, and finally near the giant planets of Jupiter, Saturn, Uranus, and Neptune. You would see sights that human eyes have never seen before.

If either *Voyager* spacecraft had a crew and you were a member, how would you respond to the sights and experiences as they unfold? Use the picture study on pages 412 and 413 to write a personal account as you and *Voyager* encounter different members of the Solar System. Your log could center on the times and events shown on these pages.

Where you're heading: the giant planets. Compare their sizes to Earth's size.

Day 1

Blast-off: How would you feel to be leaving on such a journey?

Day 215

Crossing the orbit of Mars: Examine the flight path of *Voyagers 1* and *2* on page 411. Note where Mars was on August 20, 1977—the day *Voyager 1* left Cape Kennedy. Will you be able to see Mars as you cross its orbit? Will *Voyager 2?*

Day 295

Entering the asteroid belt: What is the asteroid belt? Will this be dangerous?

Day 475

Leaving the asteroid belt: Will you feel relieved?

Jupiter from a distance of 30 million km

Day 570

Looking at Jupiter from a distance: What do you see? Describe Jupiter to someone who has not seen it before. How many different features can you describe?

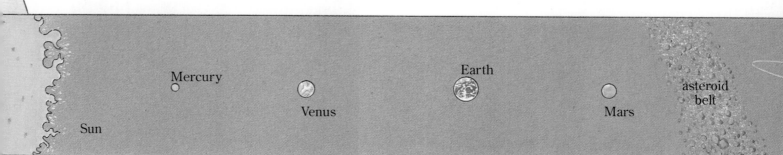

Day 630

Observing the Great Red Spot: What words could describe your view? What is the Great Red Spot? How large is it?

The Great Red Spot

Day 650

Examining some satellites of Jupiter: Jupiter has 16 satellites, one of which is about the size of our moon. What is a satellite? How are the satellites shown here different? Which one appears to be volcanically active? Which one could be covered with a thick layer of fractured ice?

Io

Europa **Ganymede**

Day 662

On the way to Saturn: What will you do for the next 905 days?

Day 1567

Encountering Saturn: How is Saturn similar to Jupiter? different from Jupiter?

Portion of Saturn's rings, in false color

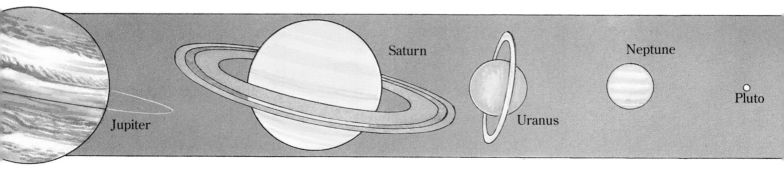

Exploring the Solar System 413

Research Activities

1. In his book, *Cosmos,* Carl Sagan wrote a similar account of what a captain of the *Voyagers* might have said if the *Voyagers* had a crew. His account makes interesting reading.

2. Before leaving the Solar System forever, *Voyager 2* made two more encounters: Uranus on January 24, 1986, and Neptune on August 24, 1989. Find out what you can about these distant planets.

Uranus seen as a crescent. Where must the Sun be in this picture?

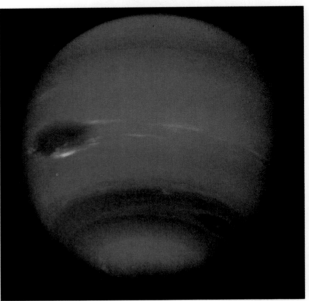

The Great Dark Spot on Neptune. This storm system is as large as Earth's diameter.

3. *Voyagers 1* and *2* will drift forever among the stars unless they are captured by a star's gravitational field. Perhaps, in the future, an alien from a distant star system may chance on the spacecraft and wonder about the world that sent the craft on its epic voyage. That alien will be able to learn something of Earth by listening to a gold-plated record that carries "The Sounds of Earth." The record begins with greetings in over 60 languages. The alien will be able to hear the roll of thunder, the roar of a volcano, crashing surf, and gurgling mud. Living sounds include that of whales, chimpanzees, and Chuck Berry singing his rock song, "Johnny B. Goode."

    Design a wordless message for another spacecraft that will eventually leave the Solar System. What would you place in the message that would give aliens some idea of what life was like on Earth?

Magellan—Mapping a Distant World

Ferdinand Magellan, a Spanish explorer, mapped the coastlines of the Americas in the early 1500s. The spacecraft *Magellan* is mapping a more distant land—a planet called Venus. As a result of the pictures sent back to Earth, more will eventually be known about the surface of Venus than that of Earth. That's because much of Earth is covered with water, forest, and ice, which hide its surface features.

The surface of Venus is hidden from observers on Earth by a thick layer of clouds. Some early astronomers suggested that rivers, streams, and perhaps even living things may lie beneath the clouds. Could this be true? Before giving your opinion, examine the information regarding Venus on page 416.

Clouds cover Venus to a depth of 60 km.

How can *Magellan* penetrate the cloud-cover to see the surface below? It uses microwave radiation, which can penetrate clouds. The reflected signals are formed into pictures by computers and then sent back to Earth.

EXPLORATION 6

Analyzing the Data

Scientists are busy interpreting the pictures sent back by *Magellan*. Using the photos on this page, be a scientist and interpret the data yourself. Perhaps you can answer these questions:

1. What surface features do you recognize on Venus that are similar to those on Earth?
2. How is the surface of Venus similar to or different from that of its near neighbor, Mercury? Can you think of reasons that would explain your observations?

The surface of Venus, photographed by *Magellan*

View of Mercury

Exploring the Solar System

Colonizing the Solar System

Project Mars

Of all the planets, Mars holds the greatest hope of sustaining a human colony. Why is this so? Why not a colony on Venus, Jupiter, or Pluto? Examine the following table of information to find some answers.

Mars	Venus	Jupiter, Saturn, Uranus, and Neptune	Pluto
• Very cold—Average temperature: −23 °C • Extremely dry • Evidence of flowing water in the past • Icecaps • Thin atmosphere of carbon dioxide • Former atmosphere may have reacted with soil and rocks	• Surface temperature: 500 °C • Very dense atmosphere of carbon dioxide • Atmospheric pressure 90 times that on Earth • Clouds of sulfuric acid	• Gaseous • Consists mainly of hydrogen and helium • Violent storms	• $1/100$ the volume of Earth • Extremely cold: −230 °C • Receives little sunlight • Surface may be covered with layer of frozen methane

By the year 2020, NASA hopes to land astronauts on Mars—the "red planet." By 2061—the next return of Halley's comet—many scientists foresee a permanent colony on Mars living in greenhouse-like structures called *biospheres*. Studies at Biosphere 2 in Arizona, shown below, may prepare scientists for a permanent space station on Mars.

There is even speculation that it may be possible to "green" Mars by releasing the gases that are locked up in the rocks and soil, and by melting some of the water in the permafrost and icecaps. If this is to happen, many more space developments must take place, such as those shown below. What do you think of these plans? Of what value is each? How would you add to or modify them?

1992	Mars orbiter launched. Its intended mission was to circle Mars, sending back measurements and images. However, in August 1993, NASA lost contact with the probe. NASA hopes to launch another probe in 1996.
1998	A permanent space station built to orbit Earth, using the space shuttle as a transporter of people and materials.
2010	Permanent space station on the Moon.
1998–2014	Numerous robot landings on Mars.
2014–2020	First astronauts land on Mars.
2020–2030	First research colony established on Mars.

Artist's conception of a permanently occupied space station

Perhaps the events may proceed as described in this hypothetical news forecast.

The first Earthlings will arrive on Mars almost exactly four centuries after the Pilgrims landed near Plymouth Rock. After a three-month flight from Earth, the Mayflower, *a nuclear-powered rocket not yet built, will deposit them on the Martian surface in a prefabricated colony designed to shelter 12 to 14 astronauts for a year.*

Would you like to be one of these Earthlings? What training would you have to undergo? This may be something that you'd like to investigate further.

EXPLORATION 7

Researching the Solar System

In this Exploration, you will choose from three different kinds of projects. Each requires you to research a part of the Solar System. In Activity 1, you can be part of a team that's planning a mission to Mars. In Activity 2, you can play the role of an advertising agency and develop a travel poster for one part of the Solar System. Or in Activity 3, you can create an imaginary alien that could survive in conditions that exist somewhere else in the Solar System.

Activity 1

Planning a Mission

Form groups of four and be part of a planning team for a mission to Mars. Your task is to create some guidelines and procedures that will ensure the success of the mission. Here are some questions you might want to consider:

1. Should vast sums of money be spent on this venture? Explain your reasoning.
2. Who will be part of the first 12-member team to spend a year on Mars?
3. What will be some of the first tasks to be done?
4. Where will their water and oxygen come from?
5. Where will they obtain food?
6. What materials will have to be recycled?
7. What essential materials must be brought from Earth?

418   Toward the Stars

Activity 2
Developing a Travel Poster

The year is 2065. The colonizing of Mars has been a success, and an increasing number of people are visiting this outpost as tourists. The usual length of stay is 1 year. Travel agencies are advertising this and other excursions throughout the Solar System.

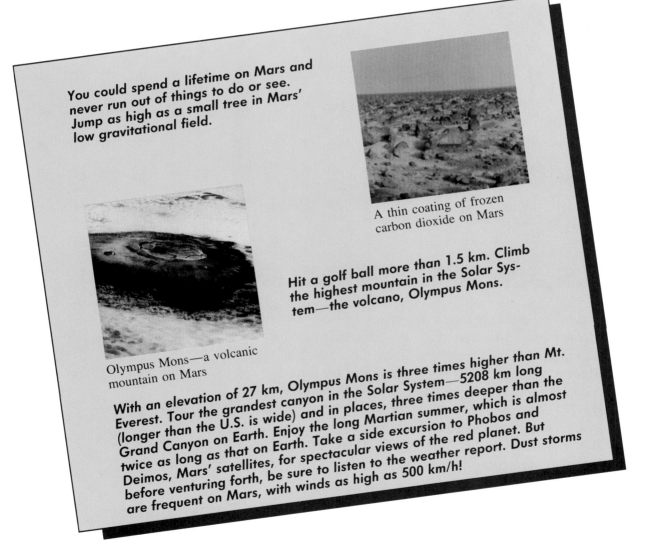

You could spend a lifetime on Mars and never run out of things to do or see. Jump as high as a small tree in Mars' low gravitational field.

A thin coating of frozen carbon dioxide on Mars

Hit a golf ball more than 1.5 km. Climb the highest mountain in the Solar System—the volcano, Olympus Mons.

Olympus Mons—a volcanic mountain on Mars

With an elevation of 27 km, Olympus Mons is three times higher than Mt. Everest. Tour the grandest canyon in the Solar System—5208 km long (longer than the U.S. is wide) and in places, three times deeper than the Grand Canyon on Earth. Enjoy the long Martian summer, which is almost twice as long as that on Earth. Take a side excursion to Phobos and Deimos, Mars' satellites, for spectacular views of the red planet. But before venturing forth, be sure to listen to the weather report. Dust storms are frequent on Mars, with winds as high as 500 km/h!

Research a component of the Solar System. You could choose a planet or one or more of its many satellites. Comets, asteroids, and the Sun would make interesting studies as well. Develop a travel poster that would encourage others to visit the site (or perhaps to keep them away).

Exploring the Solar System

EXPLORATION 7—CONTINUED

Activity 3

Create an Alien—Plant or Animal

Whisper-thin winds hiss along a dry, dusty canyon. Deadly ultraviolet radiation pours from an unshielded Sun. Nighttime cold reaches -80 °C. Perfect weather for a fellow like the Martian waterseeker. Its parasol tail can lift three meters in Mars' low gravity, shading it from ultraviolet sunburn. The long snout can probe for pockets of ice under dried-up channels. And the giant ears, needed to hear well in the thin air, also serve as blankets: In Mars' frigid nights the waterseeker stays snug by clamping its ears tightly around its whole body.

Research a part of the Solar System and create your own imaginary alien plant or animal that is adapted to the conditions there. Describe why it is well adapted to its environment. A good place to start: think of a living plant or animal that you are familiar with. Consider what adaptations would be necessary for it to survive in a more hostile world.

Challenge Your Thinking

Here are some unusual facts about the members of our Solar System. For each fact, there is a question to consider or an activity to do.

1. Venus has a most unusual day. It is longer than its year. Its year is 225 Earth days, while its day is 243 Earth days.

   Using plastic-foam balls to represent Venus and the Sun, demonstrate the year and day of Venus.

2. All planets except one rotate about an axis that is somewhat close to being perpendicular to the plane of the ecliptic. For example, you know that Earth's axis is tilted only 23½ degrees from perpendicular. But Uranus' axis is tilted 98 degrees! Perhaps Uranus was hit by a huge body in the past, and the blow changed its tilt.

   Speculate on Uranus' rather unusual seasons that result from this tilt.

Will the same pole of Uranus always point toward the Sun?

3. Pluto is the outermost planet, but not all the time. For 20 of the 248 years it takes to orbit the Sun, Pluto's orbit lies inside that of Neptune.

   Illustrate this observation by drawing a diagram of the possible orbits of Neptune and Pluto around the Sun.

4. Earth can be considered a miracle planet. It is just the right size and the right distance from the Sun to support life as we know it.

   What would be the consequences of Earth's being much larger? much smaller? closer to the Sun? farther from the Sun?

View of Earth from *Apollo 17*

Exploring the Solar System

OUR UNIVERSE

8 Messages from the Stars

Seeing the Light

Stars communicate with us through the light they emit, so you can find out a great deal about stars by studying their light. Look again at the photograph of star tracks on page 387 and think about the following questions:

- Are all stars of equal brightness? Why do you think this is so?
- Do stars differ in the color of their light? If so, what could be the reason for this?
- Exactly what are stars? What causes their glow?

You'll discover answers to these questions in the following reading and Exploration.

The Closest Star

You are very familiar with the Sun as an emitter of light and heat. Did you also know that it's a star? For those living on Earth, the Sun is a special star because it supplies the energy necessary for life. However, among the stars that you observe on a clear night, the Sun is just a typical star. In fact, it is smaller and less bright than many stars you see! On a clear night, identify several bright stars. If you were watching these stars from the same distance that separates you and the Sun, you would see that many stars are larger and brighter than the Sun. The largest stars are larger than the orbit of Jupiter.

The Sun under magnification

Hot gases rising from the Sun's surface

To get an idea of the size of the Sun, compare it with the size of Earth. How much larger than Earth is it? Is it 100 times as large? 1000 times? Actually, it is capable of holding more than a million Earths. Now think about how large that makes the brightest stars in the sky!

All stars, including the Sun, are made up chiefly of hydrogen gas, along with some helium gas. Although we haven't sampled the core of any star, scientists think that heat and light are produced in stars in the following way. When the gases in a star reach a tremendously high pressure and temperature, *nuclear reactions* occur within the star's core. In these reactions, hydrogen atoms fuse together to form another substance—helium. In doing so, a tremendous amount of energy is released. Some of that energy becomes the heat we feel and the light we see from stars. Humans have duplicated this process in a hydrogen bomb. If each star releases this much energy, imagine the amount of energy released from a constellation!

One easily recognizable constellation is Orion the Hunter, shown below. Orion's most distinguishing characteristic is a group of three stars that makes up the belt of the Hunter. Find Orion on the star chart for winter skies on page 387. Better still, locate it in your night sky. In the Northern Hemisphere, Orion is visible all winter.

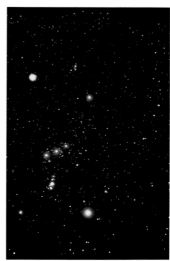

Can you locate the constellation Orion? the star Betelgeuse? Use the diagram on the right to help you.

Orion

Can you find several very bright stars in Orion? Each one is brighter and much more massive than the Sun. The reddish-colored star in Orion is called Betelgeuse (BET ul jooz). What does this red color tell you about Betelgeuse? Exploration 8 will help you answer this question.

EXPLORATION 8

The Temperature of a Star

You Will Need

- a large nail
- a Bunsen burner
- a light source with a clear bulb
- 3 D-cells (2 fresh, 1 weak)
- a flashlight bulb
- small-gauge electric wire
- a pair of tongs

What to Do

1. Using a pair of tongs, heat a nail in a Bunsen burner until it begins to glow. What color is it when it first begins to glow? Can you make it glow at any other color? How?
   Be Careful: *To avoid accidental burns, use extreme care when handling these hot materials.*

2. Now turn on the clear light bulb. What color is the filament as it glows? Which has a higher temperature, the nail or the filament in the light bulb? What can you do to make the nail glow as bright as the filament in the light bulb?

3. What is the relationship between the temperature of the glowing object and its color? If you're not sure, try this. Attach a weak dry cell to a clear flashlight bulb using conducting wires. Observe the color of the filament. Feel the temperature of the bulb. Now attach a fresh dry cell to the bulb and repeat your observations. Finally, attach two fresh dry cells and make similar observations. Note the color of the filament. Feel the temperature of the bulb. What conclusion can you draw about the relationship between the color of light and the temperature of objects emitting the light?

Questions

1. Now do you see what information we can get from the color of a star? What color are the stars with relatively high surface temperatures? with relatively low surface temperatures?
2. Polaris, the North Star, appears white in color. Rigel, a star in the constellation of Orion, appears blue. The Sun is yellow. Betelgeuse is red. Arrange these stars in order from highest to lowest surface temperature.
3. Spend some time looking at the stars. How many can you see that are bluish? red? yellow? What color are most stars?
4. Using a star chart, identify the names of some of the more distinctive stars.

History of a Star

If you've ever spent time around a campfire or warmed yourself in front of a fireplace, you know that when the fire dies down, the embers will glow for a while and the ashes will eventually cool. Is this the fate of stars as well? Do they have a life span in which they produce light and heat and then die? If so, what remains? Read on for answers to these and the questions below. Use the numbers in the right margin of "The Life of a Star" to help you in your search.

1. What is a red giant?
2. How do neutron stars and black holes begin?
3. What is a supernova?
4. How do stars produce the heat and light energy they give off?
5. What is the eventual fate of the Sun?
6. What is the relationship between the mass of a star and the way it will end?

A supernova

Our Universe 425

The Life of a Star

Can stars be seen in the daytime sky? Sometime during the next 10,000 years, a star will explode in the constellation of Orion. The exploding star, or **supernova,** will be so bright that it should be visible during the day. That star is Betelgeuse. (3)

Betelgeuse started out as a bluish and very massive star with a mass 20 times that of the Sun. It was a **supergiant.** But now, after burning for 10 million years and having exhausted its hydrogen nuclear fuel, it is nearing the end of its life span. Betelgeuse has expanded to a diameter 700 times that of the Sun. It is now a **red supergiant.** To observers on Earth, Betelgeuse appears red—the color emitted by the cooler surface gases, far removed from the core. (4) (1)

As its nuclear fuel is used up and its core cools, Betelgeuse will collapse in on itself. This is due to the extreme gravitational forces of its large mass. The resulting shock wave will send a large amount of Betelgeuse into space. This is the supernova that will be observed on Earth. (3)

What remains of Betelgeuse will be a very dense body called a **neutron star.** A neutron star contains the mass of several Suns, compressed to the size of a small (30 km) asteroid. (Ceres, the largest of the asteroids, is 1000 km across.) A thimbleful of a neutron star would have a mass of a hundred million tons! And what happens to the material thrown out into space? It mixes with the stuff of which new stars and planets are formed. (2)

The Crab nebula is the remnant of a supernova recorded on July 4, 1054.

If the mass (and therefore the gravitational forces) of the collapsing star is great enough, it may collapse even more to form a smaller and denser object called a **black hole.** The gravitational forces are so extreme that no object—not even light—can escape the black hole. Since no light can escape from this super-dense object, it cannot be seen in the usual ways—hence its name, a black hole.

(6)
(2)

Our star, the Sun, will not end its life in a giant explosion or supernova. This is a fate that is reserved for the more massive stars such as Betelgeuse. But as the Sun's hydrogen nuclear fuel runs out and the Sun starts to burn other fuels, the Sun will expand in size. Eventually, it will fill the orbits of the inner planets and perhaps even Earth's orbit. When this happens, the Sun will reach the red giant stage.

(5)
(6)

Once the Sun has become a red giant, the oceans and atmosphere of Earth will have boiled away. In time, as the Sun goes through the remainder of its nuclear fuel, it will cool and contract under its own gravitational forces until it becomes a **white dwarf**—a dim but extremely dense star. This scenario will not happen for another 5 billion years. Perhaps by then, humans will have migrated to another planet, to the moons of Jupiter, or to a star system far beyond this one—one with a quieter star.

(1)
(4)

Solar explosion

You can be thankful that the Sun is a yellow star and not a white one such as Sirius, the Dog Star. Sirius is the brightest star in the night skies. If viewed from a distance equal to that from Earth to the Sun, Sirius would be 23 times brighter than the Sun, even though it is only twice the Sun's mass. Sirius is using its energy source at such a rapid rate that its life span is estimated to be less than a billion years. The Sun, on the other hand, is already 4.5 billion years old and is expected to be around for another 5 billion years.

(6)

Checking It Out

With a partner, discuss the answers you found to the six numbered questions on page 425. In your own words, describe what you learned.

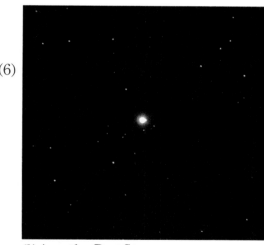
Sirius, the Dog Star

Now you're ready to do one of the following:

(a) Write a short autobiography of a star.
(b) Depict the life story of a star in the form of drawings only.

Our Universe 427

EXPLORATION 9

Simulating the Life Span of Stars

You Will Need

- 2 white balloons
- 2 small white plastic-foam balls
- red, yellow, and black felt-tip markers

What to Do

Form teams of three or four to demonstrate the history of two stars—one the size of the Sun, the other a much larger star. In your simulation, you will use balloons to represent each star at various stages of its life. Decide how to use the remaining materials listed at the left. Then figure out how to sequence the steps of the simulation. The following questions provide some hints as to how you might plan your simulation.

Questions

1. How large will each star be early in its life? How large will each be at various stages in its life? How will you present this in your simulation? The illustration below may help with this part of the simulation.

The relative sizes of stars and their remnants

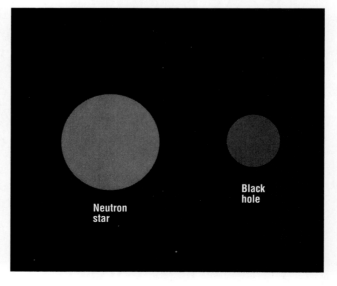

428 Toward the Stars

2. How and when will you use the markers in your simulation?
3. What does each plastic-foam ball represent in the simulation?
4. Will the star explode in a supernova? How would you represent this in your simulation?
5. How long will each star exist before its life is over? How will you represent this in your simulation?

When your team has perfected the simulation, demonstrate it to the rest of the class. Ask for feedback on your simulation. What improvements could you make? Also, give feedback on the other teams' simulations.

Earth's Place in the Universe

At the Speed of Light

The stars remind you how much larger the universe is than the Solar System. But just how large is the universe? To answer that question, you'll be taking another imaginary trip in the next Exploration. Only on this trip, you'll be traveling at the speed of light—that's 300,000 km/s (kilometers per second)! At that speed, time takes on a new dimension. Traveling at this speed, you will age much less than if you stayed on Earth.

The ticket has been bought for the trip, and you're ready to go. The brochure tells you that each time you increase your distance from Earth by 100 times, you will stop. At these stops, you'll have a chance to look back in the direction you just came from—back, in other words, toward the Solar System and Earth. You'll be looking back at Earth's place in an immense universe.

Earth

EXPLORATION 10

Heading for the Stars

Stop 1: You have traveled about 8 minutes since leaving Earth

In this time, you have traveled a distance equal to that from Earth to the Sun. This is how long it takes light to reach Earth from the Sun, when traveling at a speed of 300,000 km/s. The distance from Earth to the Sun, you'll remember, is 150 million km —one astronomical unit (AU). Looking back toward Earth, can you see the four inner planets in their orbit around the Sun? What are they?

1. How many planets are 1 AU or less from the Sun?
2. Would all planets be on the same side of the Sun as Earth?

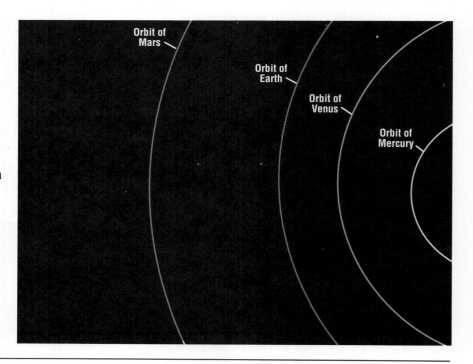

Stop 2: You have traveled 14 hours since leaving Earth

You are 100 AU from Earth. The inner planets are hidden in the brightness of the Sun. All of the outer planets are visible except for the outermost planet, which is so small that it cannot be seen. Which planet is that?

1. What does 100 AU mean?
2. If you are looking down over the north pole of the planets, in what direction would you expect them to be moving around the Sun?

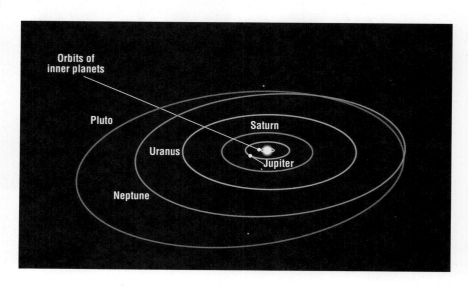

430   Toward the Stars

Stop 3: You have traveled 0.16 years (58 days) since leaving Earth

Looking back at the Sun from a distance of 10,000 AU, you see just one very bright star in a sea of blackness. Just as astronomers cannot see any planets around other stars (because any planets that may be there do not reflect enough light), so you cannot see the planets of the star called the Sun.

The distances you have traveled are now so large that kilometers and even astronomical units are becoming cumbersome. A new unit is needed—the **light-year.** A light-year is the distance light travels in one year. That distance is equal to 63,240 AU.

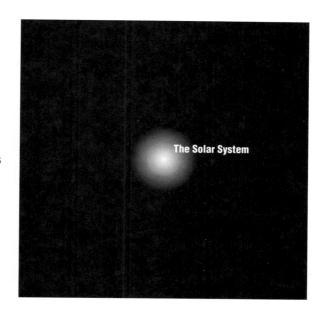

Since you have traveled 0.16 years at the speed of light, you have been traveling for 0.16 light-years.

At this stop, how far are you from Earth in kilometers?

Stop 4: You have traveled 16 years since leaving Earth

Sixteen years have passed since you left Earth. You have traveled a distance of 16 light-years. Now the view looking back toward the Sun includes other stars. You can see Sirius—the brightest star seen from Earth's Northern Hemisphere. From this distance, you notice that what you're seeing is actually two stars: Sirius and its companion white dwarf. The closest star to Earth, Alpha Centauri, can also be seen. It is only 4.3 light-years from Earth. On Earth, it can be seen anywhere south of Miami, Florida.

1. In AU's, how far are you from Earth?
2. Sirius is 8 light-years from Earth. How long would it take for light from Sirius to reach Earth?

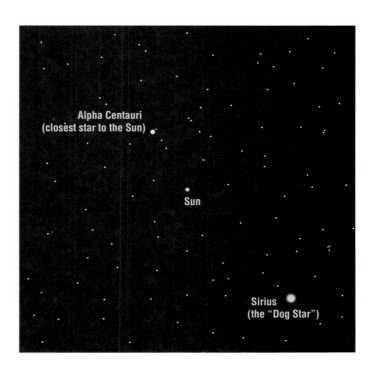

Our Universe 431

EXPLORATION 10—CONTINUED

Stop 5: You have traveled 1600 years since leaving Earth

You are now at a distance of 1600 light-years from Earth. Some of the stars that are so familiar to observers from Earth are now behind you. There is the constellation of Orion with the red star, Betelgeuse. You can also see the Pleiades, a cluster of stars in the constellation Taurus. Had you lived at the time of the dinosaurs, 150 million years ago, you would not have been able to see this group of stars. They are a cluster of young stars that formed relatively recently from clouds of dust and gas.

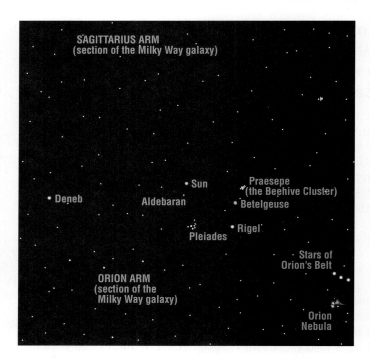

Examine the star chart on page 387. Compare the stars shown in the star chart with those you can see 1600 light-years from Earth.

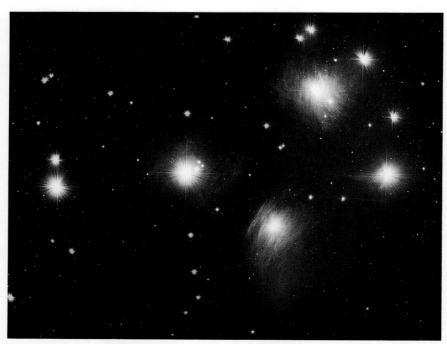

The Pleiades—still surrounded by the gases and dust from which they were formed

Stop 6: You have traveled 160,000 years since leaving Earth

At a distance of 160,000 light-years from Earth, a surprising thing has happened: As you have entered emptier space, all of the stars you have been observing have merged into a spiral-shaped structure. This is the **Milky Way** galaxy, our home galaxy. A galaxy is a huge system of stars, dust, and gases held together by gravity.

From a distance of 160,000 light-years from Earth, you see the Sun as just one of 1000 billion stars. The Sun is located near a spiral arm called the Orion Arm (named for the Orion constellation), about two-thirds from the center of the Milky Way. You now realize that on a clear night on Earth, all of the brighter stars you saw were relatively close to Earth, in the same spiral arm as the Sun.

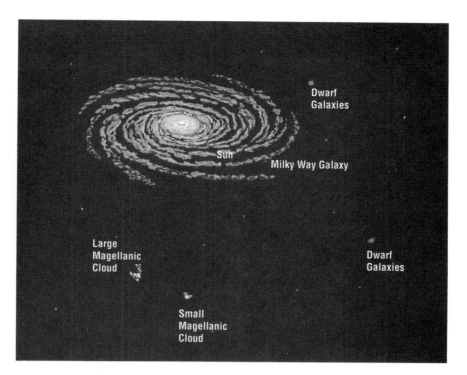

1. A galaxy has been described as an island universe. What does this mean?
2. The Milky Way galaxy is 100,000 light-years across. Approximately how far from the center is the Sun?
3. Where do most of the stars seem to be concentrated in the Milky Way galaxy?

An "island universe"

EXPLORATION 10—CONTINUED

Stop 7: You have traveled 16,000,000 years since leaving Earth

You are now 16 million light-years from Earth. The Milky Way galaxy is now seen as part of a family of galaxies called the **Local Group.** All are held together by mutual gravitational forces, just as gravitational forces in our Solar System hold the planets in orbits around the Sun. You may be familiar with one of the galaxies—the Andromeda galaxy, or M31—since it can be seen from Earth with the naked eye. It is the farthest object that can be seen from Earth without the aid of a telescope.

1. Andromeda is 2.2 million light-years from Earth. Astronomers study it through the light it emits. How long ago did that light leave Andromeda?
2. Studying distant galaxies is described as looking back into the past. What does this mean?
3. You decide to pause on your trip to write a letter back home. When will your letter arrive if it travels at the speed of light?
4. How would you address your letter? In other words, what is your cosmic address back home?
5. What are the two most important ideas this trip has conveyed to you?

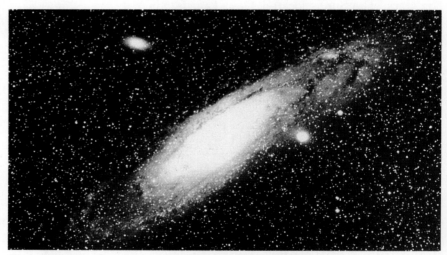

The Andromeda galaxy, a typical spiral galaxy

Stop and Think

This is where your journey into the universe stops. As you look around you from the depths of space, your view will take in 100 billion galaxies, consisting of possibly 1 billion trillion stars. In his book, *Cosmos,* Carl Sagan described it this way:

> *A handful of sand contains about 10,000 grains, more than the number of stars we can see with the naked eye on a clear night. But the number of stars we can see is only the tiniest fraction of the number of stars that are. Meanwhile the Cosmos is rich beyond measure: the total number of stars in the universe is greater than all the grains of sand on all the beaches of the planet Earth.*

A Conversation with Huck

"We had the sky, up there, all speckled with stars, and we used to lay on our backs and look up at them, and discuss about whether they was made, or only just happened."
—Huck Finn in Mark Twain's book, *Huckleberry Finn*

Around one of the stars, the Sun, you know that life exists on a planet called Earth. Only one other star has been found to have a planet orbiting around it.

Find a partner and discuss whether you think life exists elsewhere in the universe. What are your reasons? Would you like to know for sure? If we assume that the laws of nature hold true on other stars and possible planetary systems as they do here, should life forms be similar? Why?

When you look up in the sky at night, you also see it speckled with stars. But in studying this unit, you have had a "closer look" at them than Huck did. What do you suppose Huck was wondering about as he looked up at the sky? Join in a discussion with Huck. Write down what you and he might say.

There are more stars in the universe than there are grains of sand on all the beaches on Earth.

Our Universe

10 A Likely Beginning

The Big-Bang Theory

Scientists have a theory that answers part of Huckleberry Finn's question, the one you just discussed. But like all theories, it will be modified as predictions are tested by experiments and observations. Read the paragraphs below about the big-bang theory. As you read, pretend that you're a newspaper editor who has to think up headings for each paragraph. In your Journal, write down your three- or four-word headings that catch the message of each paragraph.

__?__

According to this theory, the universe began sometime between 15 and 20 billion years ago. The universe was formed in an event called the **big bang.** You can think of this as a cosmic explosion that formed the universe. The matter that was formed was hurled outward in all directions by the force of the explosion.

__?__

About 14 billion years ago, the universe cooled, and the hydrogen and helium that were formed in the big bang clumped together as a result of gravitational forces. New stars then condensed within these clumps to create what we now see as galaxies. Using a large telescope, astronomers have detected very bright energy sources that are billions of light-years from Earth. One theory is that these energy sources, called **quasars,** are galaxies in their early stages of formation.

Interacting quasars

__?__

Oxygen, carbon, and iron—all of which are found on Earth—were formed within the stars. Some of the more massive stars eventually exploded in an event called a *supernova*. These explosions sent into space material that mixed with the matter from which new stars and planets would form.

Supernova remnants

___?___

Approximately five billion years ago a new star, the Sun, and a family of planets formed from gas and dust particles in space, which mixed with heavy elements that were created inside stars. One of the planets that formed—Earth—is the home of humans. Another planet—Jupiter—never reached high enough temperatures to start nuclear reactions in its core, even though it is massive and made of the same stuff as the Sun. The Solar System almost had two suns!

Jupiter nearly became the second sun in our Solar System

___?___

There are a number of predictions suggested by this theory. If the universe began from an explosion of a super-hot fireball, some of that heat should still be detectable. By now, this radiation should have the same wavelengths as *microwaves*. In the 1960s two American physicists, Arno Penzias and Robert Wilson, were working on a way to get rid of static on their microwave antenna system. The sound seemed to be coming from the sky. After trying everything they could think of to eliminate the static (they even tried cleaning bird droppings from the antenna of their radio telescope), they realized that they may have found the radiation left over from the birth of the universe. They may have been listening to the faint echo of the big bang that occurred about 15 billion years ago! Experiments and observations have so far supported the big-bang theory. As so often happens in science, this discovery illustrates how progress is sometimes made by chance instead of by carefully planned experiments.

___?___

Since the galaxies are separating from one another, astronomers reasoned that a big bang may have occurred. In addition, the more distant galaxies are receding at a faster speed than the closer ones. In 1925 Edwin Hubble discovered evidence of an expanding universe. The universe was expanding just as it would be if a cosmic explosion had occurred.

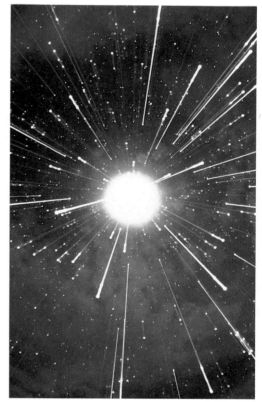

An artist's conception of how the initial explosion may have looked

EXPLORATION 11

Model of an Expanding Universe

All models are imperfect because they cannot exactly duplicate the real thing. After you construct the following model, think about its uses. Does it help you visualize a difficult idea? How is this model a good one? What are its limitations?

You Will Need

- a large round balloon
- a marker

What to Do

1. Blow up a balloon only partway. This will represent the universe.
2. Now, with a marker, place dots on the balloon, some close to each other, some far apart. The dots will represent the galaxies that are part of the universe.
3. Measure or estimate the distances between the different galaxies.
4. Now blow up the balloon some more and observe what happens to the dots.

Questions

1. Did the dots all separate by the same amount?
2. Which ones separated faster than others? slower than others?
3. Discuss this model in terms of the big-bang theory. What does the model illustrate about the theory? What predictions of the theory are illustrated by the model?
4. How is this model an imperfect picture of the universe?

438   Toward the Stars

Challenge Your Thinking

1. Some of stars that we associate with a particular constellation are not part of the same star group at all. Examine the Big Dipper as it was 100,000 years ago and as it is now. Draw a diagram of what you think it will look like in another 100,000 years. Which stars are obviously not part of the same star group? Why do they all look equally far away from Earth?

2. Examine the orbit of the white dwarf around the star Sirius.

    (a) What is a white dwarf?

    (b) What is the shape of its orbit?

    (c) Where is it moving the fastest? the slowest?

3. The following newspaper article is similar to the reading that started the unit. This time, certain terms are underlined. In your own words, describe your understanding of the underlined words.

    > Astronomers reported today the discovery of an almost inconceivably massive object within two colliding <u>galaxies</u> 300 million <u>light-years</u> from Earth. If the object is a <u>black hole</u>, it is by far the largest one ever detected. Its mass is estimated to be equivalent to all of the <u>stars</u> in the <u>Milky Way</u>. If the object isn't a <u>black hole</u>, it might be a dormant <u>quasar</u>. More observations are needed to make sense of this new <u>observation</u> in terms of current <u>theory</u>.

The Big Dipper 100,000 years ago

The Big Dipper as it appears today

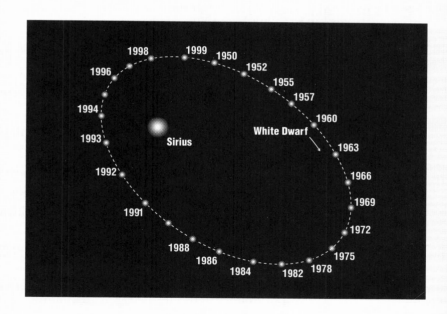

Unit 7 Making Connections

The Big Ideas

In your Journal, write a summary of this unit, using the following questions as a guide.

- What is astronomy?
- How do we learn about the Solar System and the universe?
- What are the essential differences between the heliocentric model and the geocentric model of the Solar System? Which is the better model, and why?
- Why do you see different stars in the sky at different times of the year?
- What causes the seasons?
- What objects make up the Solar System? the universe?
- What are stars, and what is their fate?
- How large is the universe?
- What is your cosmic address?
- How might the universe have begun? What evidence do we have?

Aurora Borealis (the northern lights)

Checking Your Understanding

1. Where on Earth would you be if:
   (a) The star Polaris is directly overhead?
   (b) The star Polaris never rises above the horizon?
   (c) The Sun is directly overhead once a year?
   (d) The Sun is directly overhead twice a year?
   (e) The stars at night appear to revolve clockwise over your head?
   (f) The stars at night appear to revolve counterclockwise over your head?
   (g) You have equal hours of daylight and darkness?
   (h) You have 24 hours of sunlight?

2. Earth's axis is tilted at an angle of 23½ degrees toward the plane of the ecliptic. What would the consequences be if the tilt were:
   (a) 0 degrees—no tilt at all?
   (b) 45 degrees?
   (c) 90 degrees?

440   Toward the Stars

3. Although the other planets lie near the plane of Earth and Sun (the plane of the ecliptic), they are not exactly on this plane.

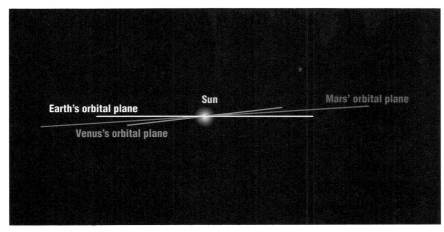

A side view of part of the Solar System, showing three of the planets' orbits

(a) What observations would you make if all of the planets did lie exactly on the same plane?
(b) Earth, the Moon, and the Sun do not all lie exactly on the plane of the ecliptic. If they did, what observations would you make on a regular basis?

4. Draw a concept map that shows the relationships between the following terms: *universe, Betelgeuse, Milky Way, Orion, constellations, stars, galaxies,* and *Polaris.*

Reading *Plus*

You have been introduced to astronomy and to human exploration of the universe. In reading pages S111–S132 in the *SourceBook,* you will take a closer look at members of the Solar System and at how humans are furthering their investigations of space.

Updating Your Journal

Reread the paragraphs you wrote for this unit's *For Your Journal* questions. Then rewrite them to reflect what you have learned in studying this unit.

Science in Action

Spotlight on Astronomy

Bob Grimm is a planetary geophysicist at a university, where he is involved in research and teaching. His educational background includes a B.A. in Physics and Geology, and a Ph.D. in Geophysics.

Q: What does your daily work involve?

Bob: Right now I spend most of my time looking at pictures of Venus sent back from the *Magellan* spacecraft. With these pictures, I trace geologic structures and piece together what the surface of the planet looks like. Using computer models and mathematical equations, I try to predict surface structures and understand the forces that create them.

Q: What sort of equipment has been developed for your use in observations?

Bob: I am the end-user of robot spacecraft that have gone out on missions and sent back photographs of the planets. These are very complicated, automatic spacecraft that can beam data down to us for years. I also program and use sophisticated computers that can store images of the planets. These computers enable us to zoom in on and roam around the planets' surface features.

Q: Do you ever have to deal with crisis situations?

Bob: When the *Magellan* spacecraft was placed into orbit, it malfunctioned just as it reached Venus. It looked like the whole project might be destroyed! The engineers in spacecraft operations who built the machine were the ones who had to deal with the crisis, but my work was directly affected by the outcome. Fortunately for all of us, the engineers were able to fix the *Magellan*.

Q: Does your job involve traveling?

Bob: Oh yes, I travel a great deal to NASA's Jet Propulsion Laboratory where all robot missions originate. This is also where data is sent after a mission has begun. Eventually, all the data coming back from space gets distributed to the universities, but I go to the center so I can see the data immediately. I also spend a lot of time traveling to the East and West coasts and to Europe, where I meet with other scientists to discuss the information we have gathered and to formulate ideas and theories.

Q: Do environmental issues ever come into play in your research or activities?

Bob: At first it may not seem so, but what we try to do is look at the "whole picture." We try to see the evolution of Earth in a bigger context—where it came from and where it's going. There are many complex issues that can be analyzed, such as how the oceans work with the atmospheres on other planets. The information we obtain from studying these types of questions can give us valuable clues to the way our *own* planet works. We can examine problems like the rising level of carbon dioxide in Earth's atmosphere by looking at Venus, whose atmosphere is mostly carbon dioxide.

Q: What activities as a young person led you to your current field?

Bob: Early space exploration in the '60s and '70s—I was greatly inspired in 1969 when *Apollo* landed on the Moon. Space exploration is all so recent, and the United States has always been a leader. That's very exciting to me!

Q: What school subjects turned out to be most important for your career?

Bob: The single most important subject for me was physics because that provided me with the tools that I use most often. Geology was also very important, and so was English. A scientist has to be able to communicate on paper and through presentations. Science is all about *ideas,* and they can only be relayed if the scientist is a good communicator.

Q: What most appeals to you about your job?

Bob: The intellectual freedom to follow my interests. It's a good feeling to know that I'm here because I'm interested in what I'm doing. I come to work each day and say to myself, "What will I think about today?"

The sense of exploration also really appeals to me. Every day, looking at new pictures of planets, I do what no one has ever done before. It's like a hunt—I try to figure something out, bring some relationships together, and soon I have a story to tell!

Q: What personal qualities do you feel are most essential for a person in your field?

Bob: Perseverance. Graduate school is hard. It took me five years, but it was worth it! Projects don't always go right. Sometimes they take so long it's easy to lose track and get "lost." But you just have to stick with it. The results are always worth it.

Creativity is also a very important quality. Sometimes you just have to break the mold and come up with new ideas. Creative talent is invaluable for science!

Q: How would you describe the importance of your work to students?

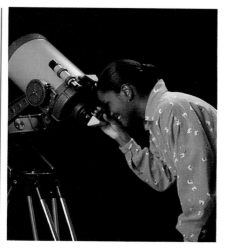

Bob: We *need* to explore. It is difficult today, when there are so many limits on budgets and resources, to advocate tremendous spending on space exploration. However, without continuously questioning our world and wondering about others, humankind has a tendency to become complacent. It is only by questioning that we will learn, and I feel that by asking questions about outer space, we will learn more about our own world.

A Project Idea

Try a planetary mapping experiment. When there is a full Moon, go out and sketch it. Be sure to distinguish the light and dark areas. Then, get a map or globe of the Moon from a map store or a library and look at structures such as lava flows and craters. How does the map compare to your drawing? Determine what the light and dark areas are made of. Does Earth have as many craters as the Moon? Why or why not? What can the number of craters on a planet tell us about its geology?

Science in Action

Science and Technology

The Most Powerful Objects in the Universe

In the 1930s astronomers first began to experiment with **radio telescopes.** These new telescopes picked up radio waves instead of visible light. By the 1950's astronomers had discovered a number of objects in the sky that give off very powerful radio signals. They called these objects *radio stars.*

The Discovery of Quasars

The nearest quasar, 2100 million light-years away

In 1964, however, two astronomers announced that the "radio stars" they had been studying are not stars at all. Actually, they are extremely distant galaxies, known as **quasars,** which are farther away than anything else that has ever been seen. Some object, which the astronomers called an "object X," is at the heart of each of these galaxies. Such objects would be sources of incredibly powerful radio waves, even more powerful than nuclear reactions can produce. These objects would be the most powerful things in the known universe.

Black Holes

Many astronomers now believe that the radio sources inside quasars are objects known as **black holes.** The existence of black holes is more or less taken for granted by many astronomers, although no one has ever seen one. Black holes, if they exist, are in fact invisible!

A black hole, according to the theory, is the result of matter that has been super-compressed. For example, if the sun were compressed from its present diameter of 1,390,000 km down to a diameter of just 6 km, it would become a black hole. The gravitational attraction of such a heavy object would be so great that nothing, *not even light,* could escape from it.

NGC 1275

Astronomers are interested in a particular galaxy known as *NGC 1275,* which is like a quasar in certain ways. It is much closer to the earth than any of the quasars, but like the quasars, it is a strong radio source. Some astronomers believe that there is

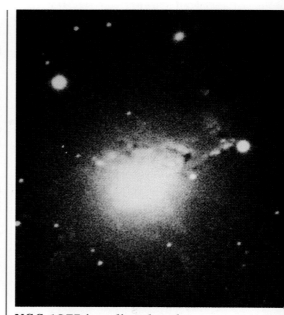

NGC 1275 is a disordered galaxy that emits powerful radio waves.

an enormous black hole, perhaps as large as the sun, at its center. If there is, all the stars and other matter in its vicinity would be attracted to it and crushed. This extreme compression of tremendous amounts of matter would give off large amounts of radio energy.

Telescope Teamwork

The largest permanent telescope installation in the world is actually a series of 27 separate radio telescopes linked by computers. When they operate together, they act like a single telescope with a dish diameter of 27 km. Such an arrangement is known as a *very large array* (VLA).

There is a way of getting an even larger system of telescopes together. This is a technique known as *very long baseline interferometry* (VLBI). Using this technique, radio telescopes in different parts of the country, perhaps even on different continents, all work together. Each telescope records data from somewhere in the sky over the same period of time. Then the data tapes are played back together. The result is a set of images from a "telescope" with an effective dish diameter that may be nearly as large as the diameter of the earth!

Colliding Galaxies?

The image on the right was taken by a VLBI team using six radio telescopes in Europe and America. The telescopes all recorded radio wave data from the NGC 1275 galaxy. The astronomers hoped to find evidence of a black hole.

What the astronomers found suggests that there may actually be two black holes at the center of NGC 1275. This information, if true, would support the hypothesis that NGC 1275 is actually the result of the collision of two different galaxies.

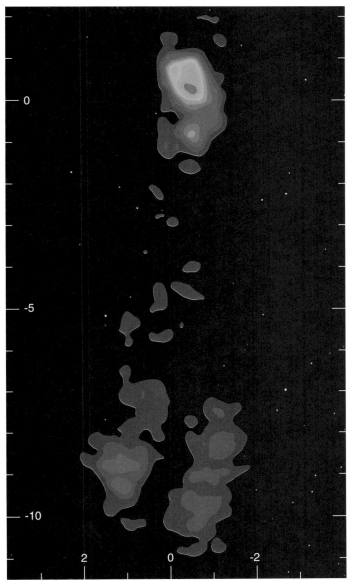

Radio image of core of galaxy NGC 1275. There may be two black holes in the region at the top. The clouds of gas below are being ejected from the galaxy's core at one-fourth the speed of light!

Find Out for Yourself

Do some research to find out about black holes. How do scientists think they might be formed? Some astronomers think there is a black hole in our own galaxy in the constellation *Cygnus*. What is the evidence for this?

There has been some science fiction written about black holes, including even taking trips into a black hole. Does such an idea seem reasonable to you? Write your own science-fiction story about entering a black hole.

Science and Technology

Unit 8: Growing Plants

GARDEN INGREDIENTS
1. A Garden in Space, p. 448
2. The Beginning of the Garden, p. 450
3. Breaking Ground, p. 454

INNER ACTIONS OF PLANTS
4. Water, Water, Everywhere! p. 463
5. Giving Plants a Hand, p. 469
6. Flowers and Pollination, p. 473
7. Plants from Plant Parts, p. 479

PLANTS IN YOUR ENVIRONMENT
8. Medicine or Poison? p. 484
9. Make a Green World! p. 486

For Your Journal

Write a paragraph summarizing your thoughts about each of the following:
1. How do plants interact with their nonliving environments?
2. What things do the plants in this photograph need in order to grow?
3. A flower and a seed look entirely different. How do you think they relate to each other?
4. How can you grow a garden in space?

GARDEN INGREDIENTS

1 A Garden in Space

By A.D. 2140, fusion-powered rockets had made it technically possible to colonize space. The expansion of Earth's population to nearly 10 billion people made emigration to a space colony an attractive possibility.

For several decades, experiments were conducted to determine how to sustain life in space. Some people favored the establishment of orbiting space colonies. The colonies would be close enough to Earth to permit regular visits between space-dwellers and their friends and relatives on Earth.

Other people favored the establishment of colonies on Mars. Travel time to Mars had been reduced, at the closest approach of Mars to Earth, to less than a month.

On January 23, 2152, the President of the United Nations, Elena Kwame Tao, announced the plans of her administration to establish a colony on the planet Mars. The announcement read:

Proposals Invited

During the next 15 years, a vast network of fully enclosed biospheres will be built on Mars. Groups interested in establishing a colony on Mars are invited to apply to the Working Committee for the Colonization of Mars, General Assembly of the United Nations.

Proposals to the Working Committee must incorporate detailed plans for the establishment of the colony. This includes the design and contents of the biosphere that will house the colony.

Could this story really happen someday, or is it only science fiction? The illustration on page 449 shows what such a **biosphere** might look like. The roof has been lifted in the drawing so that you can see the areas inside the structure. A biosphere is an area or region where conditions are suitable for living things to survive. What must living things have for survival, wherever they are? In your Journal, make a list of what you consider to be essential.

Pretend that you and a group of your friends have decided to make a biosphere proposal to the United Nations. You should consider the following questions and discuss them with others.

Think About It!

- How will you control temperatures?
- What will happen to wastes?
- Where will fresh water and fresh air come from?
- How will the plants survive?
- What foods will you need? (Make a list of ten nutritious items.)
- What is the natural source for each of the foods on your list?
- What will your Mars Biosphere have to include so that you can have a continuous supply of food?
- List everything you would need to support the food-producing plants and animals.

Write a report based on your answers. Then present your report and compare it with those of other groups. You might want to change some of your answers after listening to the other groups.

A side view

Garden Ingredients 449

2 The Beginning of the Garden

Germination of Seeds

Growing plants is a fundamental start in planning and preparing to live in an experimental biosphere. But what are the needs of one living thing—a plant—here on Earth? Will these needs be the same as in the Mars biosphere? The best way to start your investigation is to grow a plant of your own.

Julie collected seeds from flowers such as nasturtiums, zinnias, and marigolds. You may do the same or you can use seeds from a store-bought package, as Nu suggested. In either case, you and your classmates should pick one kind of seed and grow the same kind of plant.

1. Pick the first seeds that form.

2. Dry the seeds.

3. Clean and keep the heaviest, plumpest seeds.

4. Keep seeds in a cool, dry place.

Ali is uncertain whether he can rely on his seeds to develop into plants. In fact, some of the seeds you collect or buy may not **germinate.** (Germinate means to sprout, or to begin to develop.) Imagine how important it is for a farmer to know what percentage of seeds will germinate, and what size the crop will be. To find out, a farmer may test samples of seeds before planting them. You will test seeds for their germination rates, so you won't be disappointed when you begin to grow your own plant.

EXPLORATION 1

Testing for Germination Rate

You Will Need

- a package or collection of seeds
- materials of your own choice

What to Do

Plan your own procedure for testing what proportion of the seeds will germinate.

Remember, seeds need moisture and warmth to begin to grow. To get you started, here are some factors to consider as you plan your germination test:

- The kind of seeds to use (fruit, vegetables, flowers, etc.)
- The number of seeds to use in the test
- The substance to place them in (on a moist towel, in a jar, in potting soil)

Before carrying out your test, discuss with your classmates how you plan to do it. Exchange advice and ideas. Keep a record of your test in your Journal.

1. Begin by stating the kinds of seeds tested and where you got the seeds. Be sure to note how long they have been stored. Will the number of seeds you use affect your results?
2. Describe the procedure used to test the seeds and include a sketch of your seed samples.
3. Report the results.

Seed Analyst

The germination rates of the seeds that you buy in the store are tested by *seed analysts*. In the table below, the percentage of seeds that germinated in one test are listed. Seed analysts are responsible for finding out the information used in tables like this one.

Seed Germination Levels	
Percentage	Kind of Seed
90%	sweet corn
80%	beans, broccoli, cabbage, turnip, mustard, cantaloupe, cucumber, pumpkin, popcorn, watermelon, radish
75%	peas, beets, tomato, onion, asparagus, swiss chard
70%	lettuce
65%	spinach, leek, pepper, eggplant, chives
60%	carrot, parsnip
55%	celery, parsley
50%	dill, sage, summer savory, thyme
35%	watercress

Questions

1. How do the percentages shown here compare with the results from your test in Exploration 1?
2. What do you think may cause variations in the germination rate of seeds?
3. Suggest ways you could find out if a specific factor affects germination rate.
4. You must select a seed analyst to join the Mars Biosphere. Create a want-ad for a newspaper that begins, "A successful applicant . . . ."

Growing Your Own Plants

Have you ever grown plants before? In a group of three or four, make a list of plants that you have grown or have seen others grow. Group the plants under the following headings: *vegetables, flowers, indoor plants,* and *others.* Were the plants healthy? How and where did you plant them? What did you do to help the plants as they grew?

Gardeners learn from the experience of others. Talking to other gardeners is one way to become a better gardener. Another way is reading articles and books written by experienced gardeners. In the following Exploration, you will start growing your own plant from one of the germinated seeds. Keep a daily account in your Journal. Carefully record everything you do and everything you observe about the development of the plant. Do not hesitate to ask questions or to look for help in gardening books.

EXPLORATION 2

Your Plant

You Will Need

- germinated seeds from Exploration 1
- potting soil
- pots
- plant food

What to Do

Choose one of the germinated seeds. Look closely at the seed and be very gentle, since the newly formed root and the leaves are easily damaged. Plant the seed in soil that has been thoroughly broken up. Soil mixtures from a garden store would be useful. Think about the position that the root should be in. Then make a small hole in the soil. Put in the seed, and gently fill in the space around it with soil. How deep should the seed be? Leaves should be showing, as if they had just broken through the soil into the air. Lightly pack the soil so it supports the seedling in an upright position. Add enough water to make the soil moist, but not wet. Some plant food should be added as well. Remember to keep a record of what you do and what you observe as your plant grows.

Things to Observe and Record

- Record the planting conditions. Include the medium (garden soil) that the seed was planted in, the kind and size of the container, and where the container was located.
- Record when and how much you water and fertilize the seed.

- Record the date when you plant the germinated seed and sketch what it looks like.
- Record changes in the plant as it grows, including the
  —height and other dimensions, such as the circumference of the main stem;
  —number, color, and shape of the leaves;
  —number of buds and flowers;
  —characteristics of the flower;
  —signs of pests and diseases.

Special Instructions

1. If there are enough germinated seeds available, plant two or three of them. This gives you an alternative if something should happen to one of your seeds.
2. When the seeds are planted and the soil is moist, measure the mass of the container and its contents. Later you can see how growth affected the plant's mass. When you weigh the container a second time, be sure that the soil has the same amount of moisture in it.
3. Record any changes in growing conditions, such as the amount of light or heat at the location where your plant is growing.
4. When your plant is fully developed, compare your results with those of your classmates.

Other Things to Do

1. Use a series of prints or slides to show your plant's growing stages.
2. Draw a series of sketches of your plant as it grows.

Problems? You can get help from a conservatory or arboretum like this one!

Garden Ingredients

3 Breaking Ground

Soil—An Important Factor

"If you have good soil and enough rain, anything can grow."

This statement may be a slight exaggeration. However, it emphasizes the importance of two factors, soil and water, for a plant's growth.

Soil is familiar; you can see it in a garden, the park, the football field, abandoned lots, flower pots, and other places. But have you ever thought about questions such as these?

- Exactly what is soil?
- Where does it come from?
- Are there different kinds of soil?
- If there are, how do they differ from one another?
- What is the best kind of soil for growing plants?
- Is sand considered to be a soil?
- Is there any danger that our soil might disappear or be damaged? If so, how?

Discuss some of these questions with your classmates. Then consider whether soil will be important in the Mars biosphere. If so, what will the soil be like, how will it be obtained, and how will it be conserved so it will continue to be useful?

EXPLORATION 3

A Close Look at Soil

You Will Need

- a half liter of soil obtained from your backyard, a vacant lot, or a garden
- magnifying glass
- water

Keep the soil securely in a plastic bag or sealed jar. Label it with your name. You will use this soil in many of the activities that follow.

What to Do

1. Along with the particles of various minerals (the substances that rocks are made of), soil contains decayed remains of once-living things. Spread a small amount of your soil sample on a piece of paper. Use a magnifying glass to examine it closely. Do you see mineral particles—pebbles, sand, or even finer particles? What evidence can you find of things that once were living? Try to identify what these things were. Material that was once living is called **humus.** It is an important part of fertile soil.

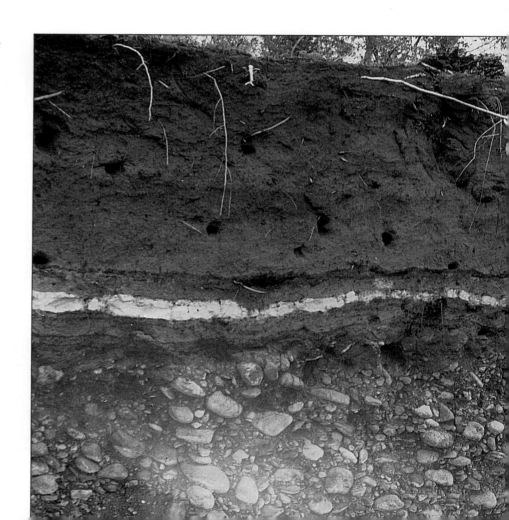

What Is Soil?

You have been chosen to write a definition of soil for a new science dictionary. Drawing on your recent experience with soil, write a definition consisting of one or two sentences. Compare it with definitions others have written and with various dictionary definitions of soil. Of all these definitions, which one do you think is best? Why?

Soil Particles

Even soils that look the same may differ in the size of the particles that they are made from. Particle size determines the texture of a soil. In the next Exploration, you will investigate particle size.

2. Add a small amount of water to your soil—just enough to make it slightly damp. Now rub some of the soil between your fingers. Does it feel smooth? gritty? What other words can you use to describe the feel of the soil? This characteristic of soil is called **texture.**

3. What color is your soil—black, gray, reddish, yellowish, or some other color? Black soils are rich in humus. Gray soils come from areas where there is poor water drainage. Gray soils do not get much exposure to the air. Red or yellow soils indicate good drainage and, consequently, good exposure to air.

4. After making close observations of your soil sample, write a complete description of it. Form a group with three others. Mix up your soil descriptions and distribute them among the members of your group. Try to match the description you receive with the appropriate sample.

Garden Ingredients

EXPLORATION 4

Sorting the Soil

You Will Need

- a jar 5 cm in diameter and 10 cm high
- water
- soil from your sample

What to Do

1. Fill the jar halfway with soil. Add water until the jar is nearly full. Put the cover on securely. Shake vigorously for one minute.
2. Place the jar on a table top or desk, and let the soil settle for 15 minutes. Do separate layers form as the soil settles? How many layers do you see in your jar?

Identify the components of your sample with this information:

- Humus: floats on top; once was living matter
- Clay: very fine particles
- Silt: slightly larger particles
- Sand: still larger particles
- Even larger particles

Compare your jar with those of your classmates. What differences do you observe?

3. Let your water-soil mixture settle for 24 hours, and then observe the layers. In your Journal, make a sketch or a graph showing the proportion of each of the three bottom layers in the jar. It is this proportion that determines the soil type.
4. Look at the Soil Classification Table. What are the sizes of the particles in your jar? You may need to magnify particles of silt and clay to see them individually.
5. Slowly pour off the water and humus from your layered mixture. Spoon off the clay and touch it to feel its texture. Observe the textures of the silt and sand particles, as well. Compare your observations with those given in the table.

Soil and Plants

How does soil interact with plants? Why is this interaction important to plants? In what other ways do plants and soils interact?

Have you ever pulled up a plant by its roots? In this case, what was the soil doing for the plant? The roots of the plant form a complex network. Water and other **nutrients** (chemicals needed for plant functioning and growth) enter the plant from the soil through the smallest roots. Thus, soil performs two important functions for plants:

- It provides support.
- It holds essential nutrients that can be taken in by the network of tiny roots.

Think About It!

A mixture of sand, silt, and clay is called **loam**. A typical mixture is 40% sand, 40% silt, and 20% clay. Mixtures with larger portions of one ingredient are called silty loam, clay loam, etc. How do you think different soil mixtures hold water and other nutrients?

Soil Classification Table

Particle	Texture
sand (0.02 mm–2 mm)	Feels gritty when rubbed between fingers. Not sticky when moist.
silt (0.002 mm–0.02 mm)	Feels smooth and powdery when rubbed between fingers. Not sticky when moist.
clay (less than 0.002 mm)	Feels smooth and sticky when moist. Forms hard clods when dry. May remain suspended in water for a long period of time.

Soil and Water

When you started your study of soils, you wrote a preliminary definition of soil. Your definition was probably more about what soil is than what it does. Now, enlarge your first definition so that it includes the uses of soil.

Do you agree with the statement, "Providing water to the roots of plants is the most important function of soil"? What are some arguments that might support this statement? Consider these questions about how soil and water interact. Use what you already know about soil to come up with the best answers.

- How much water can soil hold?
- Can some soils hold more water than others?
- Which holds more water, a soil mixture or pure sand?
- Which of these allows water to pass through at a faster rate?

Now you will have the chance to test your answers to the questions above.

EXPLORATION 5

The Amount of Water that Soil Holds

You Will Need

- 2 large plastic-foam coffee cups
- paper towels
- a glass, beaker, or wide-mouthed bottle
- your dried soil sample
- sand
- a stopwatch or clock
- a graduated cylinder

What to Do

Before starting this Exploration, let your soil sample air dry by spreading it out on a paper towel for 24 hours.

1. Using a pencil, punch three holes in the bottoms of the two coffee cups.

- coffee cup with soil or sand
- cardboard to fill space so that cup is properly suspended
- beaker

Remember to record all your measurements in your Journal. A table like this one might be helpful.

Quantities	Soil	Sand
Volume of water added	mL	mL
Volume of water that passed through	mL	mL
Volume of water held	mL	mL
Time for water to reach the bottom of the cup	seconds	seconds

Garden Ingredients 457

EXPLORATION 5—CONT.

2. Place two circles of paper towel in the bottom of each cup. The holes should be covered.
3. Fill one cup three-quarters full with dry sand and the other cup three-quarters full with dried soil from your sample. Place a glass of water on top of the sand or soil in order to pack it together. Use the same glass of water for each cup. This gets rid of air spaces.
4. Place each cup in the mouth of a beaker, bottle, or glass so that the cup is supported above the bottom of the container.
5. Add 250 mL of water to the sand. Time how long it takes for the first drop of water to drip into the beaker.
6. Repeat step 5 for the soil sample.
7. When water has stopped dripping into both containers, measure the volume of water that escaped from each cup by pouring the water into a graduated cylinder. How much water was held by the sand? by the soil?

Making Sense of Your Data

1. Compare the amount of water held by the soil and by the sand. Was the prediction you made at the start of this lesson accurate? Which do you predict would hold more water: sandy clay or silty clay? Why is the capacity to hold water an important property of soils?

2. Did your soil hold more water or less water than your classmates? Compare your results with those recorded by your class. At the same time, compare the soil types. Create and complete a table like this one in your Journal.

Name	Soil Type	Water-holding Capacity
Mike Annunzio	silty loam	mL
Susan Bates	clay loam	mL
etc.		

3. The time it takes for a certain volume of water to pass through a sample of soil is called its **percolation rate**. Which had a higher percolation rate, sand or soil? Why do you think this is so? Compare the percolation rates of the soil samples of your classmates. Which types of soils have slow percolation rates? fast percolation rates? Why is percolation rate an important characteristic of soil?

4. What could be added to a clay soil to improve it for gardening? to a sandy soil? How could the soil become better for growing plants?

5. What would you use to make a soil mixture suitable for the space biosphere? How much clay or sand would you use?

458   Growing Plants

EXPLORATION 6

Growing the Biggest and the Best

This is a long-term project. Your objective is to find out what it takes to grow the biggest and the best plant in the classroom. You will want to find out how to get the best results in each of these areas:

- Type of plant to use (must grow quickly)
- Type of soil
- Amount of water
- Amount of sunlight
- Temperature
- Distance between plants
- Size of container
- Fertilizer

As you start this project you will need to organize your activities. The class should divide into fifteen different groups. Each group will choose one investigation from those listed in the table shown. Later the groups can pool their results. Doing this will help you find out how to obtain the best results possible. Then you will all be on the way to growing the biggest and the best!

Your class is going to investigate the effect of five variables on the growth of three types of plants. How will you judge the amount of plant growth? Make a decision now. Here are some factors to consider:

- height of plant above ground
- number of leaves
- size of leaves
- mass of plant

The best idea may be to keep a record of as many observations as possible. Make these observations every second or third day, depending on how fast your plant grows. Record them in your Journal. Use several different ways to record your observations: a written description, measurements in a table, graphs, life-size or scale drawings of the plant, or some other means.

Experiment	Plant 1	Plant 2	Plant 3
Investigating different kinds of soil	Group 1	Group 2	Group 3
Investigating different amounts of humus in the soil	Group 4	Group 5	Group 6
Investigating different amounts of water	Group 7	Group 8	Group 9
Investigating different amounts of sunlight	Group 10	Group 11	Group 12
Investigating use of fertilizers	Group 13	Group 14	Group 15

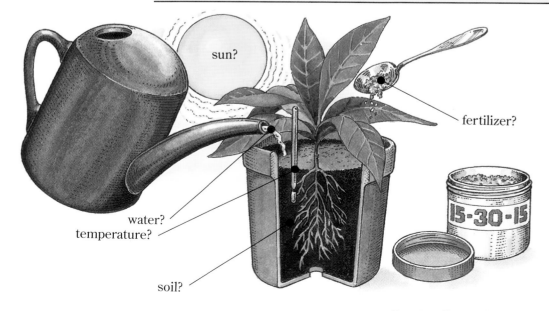

Garden Ingredients

EXPLORATION 6—CONTINUED

Wait a Minute!

Were you ready to start? Just a word of caution—you have to be careful that you don't do what Alex did! He was investigating the effect of the amount of water on plant growth. After looking at the illustration, would you say his results are reliable? How would you alter his experiment?

Alex varied the amount of water in the three containers. But he didn't keep the other variables constant. Remember to keep everything the same in each of your trials, except the variable whose effect you wish to measure.

General Hints

1. All groups should use the same kind and size of container. Large plastic-foam coffee cups are good.
2. All groups should use the same kind of garden soil, except those who are investigating soils. (Soil will be provided for you.)
3. All groups will use seedlings that have grown since the germination tests in Exploration 1. By now they may have several leaves. This is a good time to transplant them so they will have room to grow.
4. Water the plants when the soil appears to be dry. Keep a record of the amount of water added. Remember to add the same amount of water to each container, unless you are part of the watering experiment. The amount of water required may be different for different species of plants.

Hints for Each Investigation

Investigating different kinds of soil: Use sand, clay, silt, loam, or subsoil (found about 50 cm to 60 cm beneath the topsoil).

Investigating amount of humus: Add varying amounts of compost, peat moss, or bagged manure to soil that has only a small amount of natural humus in it.

Investigating amount of water: Remember that both the amount of water and the time between watering must be considered.

460 Growing Plants

Investigating amount of sunlight: Look for locations with plenty of sunlight and very little sunlight; or vary the amount of time that plants are placed in direct sunlight.

Investigating fertilizers: Use liquid fertilizer, varying either the amount used or the time between applications. Read the directions on the package before using any. The package will also tell you what types of nutrients you are adding to the soil.

Hints for Transplanting Seedlings

1. Put some of the soil that you want to use in the bottom of a container. Lift one plant from beneath the roots, taking care not to hurt the roots. (A spoon or craft stick may help.) Some soil will be removed.
2. Put the plant on the soil in the container and hold the plant upright. Fill the container with more of the same kind of soil. Make the soil firm enough to support the plant, but not so firm that there is no space for air.
3. Add enough water (room temperature) to moisten the soil. Do not pour water directly on the plant. Keep the plant at a constant temperature in a fairly dark place for a few hours. This helps it to recover from the shock of being transplanted. It can then be given the amount of light that you have decided to use in your experiment.

Judging the Results

1. Carefully listen to or read each group's report. In what ways can you learn from their experiments?
2. How might you determine exactly what is the *best* plant growth?
3. How might you improve the design of your experiment?
4. What factors were uncontrollable or difficult to control in your experiment?
5. What would you recommend to a future bean grower? radish grower? mustard plant grower?
6. Which recommendations would be useful for growing any plant?
7. "The best conditions for the growth of one kind of plant are not the best conditions for the growth of other kinds of plants." Do you agree with this statement? Support your answer with evidence from the experiments of all groups.

Garden Ingredients

Challenge Your Thinking

1. How would you complete the sentence, "Soil is . . . "? Use as many words or additional sentences as you need.

2. Prepare an article for a gardener's manual. You need to explain the instructions for raising a seedling from a seed, so a beginning gardener will easily know what to do.

3. Try your hand at the crossword puzzle. Don't write in your book; copy the puzzle into your Journal.

Clues

Across
1. A test of a hypothesis
2. A material in soil that lets the most water pass through soil
3. The time it takes for water to pass through soil (two words)
4. Once-living matter that enriches the soil
5. Chemical needed for growth and functioning

Down
1. A system that supports life
2. This is what the plant begins as
3. The chief water entrance of a plant
4. To sprout
5. Soil containing clay, silt, and sand
6. The way soil feels in your fingers
7. Fine soil particles

462 Growing Plants

INNER ACTIONS OF PLANTS

4 Water, Water, Everywhere!

You know that soil and water are important for the natural growth of plants on Earth. Consider the Mars biosphere. Will plants on Mars need soil and water just like plants on Earth? While soil may not be absolutely necessary, water must always be present for successful plant growth. How does a plant take in water? How does water travel through the plant? How is the water released again into the Earth's environment or into a closed environment such as a biosphere?

This Exploration will give you an opportunity to collect evidence about the interaction of plants and water.

EXPLORATION 7

Water and Plants Interact

Part 1
Entry

Water enters plants through their roots.

You Will Need

- germinated seeds
- a magnifying glass

What to Do

Look at the newly formed root of a seedling. Notice the root hairs, which are extensions of the root cells. Use a magnifying glass to look at them closely.
Each root hair adds to the total surface area of the root. In what way would this be useful to the plant? Imagine that the water in the soil forms a thin layer around each root hair.

Think About It!

Simone wants to plant a garden in an area of hard-packed clay soil. Ray has sandy soil in his garden. Tell these gardeners why their soils need improvement. Advise them about what to add to their soils so that the roots of their plants will have better access to the water in the soil.

Inner Actions of Plants 463

EXPLORATION 7—CONTINUED

Part 2
Which Way?

The surface of a root hair is a membrane that water can pass through. Water from outside the root hair crosses over the membrane. It enters one of the many cells that make up the root hairs. By crossing several membranes inside the root, the water moves to the other cells in the plant. But water must be able to move out of cells as well as into them. What makes water move in any particular direction through cell membranes? Try this and see.

You Will Need

- two containers
- water
- salt
- plants that pieces may be removed from

What to Do

1. Put some water into container A and a concentrated salt solution into container B.
2. Into each container, place several plant pieces, like a 5-mm thick carrot slice, a fern frond, a lettuce leaf, or a geranium leaf. Be sure to place the same kinds of things into each container.
3. After 20 minutes, observe and compare the plant pieces in the containers. Look again after 1 or 2 hours. Do they resemble the plants that they were taken from?
4. Which plant pieces remained crisp or became firmer? Which ones became spongy or limp?
5. Which direction did the water move in container A? in container B?

A Water   **B** Concentrated salt solution

In container A, water moved from the container into the plant cells. In container B, water moved out of the plant cells and into the salt solution. Such water movement is called **osmosis**. What determines the direction that the water will move? It is the concentration of solutes that determines water's movement.

Water moves through cell membranes from areas where water is more concentrated to areas where water is less concentrated. What does this mean? You know that the amount of solute in a solution determines how dilute or concentrated the solution is. Now think about a solution in this way:

Water (the solvent) and any solute dissolved in it are made of tiny particles.

Look at the diagram below. In the first section, all of the particles are water particles. In section 2, there are some particles of water and some particles of solute. That means that the concentration of water is less in the second section. In section 3, there are many particles of solute, so the concentration of water is even lower than it is in section 2. Predict what would happen if the walls between these liquids were cell membranes. Would water move through the cell membranes? In what direction?

Look at the observations that you made when you put the pieces of plants into containers A and B.

In container A, the water inside the plant cells had dissolved substances in it, such as minerals that plants need. The water in

○ water particle   ▲ solute particle

464   Growing Plants

container A did not have these dissolved substances in it. Water moved from container A into the plant cells just as it would move from section 1 to section 2 in the diagram.

In container B, there were fewer dissolved substances in the water inside the plant cells than in the water in the container that had a lot of solute (salt) dissolved in it. Water moved from the plant cells into the solution, just as it would move from section 2 into section 3 in the diagram. How did the movement of water out of the plant cells affect the plant parts? Water movement is always from areas where water particles are more concentrated to areas where they are less concentrated.

Knowledge about osmosis can be useful to gardeners, while lack of knowledge about it can cause problems. Think of some situations in which gardeners need to understand osmosis.

Beverly's Blunder

Beverly put some powdered fertilizer on top of the soil of her potted plant. Then she poured a little water over it. A few days later, the leaves of her plant wilted and turned yellow. Why did this happen? What should Beverly have done differently?

Part 3
A Quick Recovery

A plant that has been deprived of water appears limp and wilted. A wilted impatiens plant usually recovers quickly once water is provided. Explain why water moves into the root cells, and from there to all of the plant's cells. Where is the concentration of water particles greater? What direction is the water moving?

The appearance of the recovered impatiens plant shows that water has moved throughout the plant. Water has made the shape of the plant parts normal again.

Part 4
Special Routes
You Will Need

- a stalk of celery (cut at both ends)
- blue food coloring
- a clear glass
- water

What to Do

Put a few drops of blue food coloring and a piece of celery into a glass of water. Observe the celery over several hours. Describe the top end of the celery. What evidence is there that water and food coloring have moved up the celery stalk? Sketch the stalk of celery, showing any features that you think are important.

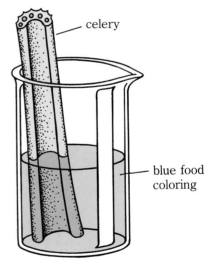

The celery plant has special cells that serve as tubes for the movement of liquids.

EXPLORATION 7—CONTINUED

Part 5
Exit

Water reaches the leaves of a plant where it is used in the food-making process called **photosynthesis.** Extra water is released from the leaves into the air. It is easy to collect some of this water.

You Will Need

- a geranium or other plant with large leaves
- a clear plastic bag
- tape

What to Do

Cover one of the leaves of the plant with a plastic bag. Tape it so that no air can enter or leave. Leave the plant in the sunlight or under a bright lamp for half an hour. Look closely at the inner surface of the bag. What do you see?

Transpiration is the term used for the movement of water out of the plant through the leaves.

Think About It!

1. A leaf has been compared to a wet towel on a clothesline.
   (a) What conditions help a towel to dry?
   (b) Would these conditions also increase the transpiration rate of a plant?
   (c) How could you help prevent a plant from drying out while it is transpiring?
   (d) What beneficial effect does transpiration have for plants on a hot day?

2. In the space biosphere, the plants' water supply would have to be maintained. How would you retrieve the water lost by plants through transpiration?

3. Eli and Tanya are farmers. They plant a field of corn with 50,000 kernels. The minimum germination rate for corn kernels is 90%. Suppose that two thirds of the germinated seeds grow into mature corn plants. One corn plant gives off about 200 L of water by transpiration while it is growing.
   (a) How much water could all of the corn plants together lose through transpiration during one growing season?
   (b) What soil type would you consider to be the best for growing corn? Why?

4. Where are the areas in the United States in which water lost from transpiration causes problems for farmers? What can they do about it?

466   Growing Plants

Water and Plants

Consider the evidence that you have gathered about water and plants. You know that water enters a plant's roots, that it moves upward through the plant, and that some of the water escapes through the leaves. Water and plants interact continuously.

In what ways is water useful to a plant?

- Water helps the plant hold its shape.
- Water carries essential dissolved substances throughout the plant.
- Water has a cooling effect as it evaporates during transpiration.
- The chief use of water by plants is for photosynthesis (making food), which takes place in the leaves. For photosynthesis to occur, plants also need carbon dioxide and sunlight. Oxygen is then released by the plant.

Investigate

You might want to investigate the process of photosynthesis by using the resources of a library. If you do, include a look at the process called *respiration* as well. Many people are surprised when they find out that plants respire. Plant respiration resembles our own respiration process; it involves taking in oxygen, which then is used to release energy from stored food. Carbon dioxide is released during respiration.

Your recent observations have shown that leaves are important to plants. The next Exploration gives you a chance to look at leaves of a variety of plants and to consider how leaf structure suits a plant's needs.

| EXPLORATION 8 |

A Look at Leaves

Part 1
Outer Looks

You Will Need

- a variety of leaves
- a magnifying glass

What to Do

1. Examine the upper and lower surfaces of each kind of leaf.
2. Describe each leaf. You may either draw the features of each leaf surface, or write a list of words to describe its features.
3. Exchange your lists or drawings with others to see if they recognize the leaves that you described.

Coleus (top), Nasturtium (bottom)

EXPLORATION 8—CONTINUED

Think About It!

1. In what ways are all the leaves alike?
2. How do you think the basic shape of leaves equips them for photosynthesis?
3. Can you tell the upper and lower surfaces of the leaves apart? How? Why do you think the surfaces are different?

Now, look for clues to answer this question: How do most substances enter and exit a leaf?

Part 2
A Closer Look

You Will Need

- a glass slide
- a cover slip
- a microscope
- a leaf (a large geranium leaf or a piece of romaine lettuce will be suitable)

What to Do

1. Carefully peel the thin layer from the lower surface of a leaf. (Hint: If the leaf is crisp, bend a piece of it until it "snaps," and then separate a piece of skin from the leaf's surface.)
2. Place the sample on a slide, with the lower surface of the leaf facing upward. Put one drop of water on top of the sample. Gently set the coverslip over your sample, and examine it under the microscope.

Compare what you observe on your slide with the picture shown here. Do you see what look like narrow openings? These are **stomata** (singular, stoma). What substances are likely to pass through these openings?

Begonia (top), Maranta (bottom)

Something to Think About

Look at the leaves of the plants you are growing. Are they all in perfect condition? Look at the leaves of other available plants.

1. Make a record of any damage you find on the leaves. This could include holes, change of color, or a wilted appearance.
2. Caterpillars and other living things find plant leaves very attractive. Explain why you think this is so.
3. What kinds of problems would you predict that your plants will have if their leaves suffer damage from other living things?

5 Giving Plants a Hand

What's in a Number?

Are the people on the other side of the fence speaking in code? It certainly looks that way! Actually, they are describing the kinds of fertilizers they use. Each set of numbers refers to the percentages of the three chemical substances most essential for plants. What does 10-52-10 mean? Such a fertilizer contains 10% nitrogen (N), 52% phosphorus (P), and 10% potassium (K). The other ingredients are inactive substances like clay or chemical substances like sulfur that make the fertilizer easy to use. What does 20-20-20 contain? 20-20-20 contains: 20% N, 20% P, and 20% K. What does 15-30-15 contain?

These number groupings are called the N-P-K ratings of fertilizers. Why did one person on the other side of the fence recommend using 10-52-10 first? This fertilizer is especially good for seedlings and for plants that have just been transplanted or repotted. Which chemical is present in the largest amount? This chemical helps plants develop strong root and stem structures. What would happen if a plant lacked this chemical?

Each mixture has its special use. For fertilizing leafy trees and shrubs, 28-14-14 is ideal. The nitrogen helps branches and leaves develop. Nitrogen must also be present for plants to have their usual green color. What symptoms would you expect to indicate a nitrogen deficiency?

Inner Actions of Plants

Potassium strengthens a plant, especially when fruits or vegetables are forming. It also increases a plant's resistance to pests or diseases. Potassium may play a part in storing or releasing nitrogen, according to the plant's needs.

Nitrogen, phosphorous, and potassium are not the only substances necessary for plants. There are many others, but these three nutrients are needed in the greatest quantities.

Increase Your Knowledge

1. Find several different kinds of fertilizers by using N-P-K ratings. Look at containers in gardening centers and read catalogues from garden supply stores. Match each fertilizer with its recommended use.

2. Investigate the composition and use of what are called "natural" fertilizers such as manure and compost.

3. Earlier, you chose a soil mixture for the space biosphere. Now consider what fertilizers you would take. Would you take:

    (a) chemical fertilizers only?

    (b) natural fertilizers only?

    (c) both types of fertilizer?

    Explain the reasons for your choice.

The composition of fertilizers is not the only chemistry important to gardeners. Plants have their own special pH needs. Some grow best in "sour" soil, that is, a soil that is acidic (pH value less than 7). Others require an alkaline soil (a pH value greater than 7). For example, cabbage grows best in slightly alkaline soil while rhododendrons thrive in acidic soil. Many plants grow best in soil with a pH of 6.5. Would this soil be acidic, neutral, or alkaline? The pH affects whether nutrients will dissolve in the water around the plant. That's why gardeners must be aware of the nutrients in a fertilizer and the pH of their soil.

EXPLORATION 9

The Litmus Test

You Will Need

- distilled water
- soil samples from different areas
- blue and red litmus paper

What to Do

Perform a simple test on your soil sample. Add a little distilled water to a small amount of soil until it makes a thick paste. Place pieces of red and blue litmus paper on the moist soil and record your observations. If the soil is acidic, blue litmus paper turns red. If the soil is alkaline, red litmus paper turns blue.

Think About It!

So far, you have been thinking about growing plants in soil. Could plants be grown without soil? Read on and see.

Plants Without Soil—The Way of the Future?

Growing a plant without soil sounds impossible! But if you replace soil with something else that can perform each function of soil, then the plant should grow. Therefore, you need to find a way to anchor the plant in an upright position, and to provide water, air, and other essential nutrients for it.

Today some plants are being grown without soil. This process is known as **hydroponics.** Instead of soil, a solution supplies the important nutrients. Perhaps hydroponics would be a useful way to grow food in your biosphere. Consider using this method for at least some of the crops when you write your proposal.

Here is an Exploration that investigates the growth of mustard plants in different nutrient solutions.

EXPLORATION 10

Which Nutrient Solutions Are Best for Growing Mustard Plants?

Before starting, form teams of three or four and choose one of the nutrient solutions listed below to investigate.

You Will Need

- rainwater or distilled water
- tap water
- rainwater or distilled water with hydroponic nutrient solution dissolved in it (See the package for directions.)
- rainwater or distilled water with quarter-strength hydroponic solution dissolved in it
- acid water: rainwater or distilled water with vinegar added (at least 10 mL of vinegar per liter of pure water)
- alkaline water: rainwater or distilled water saturated with gardener's lime
- a solution of your own choice
- baby food jars or clear plastic cups
- vermiculite or clean sand
- mustard seeds
- a dropper

What to Do

1. Fill a baby food jar or a clear plastic cup halfway with an *aggregate* such as vermiculite or well-washed sand. (An aggregate is a mixture of particles like sand or small pebbles, used here to support the plant.)
2. Slowly add your nutrient solution until the top of the aggregate is damp.
3. Place ten mustard seeds in the thin film of water on the surface of the aggregate. Put your jar in moderate sunlight. Each day, use a dropper to bring the water level in the container up to the point where the aggregate is wet on top.

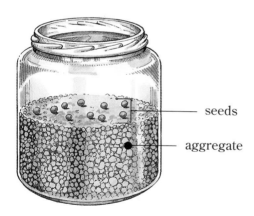

4. At the same time each day, count the number of seeds that have germinated. Record your findings in a table like this one:

Day	0	1	2	3
Number Sprouted	0	0	3	8

Inner Actions of Plants

| EXPLORATION 10—CONT. |

Analyze Your Findings

Share your results from the Exploration with the members of other teams.

1. Did the average time for germination of the seeds vary with the kind of nutrient solution used? If so, list the nutrient solutions in order according to the time it took for the seeds to germinate in each.
2. How was seedling growth affected by different types of nutrient solutions? Which solutions produced the most plant growth? Which produced the least plant growth?
3. How did the amount of hydroponic powder dissolved in the water affect plant growth?
4. Suppose your class wished to present a report of your findings to another class that is also doing this Exploration. Design a table and record the most important data found by each of the teams in your class. Will another group of students clearly understand the way you have set up and labelled parts of your table?
5. Draw some conclusions from class data that you can present to the other class.
6. Have you thought of other questions that you might investigate about hydroponics? What are some of them? Choose one of these questions and design an experiment that will help you answer it.

5. Once your seeds have sprouted, measure the height of the five largest seedlings at the same time each day. Then calculate their average height.

6. Plot a graph of the average height of a seedling against the time in days.

Tomatoes growing in a hydroponic research laboratory

6. Flowers and Pollination

New Plants From Seeds

You have been growing plants from seeds. You know that seeds are special parts of plants produced specifically to make new plants. You have worked with seeds in germination tests and have seen how plants can grow from seeds. A moment's thought will probably remind you of things you've known for years about flowers and seeds. But where did the seeds come from? A flower? How are flowers and seeds related to each other?

Look at the illustration of a flower on page 475. See if you can answer these questions.

1. What parts of a flower are often brightly colored and may attract bees or birds?
2. What parts at the base of the blossom enclose the flower bud before it opens and are often green?
3. Which parts are known as the female parts of the flower? What do they do for the plant?
4. Which parts are known as the male parts of the flower? What do they do for the plant?
5. Where do the seeds form?
6. What has to happen in order for seeds to form?

Pollination and Seed Production

Pollination is the transfer of pollen from the tip of the *stamen* of a flower to the tip of the *pistil* of the same or another flower. After this, fertilization, and later fruit and seed production, can occur. This completes the life cycle of a plant.

In nature, pollination is often aided by the action of wind and insects. Sometimes people can help make pollination successful. This is especially useful for plants grown indoors. How could you make certain that plants would be pollinated in the biosphere?

Most flowers have stamens and pistils in the same blossom. Only a slight movement is required to pollinate such plants as the tomato, pepper, eggplant, pea, or bean. Some plants, however, have separate male and female blossoms. These include cucumbers and squash. You can recognize the female blossoms by the miniature fruit *(ovary)* that appear beneath the flower.

Inner Actions of Plants

EXPLORATION 11

Parts of Flowers

You Will Need
- a variety of cut flowers
- a razor blade or sharp knife

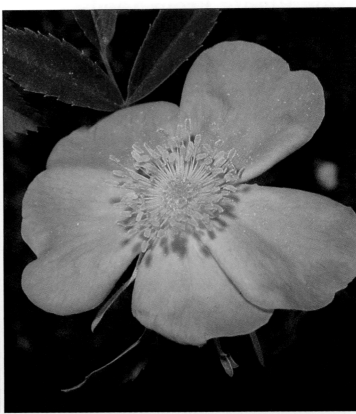

Clockwise from above, Buttercup; Carolina Rose; Oriental Lily; Zinnia

474   Growing Plants

What to Do

1. Study each flower.
   (a) Determine whether the flower is male, female, or both.
   (b) Identify the stamen, pistil, and ovary. If necessary, carefully dissect the flower with a razor blade or sharp knife.
2. Sketch and label the flower.
3. What features of each flower increase the chance that it will be pollinated in nature?

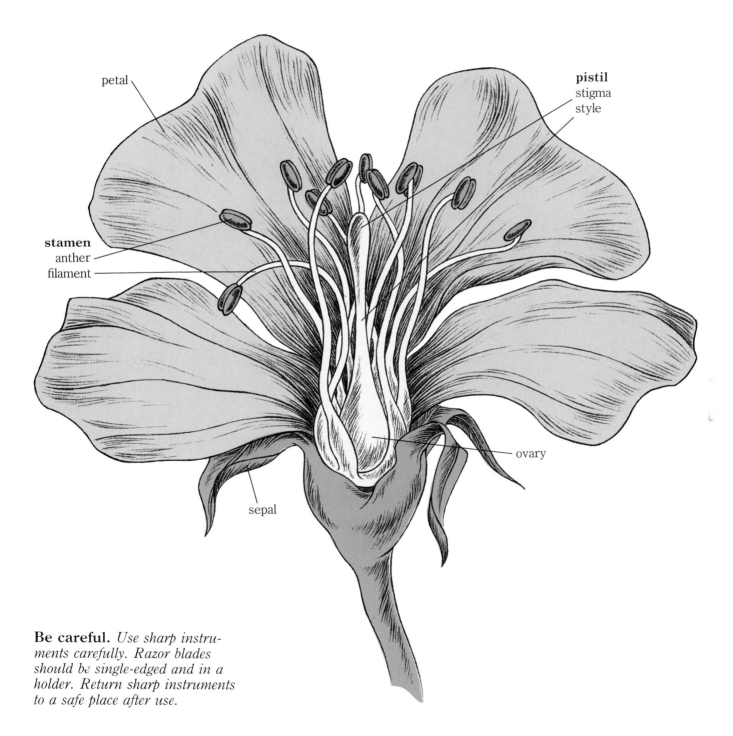

Be careful. *Use sharp instruments carefully. Razor blades should be single-edged and in a holder. Return sharp instruments to a safe place after use.*

Inner Actions of Plants 475

EXPLORATION 11—CONTINUED

A Picture Story

In your own words, describe how seeds form and develop into new plants. Use this "plant-cycle diagram" as your source of information.

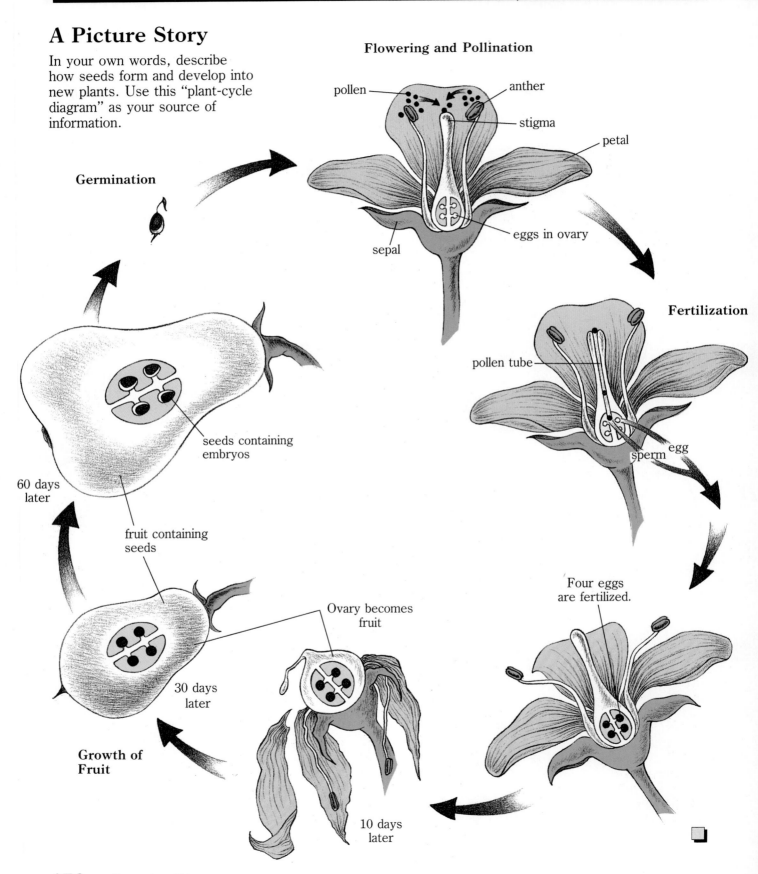

EXPLORATION 12

What Kind of Pollination?

Look at the examples of some of the ways that pollination takes place. Decide which show *self-pollination* (pollen and egg come from the same plant) and which show *cross-pollination* (pollen is transferred to a different plant).

Part 1
Peach Tree

Josh performed this experiment, which was recommended in his gardener's guidebook.

1. Pinch off petals and anthers.

2. Cover the pistil with a bag.

3. Get anthers from another tree.
4. Using a brush, transfer pollen to the stigma and cover it with the bag for a short time.

Part 2
Nasturtium Plant

Nasturtiums have perfect flowers. (They have both male and female parts.) A bee or hummingbird carries pollen from the anthers to the stigma of the same flower as it probes for nectar.

Part 3
Holly Plant

Each plant has only male flowers or only female flowers. Pollen must be transferred by the wind or an insect.

Male plant   **Female plant**

Inner Actions of Plants

| EXPLORATION 12—CONT. |

Part 4
Cucumber Plant

This plant has separate male and female flowers, called *imperfect* flowers, on the same plant.

Suggestion: Try your hand at making certain that either self-pollination or cross-pollination takes place. Save any seeds produced. Plant them, and enjoy the results of your breeding technique.

Cross-Pollination as a Plant Breeding Technique

Cross-pollination often requires a delicate touch. A scientist transfers pollen from the anthers of a flower on one plant to the stigma of a flower on another plant. Here is what might be done with two varieties of wheat, Wheat S and Wheat W.

Wheat S

- Resists disease
- Survives low temperatures
- Produces well (high yield)
- Matures in 110 days

Wheat W

- Subject to certain diseases
- Dies in low temperatures
- Has a low yield
- Matures in 90 days

Wheat S has one quality a farmer might like to change. What is it? Before you read further, propose a way to get a wheat with the desirable qualities of both S and W.

Here is one method. It is called *cross-breeding*.

1. Grow Wheat S and Wheat W.
2. When flowers appear, remove the anthers from the flowers of Wheat S. Cover the flowers with small bags.
3. Several days later, transfer pollen from the anthers of the Wheat W flowers to the stigmas of the Wheat S flowers.
4. At harvest time, collect the seeds from Wheat S.

Plants grown from these seeds will have various qualities of Wheat S and Wheat W. What combinations do you think might be produced?

7 Plants From Plant Parts

Can you grow a new plant without a seed? You can if you use roots, stems, leaves, buds, or twigs from an old plant. This way of growing new plants is called *vegetative reproduction*. Read the brief descriptions of the methods used in each example. Look at the illustrations and identify the method shown. Tell which part of the plant is being used. Remember, for each kind of plant, only some of the plant parts are suitable for vegetative reproduction.

The methods used to grow new plants from plant parts include:

1. Division—the roots of the plant are divided and planted separately.
2. Layering—plants send out new roots from points on their stems that have been left in contact with the earth. These parts are separated from the old plant and used for new plants.
3. Cuttings—a piece of a plant is cut off and replanted. This is only successful if the stem will readily grow new roots.
4. Grafting—a bud or twig from an existing plant is transferred to another plant that already has a root system.

(a)

(b)

(c)

(d)

(e)

Biosphere Decision

The members of the Biosphere Proposal Committee must decide what kinds of plants to choose and whether seeds, seedlings, or mature plants should be taken into the biosphere. Some members think that they will need plants in all three stages. What do you think?

EXPLORATION 13

Vegetative Reproduction

You Will Need

- mature plants
- scissors
- potting mix (equal volumes of perlite and vermiculite)
- a large, shallow container in which to plant the cuttings
- clear plastic wrap
- a large, deep container

What to Do

1. Water the plants the day before taking cuttings from them. This will save the plants from added stress.
2. You will root several cuttings in one container to save space and material. The potting mix should be moist, but not wet, or the cuttings might rot.
3. Fill the container with the mix. Then dig small holes about 2 cm deep with your finger or a pencil. Put the holes about 4 to 5 cm apart.
4. Take a stem or leaf cutting as described. Always hold the plants by the leaf, not by the stem, to avoid damaging the plant.

For stem cuttings, cut the stem with scissors just below the third or fourth pair of leaves. Strip off the lower pair of leaves so two or three sets of healthy leaves are left. The new roots will probably grow from the place where the leaves were removed.

For leaf cuttings, cut off a leaf with its stem attached. For a snake plant, cut off a 5-cm section from the middle of the leaf.

5. Carefully place the cutting into a hole in the mix. Fill in the hole so the cutting is held securely.
6. Put a piece of clear plastic over the whole container until the new plants have rooted.
7. When the new plants begin to grow, carefully remove them from the rooting container and transplant each into a deeper container. Use a general purpose soil mixture. Remember the precautions to take when transplanting.

Try some of these plants.

Stem Cuttings

Impatiens
Coleus
Geranium
Swedish Ivy

Leaf Cuttings

Snake plant
African violet
Begonia

Root Growth

For every method of vegetative reproduction described, what plant part must develop first (if it is not already there)? The root is the first part of the plant to emerge from the seed as it germinates. When a new plant forms from part of an old one, the roots must develop if the new plant is to live. The next Exploration will help you understand how roots grow.

Germinating barley seed

EXPLORATION 14

The Growth of a Root

You Will Need

- sprouted mung beans, radish seeds, or mustard seeds
- a fine-tipped waterproof marker
- a ruler
- paper towels
- water

What to Do

Examine mung beans, radish seeds, or mustard seeds that have already sprouted on moist paper towels. They should have roots. Gently put marks 2 mm apart on the entire length of one root. Use the waterproof marker. Watch for several days as the root lengthens. Then measure the distance between the ink marks on the root.

1. In your Journal, draw the root, showing the distances you recorded between marks.
2. Where is growth taking place in the root? What evidence leads you to think so?
3. Look for root hairs on the new root. Water enters the plant through the root hairs. What other use could root hairs have for a growing root as it moves through the soil? It may help to look at this photograph of a seedling.

Inner Actions of Plants 481

Growing Through the Soil

How does a root grow in length? Why aren't roots damaged when they push through hard soil? Examine the diagram below for answers.

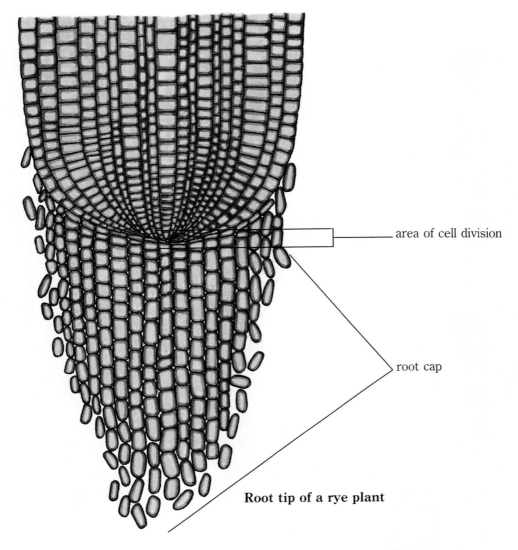

Root tip of a rye plant

The growth of the root takes place as new cells are formed in a process called *cell division*. Most of the new cells cause the root to grow longer. At the same time, however, there is a loose collection of cells forming a *root cap*. Look at the diagram of the tip of a root. These cells protect the area of cell growth. They get scraped away by soil particles as the root forces its way through the soil.

Challenge Your Thinking

1. Light affects plant development. Choose one of the following ideas and outline an experiment that would support or disprove the idea.

    (a) Does the shape of a tree have much effect on the amount of light it gets? Think about a pine tree and a maple tree.

    (b) Jeremy stated, "I think plants will grow faster in red light than in green light, and I am going to try to prove it."

    (c) Poinsettias bloom naturally in December. Dionne said, "If I can give a poinsettia plant the light conditions it would normally get in November and December, then I could make it bloom in summer or any other time I want."

2. Out of Order!

    (a) In your Journal, redraw this illustration, putting things in order. Be careful! Most, but not all, of the parts are out of place.

    (b) Label all of the stages shown. Two are labelled for you.

    (c) Why is this called a cycle?

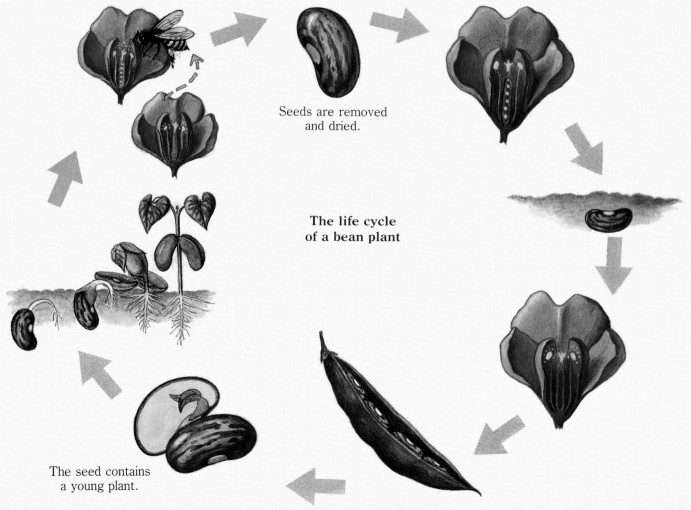

Seeds are removed and dried.

The life cycle of a bean plant

The seed contains a young plant.

Inner Actions of Plants 483

PLANTS IN YOUR ENVIRONMENT

8 Medicine or Poison?

Plant Medicines: Use With Caution!

As Mom came in the back door, three voices delivered some bad news.

"The cat's sick," said Maria.

"Poisoned!" Angela called out from the living room.

"...and the vet said it could be from something we used to kill pests in the garden!" This was from Carlos who was getting dinner ready.

Later, Mom said, "The vet says the cat will be all right this time, fortunately. But we must not let this kind of thing happen again. We'll have to be more careful about using poisons, or we'll be hurting ourselves next."

The Herrera family did something about their concerns. They began by learning about the situation. In this Exploration you will help the Herrera family identify some potential problems with pesticides and some possible alternatives that they might use.

EXPLORATION 15

Danger in the Garden

Part 1
Identifying the Issue

Sometimes plants in the Herrera garden were attacked by pests or diseases. Maria made a chart showing which chemicals had been used in the garden.

Area Number	Plants	Pest/Disease	Treatment
1	marigolds	slugs	metaldehyde bait
2	roses	aphids	dimethoate
		Japanese beetles	diazinon
3	daisies	spittlebugs	malathion
4	lilacs	leaf miners	diazinon
5	phlox	powdery mildew	benomyl

1. Find out if the pesticides (substances used to kill pests) named in the table can have effects on other living things. You might get help from people at the following locations:
   (a) Environmental Protection Agency
   (b) A garden supply store
   (c) An agricultural institute or university

2. Find out how chemicals used in agriculture can affect living things, especially if they get into drinking water or are breathed in with the air.

3. Look at the map of the Herrera's garden. Consider questions like these:
   (a) Would spray be carried by air currents? Where might it land?

484   Growing Plants

(b) How long will the chemical remain on the plants? in the ground?
(c) How will I rinse the sprayer I use? Where will I put the rinse water? Where will I put the empty poison bottle?

Add five more questions that might be asked.

The Herrera family felt they were ready to attack the problem. Mom said, "We must not spray poisons thoughtlessly, but we do want to get rid of pests and diseases from our plants. I want to find out if there are other ways of dealing with them."

Carlos said, "Janice's father has ladybugs. He said they eat pests on the roses. He doesn't use chemicals in his yard."

Mom said, "Ask him for any other ideas he has. We could find some safe alternatives."

Part 2
Identifying Alternatives

1. How can the Herrera family eliminate plant pests and diseases without using poisons? Two alternatives are illustrated here.

Find at least five other ways to help plants stay healthy without using poisons. Ask gardeners you know for alternate methods. Read a magazine devoted to organic gardening.

2. Make a collage of all the methods you discover.

Part 3
Researching

Look at gardening magazines and books. You need to answer questions like these:

- Can the use of all garden poisons be avoided?
- Are there safe ways to use some poisons to kill pests?
- Do alternatives have some disadvantages?

Part 4
Decision Making

1. Work in groups to discuss alternative treatments for the Herrera's plant problems.

2. Decide what action you would recommend for the future treatment of pests and diseases.

3. If you still find it necessary to use some poisons in your garden, name them and list the precautions that you would take to avoid possible dangers from them.

Part 5
Taking Action

The Herreras decided that they could eliminate some pesticides from their back yard. They found a collection depot where they could take these poisons, and they gladly got rid of them.

Next season the Herreras will be prepared. The cat and all of the family will be safer.

Part 6
Evaluating

Make a list of questions to ask the Herreras next year about their new ways of caring for plants.

Plants in Your Environment

9 Make a Green World!

A Design Project

Now you have a chance to do some designing on a larger scale. You will choose a project to do with other members of a team. The project requires you to plan for the growth and use of plants in one of three situations:

(a) indoors, at home or in school;

(b) outdoors, around a home or other building;

(c) in a vegetable garden, as a source of food.

Explorations 16, 17, and 18 will introduce you to these projects. Choose the one that suits you best after you have completed the Explorations.

Small Scale Designs

You have cared for plants and observed them closely. A plant's environment provides its living conditions. Plants have special needs for light, temperature, and moisture. An African violet grows well in soft, filtered light, while a cactus thrives in full sunlight. People who have learned to provide the right environment may grow both of these plants in the same house, but probably not on the same windowsill. There can be a variety of environments in one room.

What kind of environment will there be in the imaginary Mars biosphere? Consider the surroundings your biosphere plants will need.

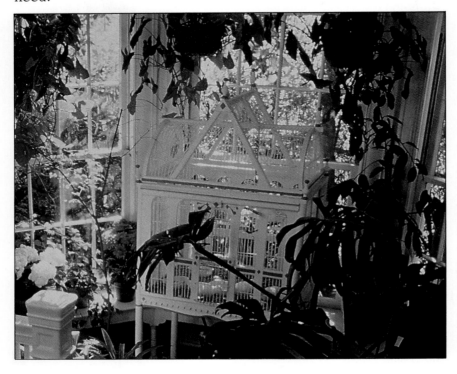

486   Growing Plants

EXPLORATION 16

A Greener Room

Han-Ling and Benjamin worked together on a plan for their classroom. They studied the conditions the room provided for green plants so that they could select plants that would grow well there.

What They Used

- graph paper and a pencil

What They Did

1. They noted where windows, doors, and furniture were located in the room.
2. They looked outside and found where trees, shrubs, and buildings partially blocked the sunlight.

Complete their study. You will begin with step 3.

3. In your Journal, draw a detailed floor plan showing what their classroom might look like. A sample sketch is shown at the right. You may add your own details.
4. Determine low, medium, high, and very high light conditions. Indicate these on the floor plan.

Low light—the light level that normally exists in a shaded window.
Medium light—the light level of an area within three meters of a sunny window that is partially blocked by buildings or trees.
High light—the light level that exists in front of an unobstructed window.
Very high light—the light level for an area that gets 3-5 hours of direct sunlight everyday.

5. Han-Ling used the following chart to choose these six plants for the room: heart leaf philodendron; red ivy; aluminum plant; prayer plant; coleus; and snake plant.
   (a) Indicate in your drawing where you would advise her to put these plants.
   (b) The classroom has space for four more plants. Help Benjamin select the next four plants. Then indicate in your Journal where he should put them.

Genus Name	Common Name	Light Level
Adiantum	maidenhair fern	medium
Aloe	aloe	very high
Asparagus	asparagus fern*	medium
Coleus	coleus*	very high
Dracaena	dragon tree	medium
Hedera	ivy*	high
Hemigraphis	red ivy	low
Maranta	prayer plant	medium
Mimosa	sensitive plant*	high
Philodendron	heart leaf philodendron*	low
Pilea	aluminum plant*	medium
Sansevieria	snake plant	low
Saxifraga	strawberry-geranium*	very high
Yucca	false agave	high

*can be used in hanging containers

Plants in Your Environment 487

EXPLORATION 17

Planning a Landscape

Think about the area around the outside of a house. How could you design a garden that would take advantage of the natural features of the landscape? The following Exploration will help you make your plans.

You Will Need

- tracing paper and a pencil

What to Do

Compare the professionally landscaped lot with the original plans.

1. What features are the same?
2. What changes were made?
3. Trace the final plan in your Journal. Use lower case letters to show where you could go to:
   (a) sit in the sun
   (b) stand talking to a friend, out of the winter wind
   (c) find a breeze in summer
4. Use capital letters to show where the designer planned:
   (a) a shady garden
   (b) a sunny garden
   (c) a decorative garden
5. What human factors, such as beauty and comfort, did the designer consider when planning the landscape?

A Base Map

The Final Plan

488 Growing Plants

EXPLORATION 18

Theo's Garden

The landscape shown in Exploration 17 includes a vegetable garden. Look again at that area of the plan. Describe the light and wind conditions that you would expect to find there. Then read how Theo planned his vegetable garden.

Theo and several of the neighbors in his apartment building rented spaces for gardens in a field outside of town. Theo planned to grow the things he likes to eat.

1. He drew an accurate diagram of his garden by letting 1 cm on the paper represent 50 cm of the actual garden plot. The diagram he drew shows the plants he intends to grow. Measure the length and width of the diagram. How large will his garden plot be? A garden this big may be large enough to provide food for four people.

2. Theo showed the diagram to his neighbor, Mr. Chandra, who always grows his own vegetables. Mr. Chandra suggested that Theo:

   (a) plant the garden well away from any large trees
   (b) plant 1 m or farther from a road or sidewalk
   (c) avoid an area at the bottom of the hill
   (d) buy young tomato, pepper, and melon plants instead of starting with seeds

   Discuss this advice with several others. Speculate on the reasons Mr. Chandra had for each suggestion.

3. Consider the advantages of having a garden that supplies you with fruit and vegetables. But remember, benefits always have costs. What are the costs of growing your own food?

Plants in Your Environment

EXPLORATION 19

Your Team Project

Now it is time to choose one of the three design projects described in Explorations 16, 17, and 18.

The following rules apply to each project.

1. Each project will be a team effort. Each member of the team will make their own contributions.
2. When the project is completed, it will be presented in an exhibit. The exhibit can include posters, photographs, illustrations, charts, diagrams, models, and other things you think will help explain what you have learned.
3. Each project must include interviews with people who have helped you. Interviews should be written up and included with the exhibit.
4. Each member of the team will obtain information for the project from at least two written sources (books, pamphlets, magazine articles) in addition to encyclopedias. This information will be included in the project report along with a list of the sources of information used.

Let's find out more about the choices you have.

Choice 1
A Greener Room or a Greener School

1. Develop a plan for your classroom or the school interior. Follow the procedure you used in Exploration 16. Be sure to share the work among the members of the group.
2. Find out more about indoor plants by visiting flower shops and nurseries where you can see the plants and ask questions about them. Talk to your parents and neighbors about the plants they have.
3. Read books, brochures, and pamphlets on indoor plants. Public libraries, flower shops, and nurseries have information to lend you.
4. In "Plants from Plant Parts," you learned how to grow plants from cuttings. Request free cuttings from parents and neighbors, or from flower shops and nurseries, and grow your own indoor plants.
5. Most indoor plants grow outdoors in tropical or semitropical parts of the world. Find out as much as you can about the origins of your favorite indoor plants.

Choice 2
Landscape Gardening

Here is your chance to redesign the school grounds. Each member of your group should participate in creating the overall plan. Then individual gardens and spaces should be assigned to members of the team who can plan in more detail.

1. Decide how you would like to make use of the school grounds. What are your goals in redesigning the landscape?
2. Measure the dimensions of the site. On graph paper, sketch any existing features.

3. Lay tracing paper over the sketch and begin to redesign the area. Make a rough outline of your own design. This will be a preliminary plan. It should be redone after you have researched plants and landscape gardening.

4. Your research should include:
   (a) visits to landscaped lots
   (b) interviews with landscape architects and gardeners (Large nurseries usually employ them.)
   (c) visits to nurseries
   (d) reading about landscape design (Borrow books from a public library, or pick up pamphlets from your local nursery.)

5. Your landscape design should include plans for each of the following:
   (a) a variety of trees and shrubs
   (b) each of these kinds of flower gardens:
      (i) a shady garden
      (ii) a sunny garden
      (iii) a cool garden (near a fountain, pond, or stream)
      (iv) a hot, dry garden

Choice 3
Growing Your Own Food

If you prefer growing edible plants to decorative ones, this project is for you. You will design a garden. Later, if you wish, you can carry out the plan and grow your own fruits and vegetables.

Each member of the group should learn about three different kinds of fruits and vegetables. The garden you plan should include the three kinds you have selected and any other vegetables and fruits that you would like to grow.

The following procedure is suggested for this project.

1. Select a location for the garden. It should get at least three hours of full sun during the growing season.

   You may select one large area to be divided among the members of your group or separate areas for each member. Each member should have an area of at least ten square meters. If your group decides to divide one large area among the members, leave a 1-meter path around each garden area.

   Draw a sketch of the garden site that you have selected. Measure its dimensions and include these on your sketch. Include nearby buildings and trees in your sketch, and note the amount of light that each part of the garden will get during the growing season.

2. Study the soil at your garden site. What kind of soil do you have? How much of your soil consists of clay? of sand? What difference does this make?

   Use a soil testing kit or ask an experienced gardener to determine if your soil is rich in nutrients.

3. Research for this project should include:
   (a) reading books on fruit and vegetable gardening
   (b) interviewing experienced gardeners
   (c) visiting nurseries and gardens

Have each member of your group research and report on one or more of the following topics:

(a) needs of different plants for light, water, and nutrients
(b) planting instructions for different kinds of plants

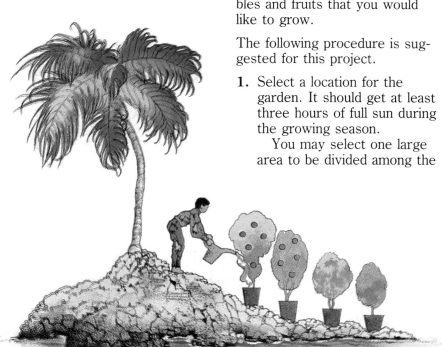

| EXPLORATION 19—CONT. |

(c) the type of soil and level of acidity (pH) needed

(d) the kinds of fruits and vegetables that grow best in your area

(e) chemical versus organic fertilizers

(f) common pests and plant diseases and how to prevent or control them

(g) the nutritional value of the various fruits and vegetables

Share your research with your group. Use the information that the group has gathered to learn more about the three plants you have selected.

4. Develop a design for your garden and sketch it. Use graph paper. In your written report, include your decisions about items (a) to (g) in step 3. At the end of the report, list all the sources of information you used including books, pamphlets, interviews, and visits.

5. The exhibit should include your plan and report, together with photographs, illustrations, samples, and other materials that will help you communicate what you have learned to other prospective gardeners.

Testing Your Design

You have completed a design. Why not find out how it works? The summer ahead will give you a perfect chance to test your design. In the fall, you can enjoy showing the results!

Design for Outer Space: Biosphere Proposal

Now it is time to submit your proposal for a space biosphere.

1. Review the biosphere report that you wrote earlier. Make any changes needed so that the report reflects what you now know about growing plants. Share your ideas with the other members of your group.

2. Together, list the problems that you would expect to face when building a biosphere.

3. Discuss how to prepare for the problems that you have listed.

4. Write your own biosphere proposal showing your plan for growing plants and providing for the people who will live in the biosphere.

Challenge Your Thinking

1. Design a newsletter for the Environmental Gardener's Group that points out specific ways to treat plant diseases and pests without poisons. Illustrate your letter with drawings of sick and healthy plants. Be sure to explain the benefits of environmentally safe pest control.

2. What's wrong with this picture? Randolph and Amy want all of their plants to grow strong and healthy. But someone planted the seeds in the wrong places! For each plant shown, describe the environment that it needs to live in. Then explain how Amy and Randolph should rearrange the plants.

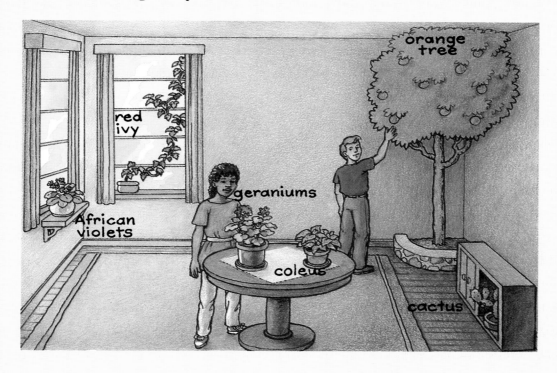

Plants in Your Environment

Unit 8 — Making Connections

The Big Ideas

In your Journal, write a summary of this unit, using the following questions as a guide.

- What must a biosphere provide for living things to survive?
- What is soil and what does it do for plants?
- How are water and other nutrients taken in and transported within plants?
- What functions do leaves perform?
- What does the N-P-K rating of fertilizer tell you?
- How can you grow plants without soil?
- What parts of flowers help pollination occur?
- What is vegetative reproduction? What role do roots play in this process?
- How can you prevent damage from pesticides?
- What factors should you consider in growing indoor plants? in outdoor landscaping? in vegetable gardening?

Checking Your Understanding

1. Megan measured the growth of a leaf over a period of time. Use her data to answer the following questions.

Day	1	3	5	7	9	11	13	15	17
Size of Leaf in Millimeters	2	3	4	6	10	16	24	35	50

(a) Plot this data on a graph.
(b) On day 19, the length of the leaf was 52 mm, and on day 21 it was 53 mm. What conclusion would you draw from these facts?
(c) What value would you expect for day 23?

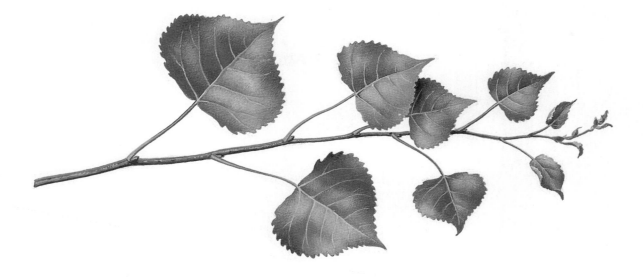

2. Describe how the following changes would affect life inside the Mars biosphere.
   (a) Eliminate the plants that are edible by humans.
   (b) Eliminate all animal life, except humans.
   (c) Eliminate the ocean.
   (d) Use chemical sprays to control pests.
3. Answer these questions about Matthew's experiment.
   (a) Matthew had two plants of the same kind. He placed one on a windowsill and the other in a closed cupboard. What hypothesis was he testing?
   (b) After two weeks, Matthew noticed that the leaves of the plant in the cupboard were not as green as the leaves of the plant on the windowsill. The plant in the cupboard was also taller and more spindly. What would be a good conclusion for this experiment?
4. Draw a concept map that connects the following words:
   flower, stomata, stamen, leaf, plant, root hair, pistil, and *root*

Reading *Plus*

You have learned about how plants grow and how to help them grow healthy and strong. Next you will explore the functions of plant parts and the many different uses humans have for plants. Read pages S133 to S144 in the *SourceBook* to increase your understanding of plants.

Updating Your Journal

Reread the paragraphs you wrote for this unit's "For Your Journal" questions. Then rewrite them to reflect what you have learned in studying this unit.

Science in Action

Spotlight on Agricultural Engineering

Christine Gorman is an agricultural engineer. We met in her office, surrounded by photographs of barns and other farm buildings.

Q: How did you become interested in agricultural engineering?

Christine: My interest in engineering probably stems from my talent for and enjoyment of mathematics. As to agriculture, I grew up on a farm and always had an interest in biology.

Q: What does your job involve?

Christine: Specifically, I am involved in the engineering of farm buildings. I give advice to farmers about designs and building materials. For example, yesterday I had to decide whether an old barn was still strong enough to withstand heavy winds and other conditions. I advised the owner that it was unsafe and should be replaced immediately. Tomorrow I will survey the site and tell the farmer what kind of barn I think he should build in that location.

Q: What are some of the benefits and drawbacks of your work?

Christine: One of the best parts is meeting a lot of people. To keep up with their questions and problems, I have to update my knowledge constantly. I am out in the field about half the time, at construction sites and in barns and places like that. One of the challenges is understanding all of the new technology. Sometimes it is hard to keep up with it.

One of the drawbacks is that it is sometimes hard to change people's minds. They hesitate to try new designs. You must try to convince them that what you have to offer may be better than what they have now.

Q: Do you ever have to deal with crisis situations?

Christine: Yes. Some of our crises involve farm fires where there is a loss of animal life. Often a fire will occur in the fall, and it is critical to rebuild immediately. Our whole staff has to go out and survey a new barn so that a building can be started the next morning. That is a real challenge to our service.

Some Project Ideas

Nowhere have inventions been more important than in agriculture. Those who live on a farm are surrounded by examples of technology. These may include a variety of machines, an irrigation system, buildings and other structures carefully designed for special purposes. There are many possibilities for interesting projects.

Your classmates might be amazed by the many inventions farmers take for granted. Pick out one example of technology on a farm. Find out what it does, how it does it, when it was invented, and whether there are any better ways of doing the same thing. Propose your own idea for improvements. The results can be presented as an exhibit. You might include diagrams to show how this technology works and how it might be improved.

Science in Action

Science and Technology

Genetic Engineering

Giant mice. Plants that glow in the dark. It may sound like science fiction, but such things have actually been created by **genetic engineering**. What are the limits of this new technology? At this point, no one can say.

The mouse on the right is a product of genetic engineering.

The World's Tiniest Code

Every living cell of a plant or animal contains enormous amounts of genetic information. This information tells the cells how to develop and what to do. When every cell follows its instructions, the result is a complete functioning organism with all its individual characteristics.

Genetic information is stored in the form of an extremely compact code. In fact, this code is "written" as small as it possibly could be—in the form of arrangements of groups of atoms! These arrangements are embedded in certain very long molecules known as DNA molecules.

Strange, But True

The individual units or "words" of the code are called **genes**. Each gene contains a specific coded instruction. For instance, a certain gene in the firefly causes the production of a substance that makes the firefly glow in the dark. If this gene could be transferred to the seed of a plant, what would the result be? A plant that glows in the dark? The answer is yes, as experiments have actually shown!

A model of a DNA molecule

Genetically engineered tobacco plant glows in the dark.

Early Results

Genetic engineering isn't just for creating oddities. It has already paid off in important ways. For example, it has made it possible to produce large amounts of *insulin,* which is needed by many diabetics. By transferring the human gene responsible for the production of insulin to certain bacteria, scientists have turned the bacteria into "insulin factories." These bacteria are grown in a laboratory, and they create large amounts of insulin. Today there is enough insulin for everyone who needs it, at lower prices than would otherwise be possible.

This illustration shows how an insulin-producing gene is transferred to bacteria.

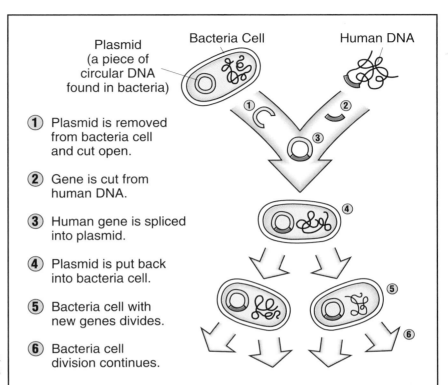

① Plasmid is removed from bacteria cell and cut open.

② Gene is cut from human DNA.

③ Human gene is spliced into plasmid.

④ Plasmid is put back into bacteria cell.

⑤ Bacteria cell with new genes divides.

⑥ Bacteria cell division continues.

Cutting and Pasting at the Molecular Level

Genetic engineers locate individual genes in a DNA molecule, cut them out, and then insert them at an appropriate place in a DNA molecule of a different organism. The procedure may sound simple, but imagine doing it on the incredibly small scale of a molecule!

The actual work, it turns out, is done by certain enzymes called *restriction enzymes.* Several hundred kinds of restriction enzymes have been identified. These enzymes have the function of cutting DNA molecules at specific points, depending on the exact nature of the enzyme.

The DNA molecules paste themselves back together, more or less automatically, because their cut ends are "sticky." When a number of fragments of cut DNA are mixed together, a certain percentage of them will splice back together in the desired way.

Find Out for Yourself

As an organism grows, many of its cells divide to form new cells. Remarkably, each of the new cells contains exactly the same complicated information as the original cell. This is because the information-bearing DNA molecules have a way of splitting into two halves and then reforming whole new DNA molecules from each half. Do some research on DNA *replication* to find out how this is done.

Write a report on some other accomplishments in the field of genetic engineering. What are future possibilities? What are some ethical considerations that might cause problems for the future of genetic engineering?

Science and Technology

CONCEPT MAPPING

A Way to Bring Ideas Together

What Is a Concept Map?

Have you ever tried to tell someone about a book or a chapter you've just read, and you find that you can remember only a few isolated words and ideas? Or maybe you've memorized facts for a test and then weeks later, you're not even sure what topic those facts are related to.

In both cases, you may have understood the ideas or concepts *by themselves,* but not in relation to one another. If you could somehow link the ideas together, you would probably understand them better and remember them longer. This is something a concept map can help you do. A concept map is a visual way of showing how ideas or concepts fit together. It can help you see the "big picture."

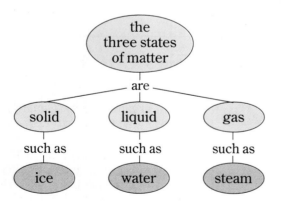

How to Make a Concept Map

1. **Make a list of the main ideas or concepts.**
   It might help to write each concept on its own slip of paper. This will make it easier to rearrange the concepts as many times as you need to before you've made sense of how the concepts are connected. After you've made a few concept maps this way, you can go directly from writing your list to actually making the map.

2. **Spread out the slips on a sheet of paper and arrange the concepts in order from the most general to the most specific.**
   Put the most general concept at the top and circle it. Ask yourself, "How does this concept relate to the remaining concepts?" As you see the relationships, arrange the concepts in order from general to specific.

3. **Connect the related concepts with lines.**

4. **On each line, write an action word or short phrase that shows how the concepts are related.**

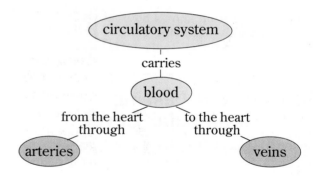

Look at the concept maps on this page and then see if you can make one for the following terms: *plants, water, photosynthesis, carbon dioxide,* and *sun's energy.* The answer is provided below, but don't look at it until you try the concept map yourself.

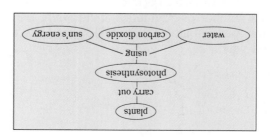

Texas EXPLORATIONS

When you think of Texas, science may not be the first thing that comes to mind. But if you take the time to look around, you may find science where you least expect it—in your own backyard! The Texas Explorations will introduce you to some of the facinating scientific happenings that occur right in your own state. Each Exploration relates to one of the units of *SciencePlus*, so you can see how the topics you study apply to Texas. But don't stop here. Texas is a large and diverse place that offers an endless variety of scientific discoveries.

Contents

UNIT 1	Fire Ants Invade the South	502
UNIT 2	The Bats Who Came to Dinner	505
UNIT 3	An Unwelcome Solution	508
UNIT 4	Nautical Notions	511
UNIT 5	Operation Ship Chip	515
UNIT 6	The Story of the Guadalupe Mountains	517
UNIT 7	Eyes on Texas Skies	520
UNIT 8	Xeriscaping	523

Colorado Canyon, Big Bend
Ranch State National Area

Texas EXPLORATIONS

UNIT 1

Fire Ants Invade the South!

It's a warm and sunny spring day. You can't wait to run barefoot through the grass. Flinging your shoes aside, you dash out into the yard, pausing for a moment to savor the feel of soft grass beneath your feet. Suddenly you feel a sharp burning sensation—like being stuck with a hot pin. This is followed a fraction of a second later by dozens more red-hot pinpricks. Help!

Does this story sound familiar? If you live in the eastern half of Texas, chances are you've had your own painful run-ins with the red imported fire ant. Unfortunately, such encounters are on the increase as these aggressive pests take over more and more of our state.

Most older Texans can remember a time when red imported fire ants weren't a problem. These pests haven't always been around. So where did they come from?

An Unwelcome Import

As its name implies, the red imported fire ant is not native to Texas. It was, in fact, an accidental immigrant to the United States. Fire ants arrived at Mobile, Alabama, in 1918, hidden in the cargo of a freighter from South America. The ants thrived in the lush, warm climate of Mobile and quickly began to spread. The rest, as they say, is history.

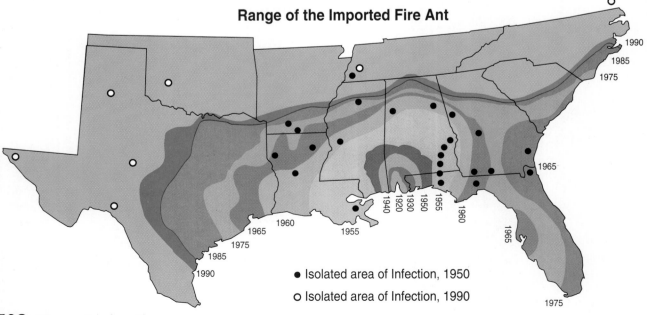

Range of the Imported Fire Ant

• Isolated area of Infection, 1950
○ Isolated area of Infection, 1990

502 Texas Explorations

By 1960 the fire ants had reached Texas. By 1990 they infested close to half the state, and they continue to spread rapidly.

An Immigrant Success Story

What is the secret to the fire ants' success? For one thing they are ferociously aggressive. It is said that fire ants "sting first and ask questions later." At the slightest provocation they swarm from their mound by the thousands, ready to attack. Fire ant venom contains a *pheromone*, a chemical messenger, that signals other fire ants to attack. Any intruder that stumbles onto a fire ant mound and does not quickly get away is likely to be stung hundreds, if not thousands, of times.

As fire ants spread, they leave a trail of damaged and disrupted ecosystems. Fire ants often wipe out native ants and other insects. They also attack and kill the helpless young of many birds and mammals.

Another key to the fire ants' success is that they breed very quickly. A fire ant queen typically lays a thousand or more eggs a day, every day, and lives four to seven years. A good-sized mound takes only days to form.

At one time, fire ants were like other ants. Each mound had one queen. The ants controlled the area immediately around the mound and defended themselves against ants from other mounds—even other fire ants. But after a massive campaign to wipe out the fire ants in the 1950s and 1960s, an extraordinary thing happened. Fire ants began forming supercolonies consisting of multiple queens scattered among many connected mounds. While a typical acre may support 2 or 3 dozen single-queen fire ant mounds, it can support 10 to 15 times as many multiple-queen mounds. The fire ant population soared.

Containing the Damage

Obviously, wiping out fire ants is not a realistic goal. Control and containment (preventing further spread) are the only viable options. How do we do this?

There are three basic methods for controlling fire ants. One method involves drenching each fire ant mound with insecticide. Another method is to use poisoned baits, which the ants pick up and feed to other ants. A promising new method uses hormone-containing baits that disrupt the fire ants' life cycle. Any successful fire ant control method must kill every queen in a mound. If even one queen survives, the mound will be reestablished.

Containing fire ants is difficult. Once a year, each mound produces hundreds of winged males and females, which fly high into the air to mate. Every female that mates becomes a queen. After mating, the female may be carried many kilometers by the wind. The male, his duty done, drops dead. When the female lands, she establishes a new colony.

Texas Explorations **503**

Three years in the life of a Texas pasture

Year 1

Year 2

Year 3

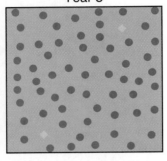

+ = Single-queen colony
◇ = Dead single-queen colony
● = Multiple-queen colony

Analysis, Please

1. Look at the map provided on page 502. Use the information in it to locate Mobile, Alabama. What clues helped you?
2. Did the fire ant population spread more quickly before 1950 or after 1950? What might account for this?
3. The figure to the left shows the changes that occurred in a 1/5-acre section of a Texas pasture over three years.
   (a) What happened between years 1 and 2?
   (b) What happened between years 2 and 3?
   (c) What effect did the arrival of the multiple-queen colonies seem to have on the single-queen colonies?
4. Of the three types of fire ant control methods mentioned, which seems the most environmentally sound? Which seems the least environmentally sound? Explain.

Go Figure!

You'll need a calculator to answer the next few questions.

5. How many offspring does a single fire ant queen produce in her lifetime, assuming that she lives for 5 years, that she produces 1000 eggs a day, and that 90 percent of the eggs mature into adults?
6. Approximately 250,000 square kilometers of Texas are currently infested with fire ants. Assuming 15,000 mounds per square kilometer, how many fire ant mounds currently infest Texas?
7. By the third year, the pasture shown at left had become heavily infested. Assuming that it was typical, about how many mounds per acre could a heavily infested area expect to have?
8. A typical fire ant mound consists of about 300,000 ants. Again, assuming that the pasture shown at left was typical, how many fire ants could you expect to find in an acre of heavily infested land?

Texas Explorations

UNIT 2

The Bats Who Came to Dinner

It's shortly before sunset on a hot August day. You're standing near the Congress Avenue bridge over the Colorado River in downtown Austin, Texas. A crowd is gathering. Everybody seems to be interested in the bridge for some reason. You sense that something is about to happen, but what?

Suddenly, there's a flurry of excitement. A young boy points and shouts "There they go." You follow his pointed finger and suddenly you see it. From underneath the bridge, swarms of fluttering, dark-brown creatures emerge. This mass exodus goes on for almost 20 minutes. "More than a million of them now," you hear someone say. So that's what a million looks like, you think.

You have just witnessed the world's largest urban bat colony departing on its nightly feeding rounds. From now until sunrise, these Mexican free-tailed bats will scour the surrounding area for insects. Each bat will consume its own weight or more in bugs. Before the night is over, the colony will have eaten 15,000 kg of insects.

As spectacular as the Congress Avenue bat flight is, it pales in comparison to that of Bracken Cave. Somewhere between 20 and 40 million Mexican free-tailed bats live in this cave near San Antonio. If that sounds like a lot, it is. The Bracken Cave colony is the largest concentration of mammals anywhere on Earth.

Blind as a Bat

For centuries, bats have been the objects of fear and loathing in Western culture. Myths and falsehoods abound about these agile aerial acrobats.

Test Your Bat IQ

How much do you know about bats? With your classmates, decide whether each of the following statements is true or false. When you're done, turn the page to find out how you did.

1. Bats are blind.
2. Bats can get tangled in your hair.
3. Many bats feed on blood.
4. Bats can eat thousands of insects a night.
5. Bats are responsible for epidemics of rabies and other diseases.
6. Bats are flying mice.
7. Bats are fierce and will attack if provoked.
8. Bats are the only flying mammals.

The Congress Avenue colony of Mexican free-tailed bats

Answers
1. False; bats generally have excellent vision.
2. False; bats are such excellent navigators that a collision with a human is unlikely.
3. False; only one species of bat—the vampire—feeds on blood. It takes only drops at a time, though, and humans are rarely its target.
4. True; bats may eat 600 mosquitoes an hour!
5. False; cases of people being infected by bats are extremely rare. Although a tiny fraction of the bat population does contract rabies, the disease so quickly kills those infected that they are not a threat unless handled.
6. False; bats are a distinctly different type of mammal belonging to their own order (a major level of biological classification). Bats may, in fact, be primates (the group of mammals that includes monkeys, apes, and humans).
7. False; Bats are quite docile and bite only in self-defense.
8. True

How did you do? Were you surprised?

Bum Rap for Bats

The truth is, bats don't deserve their bad reputation; they are very beneficial creatures. Ecosystems all over the world depend on bats. They help pollinate many useful plants and eat lots of insects. Just think about it—if the Congress Avenue colony eats 15,000 kg of insects a night, how much do you think the Bracken Cave colony eats?

A Festival of Diversity

Bats are a remarkably diverse group, about 1000 species in all. That's a quarter of all known mammal species! Texas is home to 32 species of bats—more than any other state. Why might this be? These pages show several species of bats found in Texas. Look at the pictures and then answer the questions on the following page.

Pallid bats

Silver-haired bat

How Do They Do That?

Most bats feed by chasing down and catching insects in flight, in the dark. How do they do it?

Bats use a sophisticated system called *echolocation* to precisely locate and track their prey. This method also enables them to navigate through the total darkness of caves without colliding either with the walls or with other bats. Here's how echolocation works:

- The bat emits a pulse of ultra-high-pitched sounds. At 80,000 to over 100,000 Hz (cycles per second), the sounds are well above the range of human hearing.
- These sounds bounce off the surroundings and return as echoes. The bat's brain is able to interpret the echoes to obtain information about its surroundings.

The bat's echolocation system is astonishingly sophisticated. Using echolocation, bats can tell the size and location of insects, as well as the size and placement of obstacles in their path. They can, in effect, "see" everything around them.

Long-nosed bat

Red bat

Peter's ghost-faced bat

Mexican free-tailed bat

Questions

Examine the photos on these pages to help you answer the following questions.

1. Which bat seems to have an adaptation for feeding on plant nectar? Explain.
2. Compare the bats "arms" and "hands" to your own. What similarities do you see?
3. How might the unusual shapes of some bats' faces be an adaptation?
4. Which bats show obvious signs of having acute hearing? How can you tell?

For additional information, write or call Bat Conservation International, PO Box 162603, Austin, TX 78716; (512) 327-9721

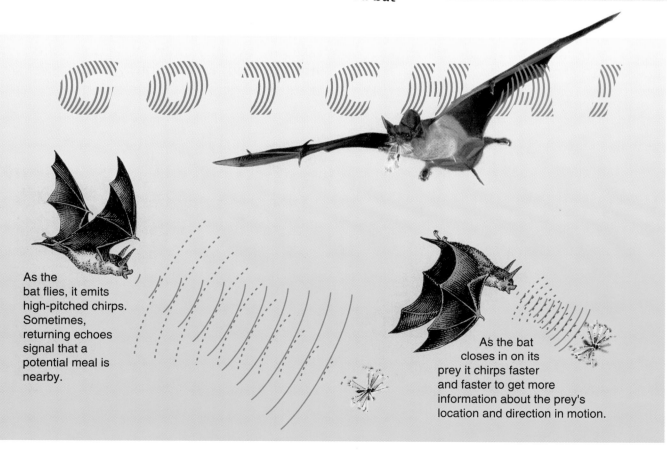

GOTCHA!

As the bat flies, it emits high-pitched chirps. Sometimes, returning echoes signal that a potential meal is nearby.

As the bat closes in on its prey it chirps faster and faster to get more information about the prey's location and direction in motion.

Texas Explorations

UNIT 3

Texas EXPLORATIONS

An Unwelcome Solution

It's 4:00 on a hot afternoon late in August. The air shimmers with heat as the day builds to a scorching climax. Slowly, imperceptibly at first, the sky begins to darken. The air grows still and heavy. From far off comes the faint rumble of thunder. Then . . . Rain—sheets of it—pouring from the sky! Streets are awash. Creeks and ravines, which moments before were dry, become rivers. All of the dust, dirt, and grime accumulated since the last rain a month and a half ago is washed away in the flood.

Everybody loves a good rain, especially after the long dry spells that our state often has. A good rain replenishes and washes things clean, and it gives people a feeling of renewal. But all of this good has a down side.

Unfortunately, there is a problem associated with rainfall—a problem we humans have created. Whenever it rains, much of the water soaks into the ground. But in a heavy rainfall, the water falls faster than it can soak in. The excess water collects and begins to flow over the surface in a process called runoff. Hard surfaces such as concrete, asphalt, and bare rock shed water as fast as it falls. As the water flows, it picks up particles and contaminants from the surface.

Pulled downward by gravity, the runoff flows into streams and creeks and from there into our rivers and lakes. Eventually, it winds up in the ocean. The runoff itself is not so much the problem. The problem is the oil and trash and chemicals that are picked up by the runoff. These pollutants form a noxious and unsightly mixture that threatens the health of our waterways and the organisms that depend on them. This is a very serious pollution problem, a problem that has a name: non-point source pollution.

Non-point source pollution comes from lots of different places. Parking lots, streets and highways, golf courses, farmland, and ranching operations are all prime sources of non-point source pollution. How do you think these places cause pollution? Think about it. Think of all the oil, grease, grime, transmission fluid, gasoline, and so on that collects on roadways and in parking lots. Think of the litter that collects around them. This stuff gets washed off every time it rains. What pollutants do you suppose are shed by lawns and by farming and ranching operations?

If you've ever seen a creek or a stream after a heavy rain, then you know how serious a problem non-point source pollution really is. Trash and crud are everywhere. The Rio Grande after a rain is literally poisonous because of all the toxins washed in from both sides of the river.

The Pollution Solution

So what can we do about this problem in Texas? First it would be a good idea to identify what kinds of non-point source pollution are getting mixed up with water in your community during a rain.

You can conduct the following experiment to make observations and analyze runoff where you live. Next time it rains, get water from different sources. Place a can or other receptacle outside to catch rainwater directly from the sky. Place another receptacle so that it collects runoff, such as that from a parking lot, golf course, or roadway. This way you can get two separate samples: one that contains only water that has fallen directly from the sky and another that has likely been affected by human activity. Be careful to secure the two receptacles so that they don't float away in the rainstorm and add to the non-point source pollution problem!

Texas Explorations **509**

Rating the Runoff

After it stops raining, collect your two receptacles and mark them: sample 1–"rainwater," sample 2–"runoff." Now, back in your classroom, use what you have learned from Unit 3 in your textbook to make a list of all the experiments that you can perform to determine what each sample is made of. Make a table to show the results of your tests.

For each sample, answer the following questions.

1. Is the sample a solution or another type of mixture? If it is a solution, is it concentrated or dilute? Are there suspended solids in either sample? How can you separate out and weigh the solids using filter paper and a triple-beam balance? What other methods, if any, can you use to separate the different components of the sample?
2. What are the soluble, not very soluble, and insoluble parts of your sample?
3. Is the sample hard water or soft water?
4. The pH of normal rainwater is 6 or higher. Use the litmus scale to determine the pH of the sample. Is it normal rainwater? What is the pH of each sample?
5. Is oil present? To find out dip a piece of filter paper in each sample. If oil is present, the filter paper will turn translucent.
6. List all the differences you find between the rainwater and the runoff.
7. Are any identifiable pollutants present?

Think About It

8. What kinds of problems might be caused when pollutants end up in our streams and lakes?
9. What impact might non-point source pollution have on aquatic life? on you?
10. A watershed is the area drained by a stream. If feasible, determine the watershed of the stream you tested (even if it's just the runoff from a parking lot.) Draw a map of the watershed, locating possible sources of pollution.
11. What are some solutions to the problem of non-point source pollution?

Going Further

An organization called Texas River Watch monitors Texas rivers for environmental quality. Find out if there is a chapter in your area. What is the major stream, river, or lake in your area? Write or call the Texas River Watch to see if there is anything you can do to help keep your local waterways clean.

Solution or not?
As you know, a solution is a special type of mixture in which a substance is completely dissolved in another. Another type of mixture is one in which particles are temporarily suspended in a liquid; after a time the particles settle out. This type of mixture is a *suspension*. There is another type of mixture in which the particles are large enough to be detected, but too small to settle out. This type of mixture is called a *colloid*.

Texas River Watch, Agriculture Division, Texas Natural Resource Conservation Commission, Agriculture Division, Community Support Section, P.O. Box 13087, Austin, TX 78711, (512) 463-8206

Texas EXPLORATIONS

UNIT 4

Nautical Notions

In the years before Texas became Texas, Spanish galleons regularly sailed the Gulf of Mexico. They traveled between Mexico, the heartland of New Spain, and the many far-flung outposts in the wilderness of North America. One frequently traveled route existed between Veracruz, Mexico, and what eventually became Galveston, Texas. Up until the 1820s, when Mexico gained its independence from Spain, Spanish soldiers, civilians, and cargo were frequently transported along this route. In honor of this rich history, every other year adventurous Texans and Mexicans sail across the Gulf of Mexico in the *Regata de Amigos*—the Regatta of Friends.

The Regatta follows the trail blazed by Spanish explorers over 400 years ago.

The Spanish galleons generally traveled well-established routes. This was not just a habit; it was a necessity. In comparison with today's sleek and maneuverable sailing craft, Spanish galleons were clumsy, lumberinghulks. To be sailed with any confidence, they required reliable winds and currents. The Galveston–Veracruz route, with its prevailing southeast breezes, met this requirement.

While times and technology may have changed, the forces of nature have not. And just like their Spanish

For more information on the Regata de Amigos, *contact the Galveston Bay Cruising Association, (713) 334-2521*

Texas Explorations **511**

counterparts, modern-day sailors must contend with these forces—wind, waves, and currents—which are sometimes cooperative and sometimes not. Amazingly, in spite of many potential obstacles, some boats complete the 1200 km Galveston–Veracruz run in as little as 80 hours. Incidentally, a Spanish galleon could be expected to make the same journey in 10 days.

Now imagine that you are one of the Texas sailors participating in the *Regata de Amigos*. What type of boat would you like to sail—a catamaran or a monohull? If you don't know what these sailing craft look like, turn to pages 195 and 196 of your textbook. The catamaran has twin hulls joined by beams; the monohull (as its name suggests) has only one hull. Both designs offer certain advantages and disadvantages. Can you think of some?

You will have the chance to design a boat for a regatta of your very own. But first, let's take a quick course to learn how sailboats work.

How Sailboats Work

Many people are puzzled by modern sailboat design. "Why," they ask, "are the sails mounted sideways?" The answer lies in the physics of sailing. Mounting the sail sideways takes advantage of physical principles to greatly enhance the sailboat's maneuverability.

It might surprise you to learn that the same principle that allows airplanes to fly also helps sailboats sail. As the wind blows into the sail, it puffs the sail outward, forming a curved forward surface. When air flows across a curved surface, it speeds up. The faster-moving air exerts less pressure than the air on the back side of the sail. The sail is "pulled" forward into the resulting partial vacuum.

The sailboat would simply drift with the wind were it not for two key control elements: the *rudder* and the *keel*. The rudder, operated either by the wheel or by a tiller, allows the craft to be steered. The keel, a large finlike panel extending from beneath the center of the boat, keeps the boat from being pushed to the side by the wind.

There are three basic sailing maneuvers: sailing with the wind, or *running*; sailing across the wind, or *reaching*; and sailing into the wind, or *tacking*. In running, the sail is held at nearly right angles to the wind and the wind simply pushes the boat along. In reaching, the sail is held at about a 45° angle to the boat and the boat travels more or less at right angles to the wind. Contrary to what you might think, reaching is actually faster than running. Reaching is, in fact, the fastest sailing maneuver. In tacking, the boat *tacks* (zigzags) back and forth as it heads upwind. Sailboats cannot sail directly into the wind. They can, however, sail to within about 45° of a headwind.

Running

Reaching

Tacking

Texas Explorations **513**

The Great Race

Now you're ready to conduct a regatta of your own. Your regatta will take place in your school on a "lake" that you construct. Each student or group of students will construct and enter a boat of original design. Boats will be raced one at a time in time-trial fashion. There will be two heats. The first heat is a running heat. The second heat is the reaching heat. If time permits, you might even attempt tacking. The boat with the fastest combined time wins the coveted "Regata de Amigos" Cup. Each boat will be permitted to make two runs for each maneuver. Only the fastest of the two times will count.

Wind will be supplied by a box fan (turned to HIGH if multi-speed). The timer will be started at the same time the boat is released and stopped when the boat reaches the opposite shore. For the reaching heat, be sure to orient the fan so that it blows from the side.

Making Your Boat

Get together with two or three of your classmates to collaborate on a design. You may use whatever materials you want to make your boat, so long as they are common classroom materials such as note cards, cloth, paper, soda straws, plastic-foam cups, craft sticks, and so on. You may also use such materials as rubber cement, string, or staples to connect the components together. You may shape the materials any way you want.

Here are a few simple guidelines:

- The mass of your boat may not exceed 15 g.
- The boat should be at least 13 cm in one direction but should not exceed 21 cm in any dimension.
- Every boat will be required to pass a seaworthiness test before racing. To pass this test, your boat must be able to float for a period of 30 seconds.

Experiment until you have what you feel is a good design. Submit an original design; don't just copy someone else's. And be creative!

Debriefing

After the race, compare the winning and losing designs. Try to figure out what gave the best designs their winning edge. Experiment with different sail configurations to see if that improves your boat's performance. Answer the following questions.

1. How are the boats affected by friction?
2. What other forces affected the boats? Draw a force diagram.
3. How could the designs have been improved?
4. How did the boats perform in each type of sailing maneuver? Were you able to get your boat to tack?

Texas Explorations

UNIT 5

Operation Ship Chip

Imagine that you are in charge of product development for South by Southwest Amalgamated Food Products, Inc., distributors of gourmet Southern and Southwestern food products. Your company wants to market a new product developed for its mail-order business. The new product, called the CowChip™, is a beef-flavored potato chip in the shape of a Texas Longhorn, the famous cattle breed. If initial tests are successful, the product line will be expanded. Each product unit will consist of a single chip in its own package. To conserve resources and to hold costs down, the packaging must be lightweight, efficient in its use of space and materials, and cheap to make. You have been put in charge of designing the packaging for the product. The packaging must keep the chip intact as it makes its trip through the U.S. mail.

Your Mission
The winning design will be the smallest and lightest package in which the potato chip actually remains intact. To demonstrate that no strengthening agent has been added to the chip, you will eat it at the conclusion of the test.

The Mail Experience
Devise a "torture test" to simulate the U.S. mail. Think about what the package will go through as it travels. Will the package be dropped? vibrated? bounced? crushed? How can you simulate these effects? Get together with your classmates to come up with a plan. After you devise your torture test, you can tackle the problem of package design. The winning package will actually be mailed to confirm its winning status.

TASTE THE FLAVOR OF THE OLD SOUTHWEST

Texas Explorations **515**

Rules

1. The minimum size (required by the U.S. Postal Service) is 3.5" × 5" × .25" (8.9 cm × 12.7 cm × 0.64 cm, or 71.7 cm^3).
2. Any materials may be used, as long as they are commonly available. But remember, you're trying to save on weight <u>and</u> cost.
3. All packages must endure exactly the same test.
4. The potato chip should be completely removed from the package unbroken.
5. The potato chip should be eaten after it is removed and its intactness is verified.
6. The formula for determining the winning design is: weight × volume. The winning design will be the one with the LOWEST score. The winning package should be mailed First Class via the U.S. Postal Service to your class, care of your school.

UNIT 6

The Story of the Guadalupe Mountains

Rugged and remote, Guadalupe Mountains National Park features almost 80,000 acres of some of the most scenic landscapes in Texas. The park includes the southernmost part of the Guadalupe Mountain range, a massive limestone block jutting out of the desert. Within the park is Guadalupe Peak, which at 2667 m is the highest mountain in Texas. Another well-known peak, El Capitan, guards the southern end of the range.

The craggy Guadalupes stand in sharp contrast to the surrounding desert, which sprawls almost 2 km below the highest peaks. Within the high valleys and deep, shady canyons of the Guadalupe Mountains, maples, pines, and aspens thrive, reminders of the cool, wet climate that prevailed thousands of years ago.

While Guadalupe Mountains National Park is notable for its scenery, it is truly remarkable for another reason: It is an almost perfectly preserved fossil remnant from the distant past.

In the Beginning

The story of the Guadalupe Mountains began about 250 million years ago along the shore of a warm sea. To the north of this sea was a large land mass. Toward the southeast, the sea bottom dropped steeply into the Delaware Basin. Lime-secreting organisms, such as algae, corals, bryzoans, trilobites, sponges, and brachiopods, found the shallow, well-oxygenated waters along the edge of the sea to be an ideal environment. Over time, their accumulated remains formed a limestone reef.

Texas Explorations **517**

The Story Continues

Over the entire area, the sea floor was slowly dropping downward. However, the lime-secreting organisms easily kept pace with this slow subsidence. As a result, over many thousands of years their skeletons continued to pile up until the reef eventually reached a thickness of several hundred meters.

As the reef grew upward, it also grew outward, advancing many kilometers into the Delaware Basin. The pounding of waves along the shore continually broke off large pieces of the reef, which tumbled down the submerged escarpment, or slope. These broken remnants, called *talus*, came to rest in a heap at the base of the reef and helped serve as the foundation for further outward growth.

During this time, the sea level rose and fell repeatedly. Occasionally, the sea level would drop just enough to expose the reef. At such times the reef organisms would die and erosion of the reef would occur. Then, after a time, the sea would rise again and life on the reef would be reestablished.

On the north side of the reef was a shallow lagoon. Sediments that washed down from the land mass settled in the lagoon's quiet waters. Periodically, the lagoon dried up, killing the marine life and causing evaporite minerals to be deposited. Halite (rock salt), gypsum, and anhydrite were the most common minerals formed in this way.

Eventually, the seas retreated altogether. Thick evaporite layers were deposited on the sea bottom as the waters dried up. In time, new sediments were deposited and began to cover the area. Over millions of years, the entire region was buried under hundreds of meters of sediment.

An Encore Appearance

Then, about 70 million years ago, tectonic forces caused an uplift to occur over the entire region. As the land rose, accumulated sediments began to wash away. Bit by bit, the former reef was exposed. The former reef had hardened into limestone, which resisted erosion while the surrounding soft sediments were stripped away. Eventually the entire sediment cover was removed, revealing a landscape that is thought to look almost exactly as it appeared 200 million years ago—only without the water! Today the limestone of the Guadalupe Mountains stands hundreds of meters above the deep-water rocks of the neighboring Delaware Basin.

Putting It All Together

Now that you have been introduced to the geologic history of the Guadalupe Mountains region, let's put your knowledge to work.

The illustration below shows the results of a geological survey of the Guadalupe Mountains area. Each dot with a letter next to it identifies a location and the type of rock found there. The rocks at each site were deposited at roughly the same time, during the Permian period. **R** stands for reef limestone, **D** stands for deep-water formation, **T** stands for talus, and **B** stands for back-reef deposit. Use this information to draw a map of the major features of the Guadalupe Mountain area. Use colored pencils to indicate these different features. (Make a copy of the data; do not write or draw in the book.) Your map should locate and identify the ancient reef, the Delaware Basin, the lagoon, and the escarpment.

Questions

1. A core drilled at the site marked **1** shows numerous erosional surfaces. Explain how this pattern could have happened.
2. The site marked **2** shows little evidence of living things, compared with site **1**. How might you explain this?
3. Gypsum deposits found at site **3** show no fossil evidence of any kind. What might be the reason for this?
4. A core drilled at the site marked **4** shows reef limestone at the surface, talus material further down, and deep-water formations at the bottom. Explain how this could have occurred.
5. Which formations mark the highlands? Which mark the lowlands? How do you know?

UNIT 7

Texas EXPLORATIONS

Eyes on Texas Skies

The stars at night are big and bright . . .
Deep in the heart of Texas!

Look skyward on any clear night and you'll see how true that is—thousands of pinpoints of light sprinkled across the blackness of space. Though the stars often seem close enough to touch, they're really quite far away. In fact, the nearest star is about 40 trillion (40,000,000,000,000) km away. The vastness of astronomical distances is simply mind-boggling. But did you ever stop to wonder just how astronomers come up with these figures? It turns out that here in Texas we have facilities dedicated to doing just that.

Astronomers work at facilities called observatories. In Texas, many astronomers spend time at the McDonald Observatory, located on top of Mount Locke, near Fort Davis. McDonald Observatory is the site of several powerful optical telescopes and one radio telescope. While McDonald Observatory does not have the most powerful telescopes, it has the good fortune to be located under the darkest skies to be found anywhere in the continental United States. This is no small matter. Light pollution is a serious threat to many observatories. Light from homes, businesses, and vehicles has nearly "blinded" observatories in more densely populated areas.

Hundreds of astronomers from all over the world utilize the McDonald Observatory. One of these astronomers, Dr. Mary Kay Hemenway of the University of Texas, is internationally recognized in the field of astrometry. *Astrometry* is the subdiscipline of astronomy that deals with measurement of stellar distances and motions. Dr. Hemenway is currently active in promoting astronomy education across the nation.

Guided tours of the observatory take place at 2 P.M. every day (summers, at 9:30 A.M.). One Wednesday a month, the observatory opens its largest telescope to the public. For more information about McDonald Observatory, write: P.O. Box 1337, Fort Davis, TX 79734; or call (915) 426-3640

AN ASTRONOMER SPEAKS

DR. MARY KAY HEMENWAY

My astronomical work involves the measurement of astronomical distances and motions. The question, "How far is it to that star?" is a lot easier to ask than it is to answer. But astronomers like me have been pretty successful in answering just such questions. I study astronomical distance indicators, called RR Lyrae Stars, which are variable stars whose motion is used to determine distances within our galaxy. The RR Lyrae method depends on a comparison between the true velocity of a star through space, called peculiar motion, and its angular velocity relative to Earth, or proper motion. We can use our findings as logical building blocks to determine the distances and velocities of even more distant objects. Our ultimate goal is to find out "how far and how fast" for everything that can be seen. Many astronomers do such field work at McDonald Observatory. It's really a wonderful facility and a real center for astronomical research.

Nowadays, it is not so unusual to meet a woman scientist, but when I began my career it most certainly was, especially in sciences such as physics. Several times during the course of my education I was the only female in the class.

One of my goals is to see more children develop a lasting interest in science. I see unmistakable signs that science is on the wane in this country, and this disturbs me greatly. It is for this reason that I currently spend most of my time involved in science education. Astronomy is such a fascinating science. It's the story of the past, present, and future of the universe. I'd like everybody to enjoy it as much as I do.

Daytime Astronomy

Astronomy probably doesn't strike you as the sort of thing that can be done during the day. But there's one star in your life that you can see only during the day—the sun. In the following activity you will use a simple method to determine the distance to the sun. This activity must be carried out on a sunny day.
Caution: *Be sure not to look directly at the sun. Looking at the sun for even short periods of time can permanently damage your eyes.*

You Will Need

- a cardboard box at least 50 cm in length
- metric graph paper
- aluminum foil (10 cm square)
- a tape measure or meter stick
- a straight pin
- scissors or razor knife

Making and Using a Pinhole Camera

1. Cut a small hole about 1 cm square in one of the small ends of the box.

2. Tape aluminum foil over the hole. Then punch a hole in the foil with the straight pin. Point this end of the box toward the sun. An image of the sun will form on the far inside end of the box. Tape a piece of graph paper to the spot where the image forms.

3. If your box has no top, you will see the image directly. If it has a top, cut a small flap in the top so that you may look inside and see the image. A box with a top will be darker inside and provide better viewing of the image.

4. Using the lines on the graph paper, measure the diameter of your image. Then measure the distance between the image and the pinhole.

5. Use the formula below to determine the distance to the sun.

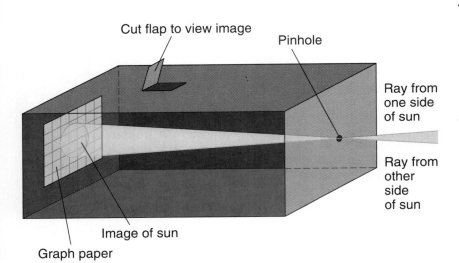

The illustration shows how the solar image is formed. If you put a piece of film on the back surface of the box, you could take a picture. For this reason the device is known as a "pinhole camera."

Taking the Sun's Measure

The geometry of a pinhole camera is such that we can use the known diameter of the sun (1,390,000 km) to solve for the distance to the sun, using a "similar-triangles" relationship. Here is the formula:

$$\frac{\text{Diameter of Image}}{\text{Diameter of Sun}} = \frac{\text{Distance to Image}}{\text{Distance to Sun}}$$

$$\text{Distance to Sun} = \frac{\text{Distance to Image} \times \text{Diameter of Sun}}{\text{Diameter of Image}}$$

Using a calculator, convert all working measurements to meters. It would be helpful to keep distances in scientific notation for ease of use.

While You're at It

Examine the image of the sun and then answer the following items.

1. Is it solid white or can you see something else? What could you do to enlarge the image?
2. Sketch what you see. Wait a while and sketch it again. Is it the same or has it changed? Do the same the following day.

Xeriscaping

Texas is a remarkably varied place. Within the borders of this state are humid woodlands, deserts, mountains, semitropical lowlands, high plains, and just about everything in between. Texas has a diverse range of climates as well. Nevertheless, much of our state is on the dry side, averaging less than 75 cm of precipitation per year. And just about all of Texas is subject to periodic droughts.

In spite of this dryness, homes in even the most arid parts of the state typically have well-tended lawns and gardens that are more suited to cooler and wetter climes. While an attractive lawn and garden enhance the value of a home and make it an inviting and more livable place, such lush greenery uses a lot of precious water. In fact, on average, 20 percent of the water used by cities goes to watering lawns. This rises to 50 percent in the summer. When water supplies are short, as in times of drought, keeping those lush lawns alive can severely tax the water supply. In the past few years, many people have realized that it doesn't make sense to maintain a lush lawn in a dry climate, and have begun to look for alternatives. Enter xeriscaping, the art of landscaping for dry climates.

Doing What Comes Naturally

The philosophy of xeriscaping (from the Greek *xeros,* or dry) is simple: Work with nature; don't fight it. Doing so saves money, resources, and time.

A sample xeriscape: a wooded hideaway

The principles of xeriscaping are:
- start with a good design
- improve the soil
- use plants that don't need a lot of water
- use mulch
- water efficiently
- limit lawn area
- practice good maintenance

Believe it or not, xeriscapes don't have to be all cactus and rocks. Examine the xeriscapes on this and the following page to see just how attractive a xeriscape can be.

Texas Explorations 523

If you were going to design a landscape for your place, how would you choose plants that could survive the local climate without a lot of care? It would depend, of course, on what part of the state you live in. If you live in Amarillo, you will probably design a very different landscape than if you live in Beaumont. The solution is simple: Use plants native to your area. Texas is full of native trees, shrubs, and grasses that are both handsome and well adapted to the local climate.

Designing Your Own Xeriscape

Now that you know something about xeriscaping, put that knowledge to work designing a xeriscape. On page 525 you will find a plan of an imaginary home with no lawn, shrubs, or trees. Get together with two or three classmates and use the accompanying data to design an ecologically sound landscape for this home. The list on page 526 provides a selection of different plants from which to choose. You may consult additional references if you like. Include plants from each of the seven categories listed. Your design should include:

1. a rock garden or brick-covered area
2. a vegetable garden
3. at least one lawn area
4. a flower bed and/or herb garden
5. an area covered by vines or ground cover

Pointers

- Work with the terrain and the lighting. For example, don't place drought-intolerant plants on a dry, sunny slope.
- Consider how conditions (and your needs) will change from season to season.
- Place tall plants behind shorter plants.
- Group flowering plants together in such a way that their colors are enhanced. For example, place a cluster of yellow flowers next to a cluster of blue ones.
- Take the type of soils you have into account.

Remember, the idea is to keep water usage to a minimum. A good xeriscape also needs less fertilizing.

Another sample xeriscape: a desert home

524 Texas Explorations

Location: North Central Texas
Precipitation: 67 cm/yr
Climate: Hot, dry, sunny, breezy summer and cool, often windy winters
Soil: Thin, alkaline topsoil with high clay content

Before you get started, you might also want to look at some additional references for more landscape examples. Or ask your teacher for suggestions.

Once More, with Feeling

Now that you've had some experience designing a xeriscape, let's do it again. Think of an area familiar to you—your schoolyard or backyard, for example. Design a xeriscape to fill that area. If circumstances permit, you might even be able to put your plan into action!

For additional information, contact:
The Texas Natural Resource Conservation Commission, P.O. Box 13231, Capitol Station, Austin, TX 78711
Your teacher has a list of additional references and sources of information that you may find helpful.

Grasses

Common Name	Growth Region	Outstanding Characteristics
Carpet Grass	1, 2	Adapted to moist sites, tolerates partial shade, abundant seedheads, produces thin turf, seeded
Buffalo Grass	3, 4, 5	Excellent drought tolerance, produces thin turf, poor shade tolerance, seeded
Bermuda Grass	6	Good drought tolerance, produces dense turf, poor shade tolerance, seed or sod
Centipede Grass	1	Low maintenance, tolerates partial shade, fair drought tolerance, seed or sod
Tall Fescue	4	Under irrigation remains green year-round, good shade tolerance, poor drought tolerance, seeded
St. Augustine Grass	1, 2, 5	Produces dense turf, good shade tolerance, poor drought tolerance, sodded
Zoysia Grass	3, 4, 5	Produces dense turf, good shade tolerance, good drought tolerance, sodded

Vines

Common Name	Growth Region	Outstanding Characteristics
Crossvine	6	Evergreen, clings to walls
Trumpetcreeper	6	Flowers
Cats-claw	1, 2, 5	Clings to walls
Coral Honeysuckle	6	Coral flowers
Honeysuckle	6	Evergreen
Virginia Creeper	6	Fall color, clings
Wisteria	6	Flowers
Mustang Grape	1, 2, 5	Fruit

Ground Covers (0.25–0.5 m)

Common Name	Growth Region	Outstanding Characteristics
Lantana	6	Flowers
Confederate Jasmine	1, 2, 5	Fragrant evergreen
Running Myrtle	6	Flowers, foliage

Small Shrubs (1–1.5 m)

Common Name	Growth Region	Outstanding Characteristics
Glossy Abelia	6	Flowers, foliage
American Beauty Berry	6	Purple or white berries
Sotol	2, 3, 5	Foliage, flowers
Hydrangea	1	Blue, pink, or white flowers
Tamarix Juniper	6	Foliage
Santa Cruz Pyracantha	6	Berries
Burkwood Viburnum	4, 5	Flowers

Medium/Large Shrubs (2–8 m)

Common Name	Grows best in	Outstanding Characteristics
Wintergreen Barberry	1, 3, 4, 5	Foliage
Pampas Grass	6	Flowers, foliage
Pineapple Guava	1, 2, 5	Evergreen
Shrub Althea	6	Summer flowers
Dwarf Crape Myrtle	6	Flowers, long season
Leatherleaf Mahonia	6	Foliage, fruit
True Myrtle	1, 2, 3, 5	Flowers, fruit, foliage
Variegated Pittosporum	1, 2, 3, 5	Foliage, form
Evergreen Sumac	3, 5	Foliage, form
Shiny Xylosma	2	Foliage
American Holly	1	Form, foliage, fruit
Glossy Privet	1, 2, 3, 5	Evergreen
Oleander	1, 2	Flowers
Chinese Photinia	1, 3, 4, 5	Red spring foliage
Yew Podocarpus	1, 2, 3, 5	Foliage, form
Arborvitae	3, 4, 5	Evergreen

Trees (8 m+)

Common Name	Growth Region	Growth Rate	Outstanding Characteristics
Pecan	6	Medium	Fruit and shade
Hackberry	6	Fast	Fast shade
Southern Magnolia	1	Slow to medium	Foliage, flowers
Loblolly	1	Fast	Evergreen
Live Oak	1, 2, 3, 5	Medium	Evergreen, dependable
Shumard Oak	1, 2, 5	Medium	Fall leaf color
Texas Oak	1, 2, 3, 5	Medium	Fall leaf color, form
Cedar Elm	6	Medium	Fall leaf color, dependability
Russian Olive	6	Fast	Shade, windbreak
Chinese Date	6	Medium	Fruit
Redbud	6	Fast	Spring flowers
Texas Persimmon	2, 3, 5	Medium	Trunks, evergreen foliage, drought tolerant
Yaupon Holly	6	Medium	Fruit, evergreen
Crape Myrtle	6	Medium	Flowers
Mesquite	1, 2, 3, 5	Fast	Fast growth rate, form and leaf color
Mexican Plum	1, 2, 3, 5	Medium	Fragrant flowers, trunks
Texas Mountain Laurel	2, 3, 5	Fast	Evergreen, fragrant spring flowers

Growth Regions
1—East Texas
2—South Texas
3—West Texas
4—North Texas
5—Central Texas
6—Adapted to most of Texas

SourceBook

This *SourceBook* is designed as a handy reference so you may learn more about many of the concepts you have discovered in your reading of Science*Plus*. For each unit of Science*Plus*, you will find a corresponding unit in the *SourceBook*. Questions on the first page of each unit will help direct your thinking as you read the material.

Contents

UNIT 1 — **S1**
- 1.1 The Biosphere — S2
- 1.2 Succession and the Biomes — S8
- 1.3 Humans and the Environment — S17

UNIT 2 — **S23**
- 2.1 The Diversity of Modern Life — S24
- 2.2 The Evolution of Diversity — S35

UNIT 3 — **S45**
- 3.1 Solutions, Suspensions, and Colloids — S46
- 3.2 Acids, Bases, and Salts — S51

UNIT 4 — **S61**
- 4.1 Motion and Force — S62
- 4.2 Laws of Motion — S68
- 4.3 Gravitation — S74

UNIT 5 — **S77**
- 5.1 Architecture in Other Cultures — S78

UNIT 6 — **S89**
- 6.1 Rocks and the Rock Cycle — S90
- 6.2 Rock-Forming Minerals — S99
- 6.3 Stories in Rocks — S105

UNIT 7 — **S111**
- 7.1 Upward and Outward — S112
- 7.2 Stars: A Universe of Suns — S118
- 7.3 One Small Step — S126

UNIT 8 — **S133**
- 8.1 Plant Structure — S134
- 8.2 Uses of Plants — S141

Unit 1

In This Unit

1.1 The Biosphere, p. S2

1.2 Succession and the Biomes, p. S8

1.3 Humans and the Environment, p. S17

Reading *Plus*

Now that you understand some of the interactions among organisms and their environment, let's take a closer look at the kinds of interactions that occur in ecosystems. Read pages S2 to S22 and write a report about your role as a human being in your ecosystem. Your report should include answers to the following questions.

1. What biome do you inhabit, and what are some of its characteristics?
2. How do some of the plant and wildlife populations interact with the human population in your community?
3. In what ways could a major environmental problem in your ecosystem be solved?

1-1 THE BIOSPHERE

Biosphere All the life-supporting environments on earth and the organisms in them.

All life exists within the **biosphere**, a thin layer that surrounds the earth. In this layer, all living organisms are born, grow, reproduce, and die. Figure 1-1 shows the vertical dimensions of the biosphere. Yet a narrow range only 200 m thick, extending 100 m above and below the earth's surface, houses almost all of the earth's millions of species of organisms. As you can imagine, the biosphere supports a wide variety of living organisms. Think about what a jellyfish needs to survive. The same biosphere that includes a warm marine environment for the jellyfish includes an icy environment for the polar bear.

Figure 1-1 The biosphere includes all the places where living things exist.

Ecosystem A group of organisms and their physical environment.

In order to better understand organisms and their environments, it is useful to divide the biosphere into **ecosystems**. An ecosystem consists of all the *biotic* and *abiotic* factors in an area. See Figure 1-2. The kinds of organisms and their interactions, the environment, the population of each species, the size of the area, and many other factors

Figure 1-2 What are the biotic and abiotic parts of this ecosystem?

contribute to an ecosystem. An ecosystem may be as small as a single decaying log in a forest—or it may be as large as the Amazon river!

Think of the different ecosystems that you have visited and interacted with. Did you include places like the park, the beach, and your own backyard? Each of these places contains biotic and abiotic factors that interact in a unique way. The interactions of an ecosystem involve many cycles and pathways that link the earth's organisms to each other and to their abiotic environment. In this section, we will examine some of these interactions.

Populations and Communities

When a coyote preys on a jackrabbit or a caterpillar eats an oak leaf, we say that these organisms are interacting. However, few ecosystems support, for example, only one coyote or one oak leaf. Instead, living organisms are members of interacting groups. Within an ecosystem, the interactions of these groups of organisms affect individual animals, the environment, and other species. Populations and communities are two of these groups.

Populations What is the population of your classroom? You would probably answer this question with a number. But how would you arrive at that number? Would you include the teacher? the potted plants? the fish in the aquarium? In an ecosystem, a **population** represents the number of organisms that belongs to the same *biological species*. So, if your classroom contains plants and fish, it may contain several populations.

Population A group of organisms of the same species.

In addition to the species of the organism, a population is also described by its location and a particular point in time. Was the population of your classroom the same two years ago? ten years ago? Consider the Alaskan brown bears in Figure 1-3. They belong to a different population than do the brown bears that are currently found in Alaska, although the species in each population is the same. Interactions among the individuals of a population contribute to the characteristics of the population and how it changes over time.

Together, the individuals in a population form a *reproductive group*. For a population to survive in an ecosystem, it must have enough individuals for reproduction to occur. If a population becomes too small, the species may disappear altogether. Organisms in small populations may not be able to find each other, or they may require a larger population for the successful growth of their young. For example, the

Figure 1-3 A family of brown bears fishing in the McNeil River of Alaska.

Limiting Factors Environmental elements that stabilize population size and keep species from producing too many offspring.

Figure 1-4 Births and deaths alone do not explain the annual changes in the size of a caribou herd. New animals periodically join the herd, while others leave.

Population Density The number of individuals of the same species in a given area.

wild populations of whooping cranes and California condors became so small that the animals could no longer reproduce in nature. You can see these birds in zoos and wildlife preserves only because people have established breeding programs that bring them together and protect their eggs and hatchlings.

Members of a population also *compete* with one another for certain resources, such as food, light, nutrients, and living space. You have seen competition among members of a population if you have ever watched alley cats fight over territorial rights or a flock of pigeons peck viciously at bread crumbs. The resources that a population needs, plus the climate of a region, are called **limiting factors**. They determine where a particular species can live and how large its population can grow.

Characteristics of Populations One important characteristic of a population is that the number of individuals in it constantly changes. A population grows as new members are born and as adults migrate in from different populations. A population may become smaller as individuals die or migrate to other populations.

Suppose a small herd of caribou, such as the one shown in Figure 1-4, grows in size from 310 to 400 members in one year. During that year, 150 animals are born and 75 die. The difference between 150 births and 75 deaths represents a *net gain* of 75 caribou. However, upon counting the caribou, you discover that the population actually grew by a total of 90 individuals. Where could the other 15 caribou have come from?

Another important characteristic to consider is the number of individuals that occupy a living area. This is called the **population density** of a species. For example, the population density of humans on Manhattan Island in New York City is about 26,000 people per square kilometer. Notice that the population density is not just the *number* of organisms in the population; it also includes the *size* of the area in which the individuals live.

Look at Figure 1-5, which represents two populations of trees that have different population densities. Can you see that the members of Population I have more living space than the members of Population II? What are the advantages of having more living space? What advantages are there to living in a dense population? If a population becomes too dense, it begins to decline because its environment cannot provide enough of its basic needs. Some

Figure 1-5 Population density is determined by calculating the number of organisms that live in a defined area.

members of the population may move elsewhere, while others may die due to lack of food and shelter.

Communities A *community* consists of all the different populations living and interacting in the same area. For example, all of the plant and animal populations in a forest belong to the same community. See Figure 1-6. Similarly, the populations of fish and marine organisms found on a coral reef form their own community. Remember that the populations of a community depend on one another for food or other nutrients and, in many cases, shelter. Studying communities allows you to learn about the interactions among many different populations that exist together in an ecosystem.

Figure 1-6 The community shown in this photograph is composed of several different populations including grizzly bears, spruce trees, aspen trees, several kinds of grasses, ducks, and many unseen organisms.

Energy Pathways

You are already familiar with two types of energy pathways—food chains and food webs. No matter how complex the food web, a community always needs more producers than consumers. Producers provide the energy necessary for themselves and the consumers. A **food pyramid** shows the energy levels of a food web. Each successive layer of a food pyramid is smaller that the one below it. The bottom of the pyramid represents the producers, which as a group

Food Pyramid A diagram that shows how some factor decreases at each feeding level in a food chain.

S5

have the most energy. Herbivores make up the next level of the pyramid. There are more herbivores in the community than carnivores, but fewer herbivores than producers. The upper levels of the pyramid consist of carnivores, some of which eat herbivores. The carnivores at the very top of the pyramid eat other carnivores. See Figure 1-7. What do you think would happen if there were more carnivores than producers?

The organisms at each level use much of their energy just to stay alive. Only 10 percent of an organism's energy can be passed up to the next level. Therefore, only a few animals are found at the top of the food pyramid. These animals—the *top carnivores*, such as lions, wolves, and tigers—exist only in small populations. They depend on large populations of other animals to survive.

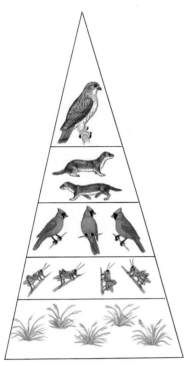

Figure 1-7 *Food pyramids show producers at the bottom, then layers of different types of consumers, with a top carnivore on top.*

Recycled Materials

Food and energy are not the only important resources an organism must have to survive. Substances from the abiotic environment, including water, carbon, oxygen, nitrogen, and mineral nutrients, are essential for the survival of the biotic factors (organisms) in an ecosystem. These materials are used and then returned to the environment by the activities of living things. In this *recycling* process, there is a constant exchange of nutrients between the biotic and abiotic factors of an ecosystem.

The Water Cycle The sun heats water on the earth's surface and causes it to evaporate into the atmosphere as a gas. This gas (water vapor) cools and forms droplets of water that fall back to the ground as rain, sleet, or snow. See Figure 1-8. How might water interact with the organisms in

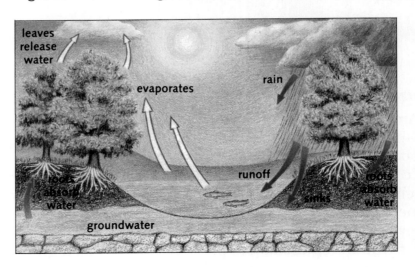

Figure 1-8 *The water cycle.*

an ecosystem? Plants use water to make food and then release that water back into the atmosphere. Animals drink water, or extract water from their food, and return it to the environment through waste products. In extremely dry areas, organisms such as cactus plants and camels have become adapted for storing water. How do you think this affects the water cycle?

The Carbon Dioxide–Oxygen Cycle The gases carbon dioxide and oxygen provide the basis of chemical interactions for all living organisms. Animals and plants need oxygen in order to release the energy stored in food. During respiration, oxygen is taken in and carbon dioxide is given off. When you breathe, you take in the oxygen that you need to get energy from your food. The waste product of this chemical reaction is the carbon dioxide that you exhale into the environment. Plants complete the cycle—they use carbon dioxide in the environment to make food for themselves and for consumers. See Figure 1-9.

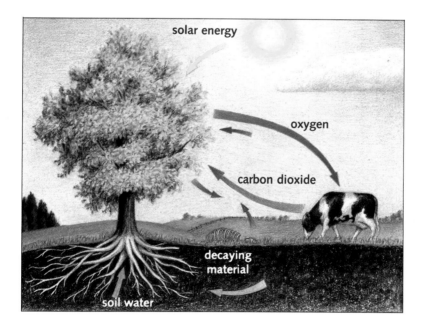

Figure 1-9 The carbon dioxide–oxygen cycle.

The Nitrogen Cycle Nitrogen gas makes up about 80 percent of the earth's atmosphere, and living organisms use it in the production of proteins. However, most organisms cannot use nitrogen directly from the atmosphere. Fortunately, *nitrogen-fixing bacteria* convert the gas into solid compounds that can be absorbed by plant roots. See Figure 1-10. Animals obtain nitrogen by eating these plants or other animals. The nitrogen is returned to the environment through waste products like ammonia and urea. Bacteria that feed

Figure 1-10 The lumps on these white clover roots contain nitrogen-fixing bacteria.

on animal wastes produce nitrogen gas that returns to the atmosphere, or nitrogen compounds that remain in the soil. See Figure 1-11.

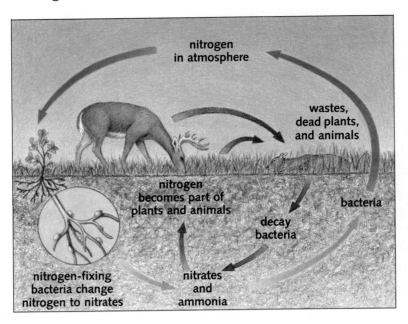

Figure 1-11 *The nitrogen cycle.*

Summary

The biosphere is a thin layer of the earth that includes all organisms and all of the places where organisms can live. It contains all of the earth's ecosystems, including both the biotic and abiotic elements and their interactions. Two interacting groups of organisms in ecosystems are populations and communities. The transfer of energy through an ecosystem can be represented by food pyramids. Water and essential nutrients, such as carbon, oxygen, and nitrogen, also circulate in ecosystems through several complex cycles.

1-2 SUCCESSION AND THE BIOMES

Have you ever wondered how a scene such as the one in Figure 1-12 came to be? Mountain-building, weathering, and erosion gradually molded the land into its present shape. The plant and animal communities that you see here developed slowly over time by the processes of evolution and succession. The makeup of these communities was determined by a variety of limiting factors such as climate, soil composition, disease, and natural disasters.

In this section, you will take a close look at succession as a process by which living communities change over time.

Figure 1-12 *The Sonoran desert of Organ Pipe National Monument in Arizona.*

You will also look at some of the geographic areas that cover major parts of the earth and share similar climates. These areas, which can be identified by their plant life, are called *biomes*.

Ecological Succession

Compare the photographs in Figure 1-13. The view on the left shows part of Glacier Bay National Monument in Alaska. Notice the land closest to the glaciers. The glacier has only recently receded, and the land is lifeless. In the photograph on the right, the land near the mouth of the bay has been exposed for hundreds of years. You can see the forest of spruce trees and can imagine the kind of animal life that inhabits the forest.

Figure 1-13 The pioneer stage of succession (left) *as well as a climax community* (right) *can be seen in Glacier Bay, Alaska.*

The glaciers of Alaska's Glacier Bay formed during the last Ice Age. Since that time, a slow, steady melting has taken place, exposing land that was covered by ice for thousands of years. During the next few centuries, a gradual process of change will occur in Glacier Bay as living communities become established on the bare rock. Over time, one community will gradually replace another in an ecological process called **succession**.

Primary Succession Imagine an underwater volcano erupting violently—huge rivers of lava pour out of the ocean floor until a new island appears in the water. This brand new island is barren of plant and animal life. But not for long! Living organisms will invade this new environment as the process of succession begins. Succession that starts in an area where the land has never had organisms living on it is called *primary succession*. In addition to land formed by lava flows and exposed by the retreat of glaciers, sand dunes are also places where primary succession occurs. The types of communities that come and go during primary succession are different in each of these areas, but they all follow similar trends.

Let's take a closer look at primary succession as it proceeds around Glacier Bay, Alaska. Traveling up Glacier Bay

Succession A series of changes that occur in the plant and animal communities of an ecosystem.

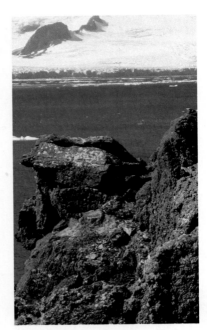

Figure 1-14 *Lichens completely encrust these rocks to begin the process of primary succession.*

Figure 1-15 *Soil is formed during the pioneer stage of succession. Soil is necessary for the next stage of succession to take place.*

is like traveling backward in time. By retracing your steps back toward the mouth of the bay, you can investigate the changes in the land as if they were occurring by succession. Communities representing each stage of succession thrive in different parts of the bay. Standing at the edge of the glacier, you will notice that primary succession begins soon after bare, lifeless rock is uncovered by the melting of glacial ice.

Right below your feet, you can see succession beginning at the base of the glacier. *Pioneer species* are the first living things to appear in a previously lifeless area. The harsh habitat of the pioneers is characterized by high winds, extreme cold, bright sun, and heavy rain. There is no soil at this stage. *Lichens*, which consist of a fungus and an alga growing together, are one of the first pioneer species you find. They often completely cover the surface of rocks. See Figure 1-14. In order to get necessary nutrients, lichens produce an acid that breaks down the hard rock. As dead lichens collect and decay, they mix with small pieces of the rock to form a thin crust—the beginning of soil.

A short distance away, *mosses* (another pioneer species) grow in the new soil formed by lichens. Mosses are simple plants that form thick mats. Their decaying parts add to the mixture of dead lichens and rock particles. Mosses also trap bits of dust and dead matter brought in by the wind and waves. More organic matter is added by a variety of animal species, such as insects and birds, that are a part of the moss community. This organic matter enriches the developing soil. See Figure 1-15.

If you climb down from the glacier and cross over the areas of lichen and moss, you will see that the soil becomes richer. The land no longer looks as barren as it did while you were standing on top of the ice. Meadows of grasses and wild flowers have replaced the pioneer species. Wind, birds, and small mammals bring seeds for new plants into the area. The roots of these plants break up rock and trap decaying organic matter that helps form richer layers of soil. Eventually, the soil can support the growth of shrubs such as willow and alder, which thrive in the bright sunlight of open meadows.

If you leave behind the grassy meadows and shrubs, you find that these communities are replaced by forests of white spruce. Spruce seedlings grow well in the shade that the shrubs provide, so the seedlings eventually replace the mature willows and alders. See Figure 1-16. The spruce trees represent the final stage of succession. The community that exists at this stage is called the **climax community**. It includes the population of white spruce, some small shrubs, and the animals that find shelter and food in such a forest.

Secondary Succession Natural disasters such as fires, floods, diseases, and swarms of insects can disturb a climax community. Agriculture and forestry can destroy these living communities as well. Once the disturbance is over, and if no other disturbances occur, the climax community will slowly re-establish itself. This type of succession, which occurs after an original community has been disturbed or destroyed, is called *secondary succession*. See Figure 1-17.

The pioneers for secondary succession are usually grasses and other weed-like species of flowering plants that are brought in from the surrounding area by wind, mammals, or birds. Grasses and weeds grow quickly, and most have root systems that protect the soil of a disturbed area from erosion. Often, seeds from the plants that originally lived in the area are lying undisturbed beneath the ground. When the environment is ready to support the seedlings, these seeds will germinate and grow. As the plant communities are re-established, the animals that once lived there return as well.

Climax Community The final stage in the succession of a community.

Figure 1-16 In this part of Glacier Bay, spruce trees are beginning to take over the community of sun-loving willow and alder trees.

Figure 1-17 Succession in an old New England field.

Since soil is already present, the process of secondary succession takes place more quickly than primary succession. In many cases, the original climax community can be regained by secondary succession. In some cases, however, it cannot. For example, the eastern forests of the United States were dominated by chestnut and oak trees until the

S11

early 1900s. At that time, the chestnut trees were struck by a fungal disease that destroyed all of them by the 1930s. The climax community now consists of an oak and hickory forest.

Terrestrial Biomes

Biomes Large geographic areas of similar climate and life forms.

Large areas of the earth's continents share similar climates and soil types. These geographic regions are called **biomes**. Different plants and animals are found in the various biomes. Scientists usually identify these regions by their climax plant species. The six major terrestrial (land) biomes are indicated on the map in Figure 1-18.

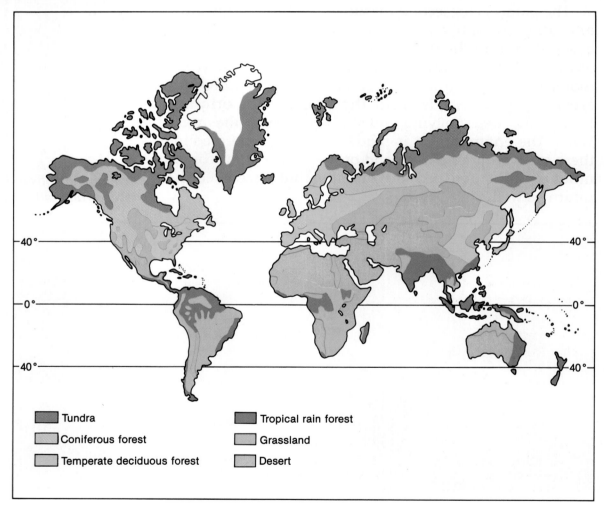

Figure 1-18 The major biomes of the world.

The Tundra Just south of the arctic icecap is the freezing, harsh biome known as the *tundra*. A land of extreme cold, high winds, and very little (25 cm or less) rain or snow, the tundra has winters where temperatures may drop as low as −40 °C! The tundra receives the least amount of sunlight of

Figure 1-19 Typical plant growth on the tundra (left). The light-colored plants are dwarf willows, which appear during the brief summer on the tundra in Alaska. Caribou (middle) *and arctic fox* (right) *are both adapted to tundra life.*

any of the earth's terrestrial biomes. Life on the tundra is limited to hardy species that are adapted to the cold, harsh climate or are able to migrate to warmer regions when necessary. See Figure 1-19. Plant life includes low-growing species of grasses, lichens, mosses, and flowering herbs that can reproduce during the short growing season. Small, woody shrubs of willow and birch grow in some areas. Animal inhabitants include migratory birds, musk-oxen, wolves, grizzly and polar bears, lemmings, and arctic hares and foxes.

The Forests Three kinds of forest biomes cover a large portion of the earth. They differ in climate, amount of rain, temperature range, and growing season. See Table 1-1. These and other limiting factors, such as soil type and humidity, determine the kinds of communities that inhabit each forest biome.

Forest Type	Average Rainfall	Average Temperature	Length of Growing Season
Rain forest	212 cm	High: 35 °C Low: 25 °C	9–12 months
Deciduous forest	75–125 cm	High: 27 °C Low: −10 °C	6 months
Coniferous forest	25–125 cm	High: 20 °C Low: −30 °C	2–5 months

Table 1-1

The *coniferous forest* biome lies just south of the tundra. The dominant plants of this biome are evergreen *conifers*, or cone-bearing trees, such as spruce, fir, pine, and hemlock. The smooth, slender needles and bendable branches of the trees allow the conifers to shed heavy snows before the trees can be damaged. See Figure 1-20. Animals adapted to this

Figure 1-20 Elk in a coniferous forest in Yellowstone National Park. Note the absence of snow on the branches of the conifers.

Figure 1-21 Typical layers of vegetation in a tropical rain forest.

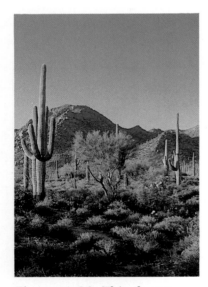

Figure 1-22 This desert scene is *typical of the southwestern United States and northern Mexico.* Shown here is the saguaro cactus, which may reach 15 m in height.

cold biome include snowshoe hares, moose, elk, bears, beavers, and many species of birds.

Moving from the coniferous forest toward the equator, the *deciduous forest* biome is found. This biome has a more humid climate marked by distinct seasonal changes. *Deciduous trees*, like oak, hickory, beech, maple, and elm, are the dominant plant species of this biome. Animal life includes deer, bears, snakes, rabbits, squirrels, and many birds and insects.

The third type of forest biome is the *tropical rain forest* which is found along and near the equator. The hot, wet climate of the tropical rain forest supports the greatest variety of life found anywhere on earth. New species of plants and animals are discovered each year. Some may never be found by humans because they live up to 60 meters above the ground, supported by the ecosystem found in the *canopy* of the tree branches. See Figure 1-21. The animal species include: monkeys, squirrels, bats, snakes, deer, many species of rodents, numerous species of birds, and a vast number of insects.

Grasslands The grassland biome is usually found in the same general areas as the deciduous forest biome, except that the grasses require less rainfall than the amount necessary for trees. In North America the grasslands are called *prairies*. Great herds of bison once roamed these prairies. Because grassland soils are rich and well developed, much of the world's best farmland is found in this biome. Today, much of the original vegetation has been replaced by fields of grain such as wheat and corn, while great herds of cattle and sheep graze in place of the herds of bison. Still, a variety of small animals such as jack rabbits, ground squirrels, prairie dogs, and mice inhabit the grasslands. They are prey for grassland predators such as coyotes, foxes, snakes, and hawks.

The Desert Deserts occur in geographic areas that receive less than 25 cm of rainfall per year. Rainfall in most deserts is rare; some deserts do not receive any rain for several years at a time. Although not all deserts are hot, the hottest places on earth are found in deserts. For example, Death Valley, California, a part of the Mojave Desert of the southwestern United States, has a record high temperature of 57 °C—in the shade! Desert plant communities contain cacti, small trees, woody shrubs, and many wildflowers. See Figure 1-22. Since moisture is the primary limiting factor for survival in the desert, the plants that grow there have many adaptations for getting and conserving water. Many desert animals are *nocturnal*, meaning they are only active at night. Kangaroo rats and other rodents, bats, snakes, toads, birds, ants and other insects populate the desert.

Aquatic Biomes

Oceans and seas contain saltwater ecosystems. Lakes, ponds, rivers, and streams contain freshwater ecosystems. Along coastlines, saltwater and freshwater come together to form a special ecosystem called an *estuary*. All of these ecosystems make up the aquatic (water) biomes of the earth. Together, the aquatic biomes cover more than three quarters of the earth's surface. Temperature, sunlight, salinity, water flow, and the amount of available nutrients are important variables that determine how organisms are distributed in the aquatic biomes.

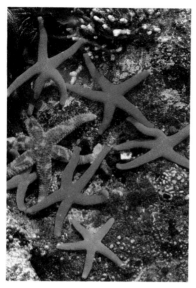

Figure 1-23 Sea stars have adaptations for life in the intertidal zone.

The Marine Biome Covering about 70 percent of the earth's surface, the marine biome is the largest biome on earth. Since it is so large, factors such as temperature, light, and depth of water vary from region to region. The *intertidal zone*, which is closest to the shore, is rich in nutrients that have washed down from shoreline communities. During low tide, the region is exposed to air. Therefore, organisms living in the intertidal zone must be adapted for life underwater as well as in the air. See Figure 1-23. One of the most productive intertidal ecosystems, the estuary, occurs in bays, salt marshes, or swamps where the salinity of the water is between seawater and freshwater. Many marine organisms begin their lives in an estuary, which supports large and diverse communities of organisms such as clams, snails, oysters, and crabs, as well as many plant species.

Beyond the intertidal zone is the *neritic zone*, which extends from the low-tide line out to a depth of 200 m. Floating microorganisms, plants, and animals called **plankton** are the most abundant life forms. See Figure 1-24. Other organisms found in this zone are dolphins, sea turtles, fish, and bottom-dwelling crabs and oysters.

The *oceanic zone*, which extends outward from the neritic zone and downward to the bottom of the ocean, is relatively barren of life because there is no light and few nutrients. In this zone, at depths of 2000 m, are some of the strangest-looking organisms on earth. They feed on each

Figure 1-24 Plant and animal plankton are found floating in the neritic zone.

Plankton Small organisms that float near the surface of the ocean.

Figure 1-25 A predator of the deep ocean.

other or on dead organisms that fall from the water above. See Figure 1-25.

The Freshwater Biome In freshwater biomes, limiting factors include rate of water flow, oxygen, available nutrients, light, and temperature. These factors vary greatly between streams, ponds, and lakes. Lakes and ponds have nutrient-rich water that supports the growth of surface algae and floating plants. Many species of fish, insects, reptiles, birds, and mammals live on or near the surface of lakes and ponds. See Figure 1-26.

Figure 1-26 A lake is a freshwater ecosystem that contains many different populations.

Life is sparse, however, in swiftly flowing streams, such as the one in Figure 1-27. Cold water and the currents that carry away both the nutrients and organisms are the main

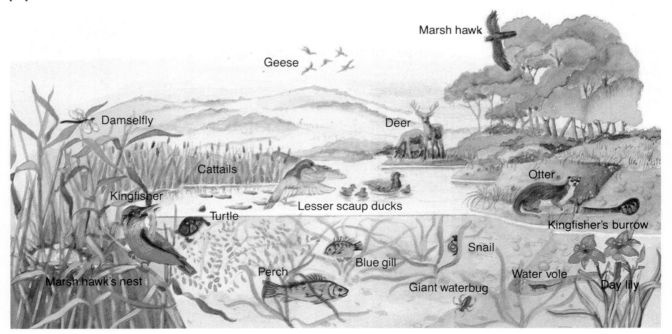

limiting factors. Green plants that attach to rocks live in swift streams, but floating plants and algae merely pass through with the current. The animals of this ecosystem are either excellent swimmers, like trout and bass, or are able attach to rocks and filter nutrients and oxygen from the passing water.

Summary

Succession is the process by which one community is replaced by another over time. Primary succession occurs in areas that have not previously supported life. Secondary succession occurs in areas where an existing community was disturbed or destroyed. The biosphere includes several large geographic areas that are characterized by their climates and the types of plants and animals that live there. The six major terrestrial biomes are tundra, grassland, desert, and coniferous, deciduous, and tropical rain forests. The major aquatic regions include the marine and freshwater biomes.

1-3 HUMANS AND THE ENVIRONMENT

Until it died in 1914, the last known passenger pigeon lived in the Cincinnati Zoological Gardens—its calls unanswered by others of its kind. Prior to the mid 1800s, this beautiful bird, shown in Figure 1-28, inhabited the forests of eastern North America in numbers of 3 to 5 billion. How could so large a population become **extinct** so quickly?

In this case, the cause was a migration of an animal species into the birds' ecosystem. This species consumed the oak and beech forests in which the birds nested. The invading species also preyed upon the birds. As the population of passenger pigeons declined, they became unable to reproduce as quickly as they once had. Soon, all of the passenger pigeons were gone—their species extinct. Who were the invaders that caused this extinction? They were human settlers that used the wood of the forests to build their homes and sold the birds to city dwellers to eat as delicacies. Though the extinction of a species is a natural process, human activities in recent years have greatly increased the rate at which species become extinct.

The Use and Misuse of Natural Resources

Human societies need a steady supply of food and materials from the earth and its biosphere. These organisms and

Figure 1-27 *The mountain stream shown here is an example of a moving freshwater ecosystem.*

Extinct Refers to a species that no longer exists.

Figure 1-28 *Male and female passenger pigeons, as painted by John James Audubon.*

S17

Natural Resources All natural substances that humans remove from the environment for their own use.

Pollutants Unwanted materials, such as gases, particles, and chemicals, that are released into the environment.

Figure 1-29 Smoke billowing from factory smokestacks was seen as a sign of progress during the Industrial Revolution. Today, however, people are aware that many materials in this smoke are harmful to plants, animals, and people.

Conservation The protection and wise use of natural resources.

substances are called **natural resources**. Some of our natural resources can be replaced, but others cannot. *Renewable resources* can be replaced at about the same speed we use them. These include trees, crops, livestock, fish, wildlife, water, oxygen, nitrogen, and carbon. *Nonrenewable resources* cannot be replaced at the speed at which we use them. They include soil, minerals, metal ores, and fossil fuels.

Pollution is the process of making the environment unclean with waste products, such as poisonous gases, liquid chemicals, radiation, heat, garbage, and even noise. All waste materials are referred to as **pollutants**. You find them in the air you breathe, the water you drink, and the soil where your food grows. A major factor in the pollution of earth's environment is the amount of waste produced by the people that live on the planet. Americans alone produce about 4 trillion kilograms of wastes each year. In 1989, over a billion kilograms of toxic materials were released into the air by American industries. This amount is equal to about 4.5 kilograms of poison for every person in the United States.

As our human population grows, we require more and more living space, fuel, food, and clothing. Manufacturing, agriculture, mining, and transportation help to meet these needs; they also improve the quality of our lives. However, these processes use up certain resources while their waste products spoil the environment. See Figure 1-29. By practicing **conservation**, the rapid depletion of the earth's natural resources and the pollution of its environment could be slowed down. As members of a world-wide biological community, we must make informed decisions about pollution and the conservation of our resources.

Air Pollution Burning fuels and burning trash are the major causes of *air pollution*. Automobiles, factories, power plants, and other industries are the largest contributors to this pollution. Most coal and petroleum fuels contain sulfur, which when burned, combines with oxygen to put deadly sulfur dioxide gas into the air. Breathing air containing sulfur dioxide gas can cause health problems such as lung diseases. It also causes acid rain.

Water Pollution One of the most dangerous kinds of *water pollution* occurs when *industrial wastes* and untreated *sewage* are dumped directly into streams, lakes, and oceans. See Figure 1-30. Sewage can add bacteria and viruses to water supplies. It also carries household chemicals, such as phosphates from detergents, that cause algae and water plants to grow out of control. Another source of water pollution is the poisonous chemicals used to control pests on crops and lawns. These chemicals frequently wash off the land and into water supplies where they harm both aquatic organisms and the humans who eat these species.

Soil Pollution A third form of pollution is the buildup of harmful substances in the soil. Some *soil pollution* comes from pollutants that fall out of the air. Most, however, is caused by chemicals used to control pests and plant diseases. Poisonous chemicals are often buried in the ground in containers that are supposed to remain sealed. However, many of these chemicals, including nuclear wastes, have leaked into groundwater. Groundwater provides drinking water for more than half of the population of the United States.

Figure 1-30 Wastes from industry may be one of the major sources of water pollution.

The Consequences of Pollution

From a distance, the picture in Figure 1-31 looks perfect. The dark green of tall trees surrounds the deep blue of a crystal-clear lake. But everything is not perfect. There are few birds or frogs around, and the fish are almost gone. This lake, like thousands of others in eastern North America, is dying because its waters have become an unfit habitat for many of the organisms that once lived there. What caused this destruction? Acid rain, a consequence of industrial pollutants, has contaminated the lake.

Figure 1-31 Many lakes in the Adirondack Mountains of New York are suffering from acid rain.

Since the 1700s, over 60 animal species have disappeared from the United States alone. In those days, the extinction rate was about one species every fifty years—it is now one species every year! In addition, there are presently over 400 species in the United States that are **endangered**; the number worldwide is nearly 900 species, according to the U.S. Fish and Wildlife Service.

Why should you be concerned if organisms become extinct? After all, species become extinct naturally. However,

Endangered In danger of becoming extinct.

Figure 1-32 *There will never be another Tasmanian wolf (left). Once it was an important predator in its native Tasmania, an island south of Australia. The giant panda (right) is endangered and may soon follow the Tasmanian wolf into extinction.*

as the rate of extinction increases, the variety of life on earth decreases. See Figure 1-32. New species will develop, but the rate of that development is relatively slow. Every organism occupies a niche in a food chain. If we eliminate a species, could an important food chain collapse and threaten the food web of an entire ecosystem? Remember, humans are at the top of many food chains, so we may be endangering ourselves. It has been said that each time we eliminate a species, we lose a little of ourselves. After all, we are all a part of the same biosphere.

What Can We Do?

As humans, we have the knowledge and the ability to change the environment to meet our needs. Along with this ability comes a responsibility to all other life on the planet. What can humans do to make the future better for all forms of life on earth?

Citizenship and Participation Individual and corporate efforts over the last 20 years have cleaned up many polluted ponds, lakes, and streams and improved the quality of the air. In addition, *environmentalists* helped pass many laws—at local, state, and national levels—that control the emission of all types of pollutants. See Figure 1-33. Environmentalists are not just scientists and government officials; they are people, just like you and your parents, who are concerned about issues such as pollution and its effects. They teach others, volunteer for clean-up projects, and participate in the law-making process.

Figure 1-33 *Many states require the exhaust of motor vehicles to be tested for pollution levels. Those that give off too many pollutants must be repaired.*

One thing you can do is to learn about the causes and consequences of pollution in your own community. Tell your friends and family about what you find out and participate in whatever way you can. For example, picking up litter, such as plastic bags, may help to save the lives of local wildlife, which often die when they become entangled in or accidently eat the plastic. Practicing conservation by recycling and using less water saves community resources. What other ways can you help the environment in your area?

Solid Waste Management One of the most difficult problems facing humans today is the disposal of solid wastes. Some communities still dispose of these wastes, also called refuse or garbage, in *open dumps* where they are left on the ground. Open dumps are often homes to disease-carrying rats, insects, bacteria, and fungi. In addition, rain carries pollutants from these dumps directly into water supplies. Other communities get garbage out of their sight by loading it onto barges that carry it out to sea where it is dumped into the ocean. This practice harms ocean life and causes problems for marine communities. Trash has even been found on the most remote, uninhabited islands of the Pacific Ocean, carried there by waves and currents. These practices must be stopped.

Figure 1-34 The sanitary landfill shown here provides a less harmful way to dispose of solid wastes than an open dump.

Many communities now dispose of solid waste in *sanitary landfills*, like the one shown in Figure 1-34. Others burn wastes in high-temperature furnaces that heat water for use in generating electricity. In landfills, solid wastes are covered with a layer of clay on all sides. The clay helps prevent rainwater from washing through the garbage and carrying pollutants into water supplies. It also keeps disease-carrying rodents and insects away. Many communities recycle materials such as metals, glass, and paper, then sell them to companies that use them to make new items. See Figure 1-35. This practice reduces the amount of garbage that a community must dispose of, and slows down the depletion of natural resources.

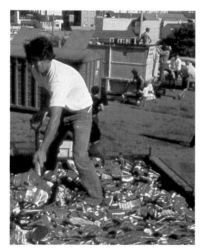

Figure 1-35 Community recycling centers provide one way for citizens to recycle refuse such as aluminum cans and newspapers.

Figure 1-36 Using water foolishly is a waste of an important natural resource.

Urban Development The building of cities, towns, subdivisions, and factories.

Figure 1-37 Peregrine falcons are endangered North American predators. These chicks have been incubated and hatched in a breeding program.

Water Conservation The goal of water conservation is to preserve our limited supply of clean, fresh water. Humans use half of the earth's freshwater supply to irrigate agricultural plants. We use water to manufacture products and to generate electricity. In addition, an average person in the United States uses over 700 liters of water daily! What does a person do with so much water each day? Only about two liters is used for drinking. The rest is used for such things as flushing toilets, bathing, food preparation, watering lawns, washing cars, and filling swimming pools.

There are many ways that people can conserve water. One way is to use less water by reducing waste. See Figure 1-36. Leaking faucets and long showers are common ways that people waste water. Another way to conserve water is to protect watersheds where water either enters the ground or is captured by streams. Clearing forests and grasslands for farming and for **urban development** allows water to runoff instead of soaking into the ground. This not only reduces the supply of groundwater, but it causes flooding and degrades the quality of the water with pollutants and fine soil particles.

Wildlife Conservation The goal of wildlife conservation is to protect the wild animals and plants that live on the earth. See Figure 1-37. Setting aside natural habitats is one way to protect wildlife. Wildlife refuges, national forests and parks, and state forests and wilderness areas all provide natural habitats for many kinds of wildlife. Human access is limited in many of these areas in order to reduce the potential for pollution. Laws that permit hunting and fishing only at certain times of the year also protect wildlife. Such laws limit the number, and often the size and age, of animals that can be killed in order to keep those populations healthy. Good conservation practices reduce the effect of humans on wildlife. This allows ecosystems to continue to exist in a more natural way.

Summary

Humans require a variety of materials from the environment for their activities. Many of these activities cause pollution of the air, water, and soil that poisons both wildlife and humans. The result has been loss of habitat for wildlife and the extinction of many species. Some solutions to these problems include community involvement, proper waste management, and water and wildlife conservation.

Unit 2

Reading *Plus*

Now that you have been introduced to some of the many different kinds of organisms and how they are classified, let's take a closer look at this diversity and some ways that it developed. Read pages S24 to S44, and then pretend you are a modern-day Darwin and write an essay about evolution as it relates to humans. Your essay should include answers to the following questions.

1. Why are you classified as a mammal? Be specific and complete. (Remember, to be a mammal, you must first be an animal.)
2. How might humans change over the next few millennia by the process of natural selection?
3. In what other ways, besides selective breeding and resistance to pesticides and antibiotics, are humans affecting the evolution of living things?

In This Unit

2.1 The Diversity of Modern Life, p. S24

2.2 The Evolution of Diversity, p. S35

2-1 THE DIVERSITY OF MODERN LIFE

Did you know that the inside and outside of your body is occupied by many living things that are so small they can be seen only with powerful microscopes? Have you ever imagined that a forest of giants over 30 m tall could consist of algae instead of trees—or that you may someday live on the moon? Even though our own species has the size, shape, and intelligence necessary to send rockets into space, most living things go unnoticed. Already close to 3 million species have been identified and named, though some scientists estimate there may be as many as 30 million different kinds of organisms living on earth today! Let's look at some of the variety of earth's organisms. They are grouped into large categories, called *kingdoms*, which are based on some of the fundamental similarities they share. See Figure 2-1.

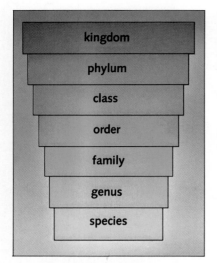

Figure 2-1 Each of the five kingdoms contains subcategories by which all living organisms can be classified.

The Monerans

The smallest and simplest life forms are placed in the Kingdom Monera. Monerans are also some of the most important organisms in your everyday life. They are different from all other living things because of their simple cell structure, which has no nucleus or cell organelles. All monerans are single-celled organisms that often live in chains or clusters of cells. Bacteria and blue-green bacteria are the two major groups (phyla) in this kingdom. See Figure 2-2.

Figure 2-2 Two examples of bacteria (top) and two examples of blue-green bacteria (bottom).

S24

Bacteria These monerans are single-celled organisms that absorb food molecules from their surroundings, such as the surface of your skin or the inside of your intestines. Their cells take three basic shapes—round (cocci), rod-shaped (bacilli), and spiral-shaped (spirilli). Many live in pairs, chains, or clusters. Bacteria live in almost every habitat on earth. You may already be familiar with the bacteria that live on or in other organisms and cause disease. However, most bacteria are beneficial by helping animals with digestion, by decaying dead organisms, or by putting nitrogen back into the soil.

Blue-Green Bacteria Blue-green bacteria contain chlorophyll and produce their own food by photosynthesis. They also have different shapes and usually attach to one another in long chains or clusters. Most blue-green bacteria live in fresh water and form the smelly, green scum that you may see on polluted ponds and streams. Some blue-green bacteria form the food-producing member of a lichen.

The Protists

The more advanced single-celled organisms are placed in the Kingdom Protista along with some simple, multicellular life forms. Unlike the monerans, however, protists have much larger and more complex cells that contain nuclei and cell organelles. Although some multicellular forms are quite large, their cells do not specialize into tissues or organs. The protists are divided into several phyla that share similar characteristics. Some of these phyla contain the animal-like protists we call *protozoans*. The others contain the plant-like protists we call *algae*.

Protozoans These protists are all single-celled organisms with no cell walls or chlorophyll. Protozoans *ingest* their food (take it in whole), as you do, and *digest* it inside their cells. Since they must find their food, most protozoans move about in one of three ways—by a flowing motion, such as that used by an amoeba, or by using special structures such as whip-like *flagella* or hair-like *cilia*. Some protozoans, however, have no means of self-locomotion. These protists, many of which cause diseases, simply attach to their host's cells. Protozoans are divided into four phyla based on the way they move. See Figure 2-3.

Algae The algae are protists that have cell walls and chlorophyll, and they make their own food by photosynthesis. Algae may be single-celled, or they may form long chains,

Figure 2-3 Some examples of protozoans: radiolarians (top), Stentor (middle), *and the protozoans responsible for African sleeping sickness* (bottom).

complex colonies, or large masses of unspecialized cells. In some of the largest algae, such as the 30-m giant kelp, the body shape is similar to a plant's. Most algae live in either salt water or fresh water; but, some may be living near you on moist, shady tree bark, on wet cement and rocks, or with a fungus as part of a lichen. Algae are divided into five major phyla based on their color and structure. A sixth phylum contains the *euglena,* which resembles both protozoans and green algae. See Figure 2-4.

Figure 2-4 Volvox (left) *is a colonial green alga. Kelp* (middle) *is an example of a marine brown alga. This example of a red alga* (right) *lives in very deep water.*

The Fungi

Some make us sneeze, some live in our bathtubs, and many kinds spoil our food. This is one answer to the question: "What are fungi?". Most members of the Kingdom Fungi are multicellular organisms with a complex cell structure, cell walls, and no chlorophyll. You can find them growing both inside and on top of their food. Fungi eat by releasing digestive chemicals into their food and absorbing the digested food back into their bodies. The fungi are divided into phyla based on their body structure and how they produce **spores**.

Spore A special reproductive cell formed by fungi and other simple organisms.

Thread-like Fungi

Have you ever noticed a black or gray powdery substance growing on bread? These are the spore

Figure 2-5 *The common black bread mold.*

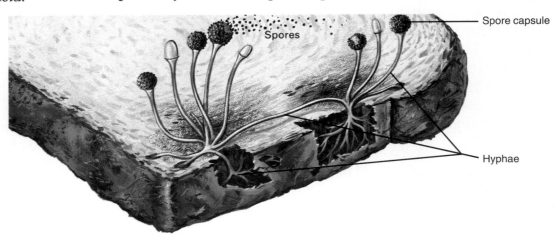

capsules of bread mold, an example of the thread-like fungi. If you look closely at the surface of the bread, you may be able to see the tiny, thread-like hairs or filaments called *hyphae* that make up the bodies of all fungi. Many other hyphae grow inside the bread as well. See Figure 2-5.

Club Fungi Mushrooms, puffballs, and bracket fungi are members of this group. These familiar structures are made up of closely packed hyphae that grow on the surface of a food source and come in a variety of shapes and colors. See Figure 2-6 *(left)*. In mushrooms, spores are found along the *gills* on the underside of the mushroom's cap. You may be able to see these tiny spores, which are often quite colorful, by first cutting off the cap of an opened mushroom. Then place it lower-side down on a white piece of paper and check it after a day or so. The pattern that may appear will be caused by the spores.

Sac Fungi Yeasts, mildews, and morels (a type of edible fungus that resembles a wrinkled mushroom) are members of this group. Many of these fungi are parasites that attack and cause diseases in several kinds of trees, such as elm and chestnut. Some yeasts even cause diseases in humans. The spores of these fungi are found inside tiny sacs or cups that form on the outside of the host. See Figure 2-6 *(right)*.

Figure 2-6 A bracket fungus (left) *on a rotting log. The fruiting bodies of a sac fungus* (right) *resemble small cups.*

The Plants

One of the two most advanced groups of organisms, the Plant Kingdom, includes multicellular organisms that have chlorophyll, complex cell structure, and cell walls. Most plants live on land, anchored in soil or attached to something else. Unlike the multicellular algae and the fungi, plant cells specialize to form many different types of tissues, organ structures, and systems. Both the largest and longest-living things on earth are plants. Plants are divided into two main groups based on whether or not they have water-conducting (vascular) tissue.

Figure 2-7 *Hair cap mosses (top) and a liverwort (bottom) called* Marcantia.

Figure 2-8 *The club moss* Lycopodium *(left), horsetails (middle), and ferns (right).*

Nonvascular Plants Have you ever noticed a bright green carpet covering a forest floor or growing on a rotten log? This carpet is made of plants that do not have water-conducting tissue. The nonvascular plants must live in moist environments, such as a shaded forest floor, and must also stay very small. The mosses and the liverworts, shown in Figure 2-7, are the two types of plants in this group. Notice that individual mosses and liverworts grow together in tightly packed mats. This helps them to conserve water and also to reproduce. The tiny capsules that appear at the top of a moss plant contain spores.

Vascular Plants Most plants have *vascular tissue* that conducts water, minerals, and food molecules throughout their bodies, which are composed of roots, stems, and leaves. Vascular tissue not only allows these plants to live in drier environments, but also serves as a support system for holding up their stems and leaves. Without vascular tissue, plants like the giant redwoods could not grow to be over 100 meters tall. Five phyla (or *divisions*, as botanists call them) of vascular plants are listed below. See Figure 2-8 for examples of each of the first three groups.

1. The *club mosses* are vascular plants that resemble the mosses. However, their stems are much sturdier due to their vascular tissue, and they grow larger than the mosses. Club mosses can also be found carpeting a forest floor.

2. The *horsetails* are another kind of vascular plant. They have hollow stems with many vertical ridges and joints. A whorl of needle-like leaves appears at each joint. Their stems also feel gritty due to the presence of silica, the material that makes sand. Because they are gritty and coarse, early settlers used them to scour pots and pans.

3. The *ferns* are probably a more familiar kind of vascular plant, since you may have ferns in your house or your garden. Ferns have distinctive leaves, called *fronds*, which consist of a strong central vein and many pairs of leaflets. Their stems remain underground. Contrary to what many people think, the horizontal rows of bumps that appear on the under side of fern fronds are not a disease or bug—they are structures that contain spores.

4. The *gymnosperms*, whose name means "naked seed," are one of the two types of *seed plants*. Seed plants are by far the most common and most successful kind of plants. Gymnosperms produce seeds on the dry scales of cones. Most gymnosperms are *conifers* (cone-bearing plants), such as pine, redwood, juniper, fir, spruce, hemlock, and

Figure 2-9 Cycads (left), ginkos (middle), and the larch (right), are all examples of gymnosperms.

cypress trees. All have small needle-like leaves and most stay green all year. Some conifers are believed to be over 5000 years old. Other examples of gymnosperms include cycads and ginkgo trees. See Figure 2-9.

5. The *angiosperms*, whose name means "covered seed," are also seed plants. Usually called *flowering plants*, angiosperms are the largest, most diverse group of modern plants. Their seeds are produced inside of a special structure, called an *ovary*, which is part of a flower. See Figure 2-10. An ovary ripens into a **fruit** that surrounds and protects maturing seeds. Fruits also help to *disperse* seeds (spread them around) once they are mature. You have eaten the fruit of many angiosperms including apples, oranges, peaches, tomatoes, eggplant, squash, cucumbers, grains, peanuts, walnuts, almonds, and coconuts. Although their flowers are not as obvious or showy, plants such as grasses, palms, all crop plants, and fruit, nut, and shade trees are flowering plants, too.

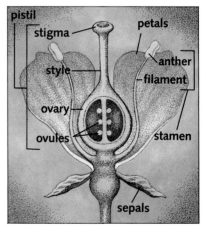

Figure 2-10 The structure of a typical flower showing the ovary and ovules, which will become seeds.

Fruit A ripened ovary that contains the seeds of a flowering plant.

The Animals

You are a member of the Animal Kingdom. How can you tell? You are not a moneran, because you are made of many cells and most of your cells have a nucleus. You are not a protist, because you are multicellular and have complex tissues and organ systems. You are not a plant, because you do not have cell walls and you cannot make your own food. You are not a fungus, because you digest your food inside of your body with a complex digestive system. Therefore, since there is only one kingdom left, you must be an animal.

Animals are divided into nine major phyla and several minor phyla. Eight of these groups contain soft-bodied organisms that are referred to as *invertebrates*. Humans belong to the ninth group of animals, which includes fish, amphibians, reptiles, birds, and mammals—the *vertebrates*.

Invertebrates Invertebrates are soft-bodied animals that have no backbones for internal support. Instead, their bodies stay relatively small, have an outside covering for support, or are supported by the water in which they live. The members of each phylum share characteristic body plans. As you read through the list of invertebrates below, notice how each group is more complex than the previous one. Figures 2-11 to 2-14 show examples of these animal groups.

1. *Sponges* are the simplest of the animal groups. Adult sponges live in water, attached to some object, and are composed of two layers of body cells. These body cells cling to a network of tiny spikes or fibers that surrounds a hollow central cavity. The walls of a sponge have many tiny pores through which water and tiny bits of food are drawn into the central cavity. Water and wastes leave through a large opening on top.

2. *Coelenterates* are simple animals composed of two specialized cell layers (tissues) that are separated by a jelly-like substance. Hydras, jellyfish, sea anemones, and corals are all types of coelenterates. All live in water and have a hollow, sac-like body with a single opening. This opening, through which food enters and wastes are expelled, is surrounded by tentacles that are lined with stinging cells for capturing and paralyzing live food.

3. *Flatworms* have a flattened body shape, with one body opening, a body cavity with many lobes, and a simple nervous system. Most flatworms are parasites that live

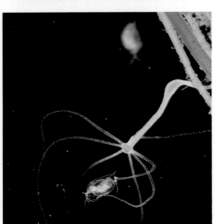

Figure 2-11 Two simple invertebrates: sponges, (top) and a coelenterate (hydra) (bottom).

inside of other animals. Blood flukes, liver flukes, and tapeworms are some flatworm parasites of humans. Turbellarians are free-living flatworms, the majority of which live in marine environments. Freshwater turbellarians—the planarians—are distinctive because they have two *eyespots* that are used to detect light.

Figure 2-12 The tapered body shape of this turbellarian is an adaptation that allows it to move freely and to seek its own food.

4. *Roundworms* have a rounded body shape, a tube in which food is digested, and two body openings—a mouth for taking in food and another for expelling wastes. Most roundworms are free-living, aquatic organisms that are easily visible in pond water. Many others live in moist soil. Some roundworms are parasites of animals or plants, for example ascaris worms, which live in the intestines of humans and other mammals, and root-knot nematodes, which damage the roots of plants.

5. *Segmented worms* have rounded bodies with two body openings, but their bodies are divided into a series of segments. They have several well-developed body systems that include organs such as a simple "brain" that connects to a nerve cord and a set of five primitive "hearts." The earthworm is a common example of the segmented worms, which also include leeches and marine tube worms.

Figure 2-13 A roundworm (left) *and a segmented worm, the familiar earthworm* (right).

6. *Mollusks* are soft-bodied animals with well-developed organ systems. Most mollusks make hard, calcium-containing shells that completely cover the body for

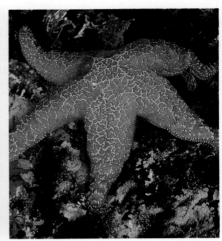

Figure 2-14 Higher-order invertebrates (left to right): *a mollusk (octopus), an arthropod (crab), and an echinoderm (starfish).*

Exoskeleton A hard outer covering for protection that is typical of arthropods.

Regeneration The ability of some organisms to grow new body parts.

Endoskeleton An internal system that functions for support of the body in vertebrates.

protection. Clams and oysters are covered by a two-part, external shell, while snails secrete a single, spiralling shell. Octopuses and squids, which are mollusks that have tentacles, do not make a shell for protection. Instead, they swim very fast and shoot out an inky substance to confuse their predators.

7. *Arthropods* are by far the largest group of animals. These organisms are characterized by two, three, or many body segments and numerous jointed *appendages*. They also have a hardened **exoskeleton** for protecting their many, well developed organ systems. Arthropods live successfully on dry land and in water, and many have a complex life cycle. Among the arthropods are lobsters, crabs, shrimp, insects, spiders, centipedes, and millipedes.

8. *Echinoderms*, also known as the "spiny-skinned" animals, include starfish, sea urchins, sand dollars, and sea cucumbers. Echinoderms live in the ocean and have bodies protected by a spiny skeleton that lies under the skin. All echinoderms move using special structures that resemble suction cups called *tube feet*. Many echinoderms have several flexible body branches, or arms, on which the tube feet are found in rows on the underside. They are famous for their ability to replace missing arms by a process called **regeneration**.

Vertebrates The vertebrates are animals that have a backbone. This backbone is part of an internal skeleton, called an **endoskeleton**, which provides support for the body and aids in body movements. All vertebrates have a body structure that includes a head, a neck, and a body with appendages. They also have the most highly-specialized organ systems in the animal kingdom. Seven kinds of vertebrates are included in the last major phylum of the Kingdom Animalia. They are:

1. *Jawless Fishes.* Represented by the sea lamprey in Figure 2-15, these vertebrates have a jawless mouth adapted for sucking body fluids from other fishes, an elongated, snake-like body with no appendages, gills for obtaining oxygen, and flexible skeletons of *cartilage*. Like all fishes, they are **cold-blooded** and live only in water.

2. *Cartilaginous Fishes.* The shark in Figure 2-15 represents this group of vertebrates, which have skeletons made of cartilage, two pairs of fleshy fins, gills, and strong jaws with many rows of teeth for tearing and eating flesh. Stingrays and skates are other members of this group, which, like jawless fishes, lives only in the ocean.

Cold-blooded Having a body temperature that changes according to the temperature of the environment.

Figure 2-15 The sea lamprey (left) is a jawless fish, a shark (middle) is a cartilaginous fish, and the angel fish (right) is a bony fish.

3. *Bony Fishes.* Catfish, trout, goldfish, flounder, and eels are examples of bony fishes. These fishes have gills, and they have skeletons made of bone. Most have a streamlined body that is tapered at both ends, and two pairs of fan-like fins. The fins and the body shape of the bony fishes allow them to move easily through the water. They are the most numerous, varied, and successful kind of fishes.

4. *Amphibians.* Frogs, toads, and salamanders are examples of amphibians. The name *amphibian* means "double life." Most amphibians live part of their lives in water and part on land. They have smooth skin that must be kept moist, gills when they are young and lungs when they mature, and two pairs of legs as adults so they can move on land. Like the fishes, their eggs are laid in water, and all forms are cold-blooded.

5. *Reptiles.* Turtles, snakes, lizards, crocodiles, and alligators are the living members of this group, which was once dominated by dinosaurs. The reptiles are truly adapted to life on land. For this purpose, they are covered by hard

plates or scales that prevent water loss by evaporation. Reptiles also have lungs; two pairs of strong legs with clawed toes for digging, climbing, and moving on land; and they lay eggs that are covered with a tough, leathery shell that prevents them from drying out. However, they are still cold-blooded.

Figure 2-16 *Amphibians (left) and a reptile (right).*

6. *Birds.* Birds are recognized by a body covering of feathers and a pair of wings, which gives most of them the ability to fly. In addition, birds lay eggs covered by a hard shell, have a beak for a mouth, scaly legs, and feet with clawed toes. Birds also have light, hollow bones, and enlarged lungs as an adaption for flying. Unlike the fish, amphibians, and reptiles, birds are **warm-blooded**. This allows birds to occupy many different habitats on earth.

7. *Mammals.* Mammals are the most complex kind of vertebrates. They feed their young with milk from mammary glands, have hair somewhere on their bodies, breathe with lungs, and are warm-blooded. While one category of mammals lays eggs, almost all give birth to live young. Mammals are also known for their very advanced nervous system, which includes keen senses and a highly developed brain. Like the birds, mammals also live on land, in the air, and in water.

Warm-blooded Having a body temperature that remains constant despite temperature changes in the environment.

Figure 2-17 *Warm-blooded vertebrates include birds (left) and mammals (right).*

Summary

According to the current biological classification system, the diversity of life on earth today can be separated into five major groups called kingdoms. The members of each kingdom have a few common characteristics. The organisms of each kingdom are further divided into smaller groups that have even more characteristics in common.

2-2 THE EVOLUTION OF DIVERSITY

Imagine going on a long trip and arriving in a strange new place where even the animals and plants look strange. You find giant tortoises grazing like cattle, seafaring lizards, and unusual, flightless birds. Yet, other life forms look much like organisms with which you are familiar. This was the experience of Charles Darwin when he visited the Galapagos Islands near South America. See Figure 2-18. Darwin's voyage inspired him to develop the theory that life has evolved by a process called *natural selection*.

Since Darwin's time, scientists have gathered much information from **fossils** about the life forms that once existed. From this information, they have developed explanations as to how life changes. They have even performed experiments that show how life could have originated from substances that occur naturally on earth. Today the concept that all life has evolved from a few simple forms is a fundamental theme in biology.

Figure 2-18 *The location of the Galapagos Islands.*

Fossil The remains or imprint of an organism that once lived.

A History of Life on Earth

If you could travel back to the time before history began, what kinds of organisms would you see? Depending on the time of your arrival, you might see horses the size of dogs, flying reptiles, forests of giant ferns, and even giant insects. Biologists tell us that all the species alive today make up only one percent of all the life forms that ever lived—99 percent are extinct! Some are preserved in the rocks of the crust as fossils. To get the most out of your trip you would need to move quickly, since the earth's history covers an unimaginably long period of time.

Earth's Calendar During our lives, we mark the passing of each year on a calendar. The study of human history, however, is broken into longer units of time, such as decades, centuries, and *millennia*. To us, a millennium, or 1000 years, is a very long period of time, since very few people live more

Figure 2-19 This diagram shows the relative lengths of the geologic eras. The Precambrian era is by far the longest.

Figure 2-20 Stromatolites, fossil structures formed by mats of algae, have been found in rocks that are over 3.5 billion years old.

than a century. The earth, however, is estimated to be over 4.5 billion years old, which is more than forty-five million (45,000,000) centuries of time! How much time is this? If we could squeeze the entire history of the earth into one calendar year, we would be living in the last split-second of December 31 of that year!

Earth's history has been divided into several time periods called *eras*, as shown in Figure 2-19. Each era is further divided into *periods* and then *epochs*. Eras are named for the kinds of life forms that were abundant during that time. Each one ends with a major change in the life forms that existed.

We live in the *Cenozoic Era* (recent life), which began about 65 million years ago. The era before that is called the *Mesozoic Era* (middle life), which lasted about 160 million years. The *Paleozoic Era* (ancient life) came before the Mesozoic and lasted about 345 million years. An abundance of fossils is found in the rocks that formed during these three eras. However, they cover only 13 percent of the earth's history.

The remaining 87 percent of earth's history is often called the *Precambrian Era*. This name comes from the Cambrian Period, which is the oldest period of the Paleozoic Era. The life forms of this era were the first to inhabit the earth. Unfortunately, there are not many fossils of these organisms because most were single-celled or had soft bodies. These kinds of organisms do not make good fossils. However, microscopic examination of fossils, such as the one in Figure 2-20, finally revealed the presence of tiny organisms that lived over 3.5 billion years ago.

How Life Has Changed The earth's history, as recorded in its rocks, reveals that the diversity of life today has evolved from simple to complex forms. This gradual progression is illustrated in Table 2-1. As you study the table, think back to the overview of the five kingdoms that you have just read.

Evolution by Natural Selection

Charles Darwin became convinced that life had changed over time while serving as ship's naturalist on the HMS *Beagle*, a British survey and mapping ship. His job on the ship was to study and collect the different kinds of plants and animals he found in the many places where the ship stopped. See Figure 2-21. In addition, he collected many fossils and made notes of everything he saw. During the long, five-year voyage, Darwin carefully studied these notes and collections.

Figure 2-21 *The marine iguana is one species that Darwin found living on the Galapagos Islands.*

Developing a Theory When he returned home, Darwin began to classify his collections and wondered how organisms could change over millions of years. He knew that farmers selected and bred animals and plants with certain traits, but he did not know what could cause such selection in nature. Then, Darwin happened to read an essay by the noted clergyman Thomas Malthus. The essay discussed Malthus' observations that the number of people in a population gets larger when there is enough food and other necessities, and gets

Era	Period	Million Years Ago	Organisms
Cenozoic	Quaternary	0.025	Complex human societies develop.
		2	Many extinctions occur.
	Tertiary	5	Modern mammals, birds, and sea life appear; early hominids appear.
		24	Grasslands spread.
		37	Primitive apes, elephants, horses, and camels develop. Forests of gymnosperms and angiosperms spread.
		58	Mammals spread rapidly. Fruit-bearing trees become common.
		65	Primates appear.
Mesozoic	Cretaceous	144	Flowering plants and trees appear. Dinosaurs are extinct by the end.
	Jurassic	208	First feathered birds and large dinosaurs appear.
	Triassic	225	Dinosaurs and first mammals appear.
Paleozoic	Permian	286	Conifers and modern insects appear. Many invertebrates become extinct.
	Carboniferous	360	First reptiles appear. Amphibians are common. Coral reefs form. Forests of large ferns develop.
	Devonian	408	Amphibians and insects appear. Gymnosperms appear.
	Silurian	438	Primitive vascular plants, fish, and shell-forming sea animals appear.
	Ordovician	505	First vertebrates appear in the sea. First plants appear on land.
	Cambrian	570	Clams and snails appear. Algae are common.
Precambrian		4,500	Marine invertebrates appear. Bacteria, blue-green bacteria, and algae predominate.

Table 2-1

Figure 2-22 What might the earth be like if human population were not regulated?

Natural Selection The process by which only the organisms that are best suited to their environment survive, passing their traits on to their offspring.

smaller when there is not. Malthus predicted that people would overrun the earth if starvation, disease, crime, and war did not keep their numbers down. See Figure 2-22.

Darwin realized that *all* organisms must compete for food, water, and other necessities in order to survive in what the essay had called a *struggle for existence*. This idea became the key to Darwin's theory of evolution by **natural selection**. He reasoned that the organisms that are better able to compete are the ones most likely to survive and leave more offspring. Those that are not, are less likely to live long enough to produce offspring.

Darwin worked on his theory for twenty years. But, before he finished, he received an essay from a man named Alfred Russel Wallace that stated the main points of the theory that Darwin had worked on for so long. Wallace was also a naturalist and had worked in the jungles of Indonesia. He and Darwin had been writing to each other about their research and their ideas. Darwin quickly finished a paper on his theory and, in 1858, both papers were presented jointly at a scientific meeting in London. The next year, Darwin published his famous book, *On the Origin of Species by Means of Natural Selection*.

Parts of the Theory   There are five main parts to the theory of evolution by natural selection. The parts presented here are a bit different from the way Darwin stated them. Our knowledge of genes explains how variations can be transmitted—something Darwin did not know.

1. *Variation.* All members of the same species are somewhat different from one another, or in other words, individuals are not exactly alike in all of their traits. See Figure 2-23. Some differences, like color, may be very noticeable; but many less noticeable traits, such as size, weight, and running speed, can be different as well. These differences

Figure 2-23 Two of these Bengal tigers have a mutation that causes albinism. Would this mutation be helpful or harmful to these animals in the wild?

are called *variations*. Darwin believed that variations were caused by some agent that could be passed from parent to offspring, but he did not know what that agent was. Today, we know that variations result from sections of DNA called *genes*. New variations appear when genes change slightly by the process of **mutation**.

Mutation A change in the DNA of a gene.

2. *Overproduction of Offspring.* Each species produces many more offspring than can survive and reproduce. For example, female fishes lay enormous numbers of eggs. If all of these eggs hatched and the young survived, the waters of the earth would quickly be overrun with fishes. Plants also produce large numbers of seeds, but not all of them grow into adult plants.

3. *Struggle for Existence.* The overproduction of offspring leads to a struggle for survival, since all organisms must compete for a limited amount of food, water, and living space. During this struggle, some individuals will get enough of these necessities to survive and reproduce, and some will not.

4. *Natural Selection.* Individuals with certain variations compete more successfully than individuals that lack those variations. For example, in a jungle, which of the tigers in Figure 2-23 would be more likely to stalk its prey undetected? In other words, which tiger would be the most successful hunter? Would the same be true if the tigers lived on the tundra? As you can see, the environment determines which variations are successful. Individuals with traits that make them successful in their environments have the best chance to survive and reproduce. Offspring that inherit successful variations also have a survival advantage, as long as the environment remains the same. Thus in a population, the percentage of individuals with successful variations increases over time.

Adaptation Due to variation and natural selection, populations and species may *adapt* (change to become better suited) to new conditions in their environments. For example, recall what happened to the peppered moths in England during the Industrial Revolution. Where tree trunks were darkened by pollution, populations of mostly light-colored moths were gradually replaced with ones dominated by dark-colored moths. Because of the variation for dark color, this species was able to adapt to and survive a change in its environment. As many slight, yet beneficial changes accumulate over millions of years, significant changes in structure

Figure 2-24 How did color affect the survival of the peppered moth?

and function can occur. See Figure 2-25. In this way, species that are adapted to particular environments develop.

Figure 2-25 The shape of a mammal's foot shows it has adapted to the way it moves.

Diversity Directed by Humans

If you could put together the perfect pet, what would it be like? If you could design the perfect vegetable how would it taste? Believe it or not, people have been able to do just that by a process of *artificial selection*.

Selective Breeding The process of making hereditary improvements in animals and plants.

Selective Breeding Humans first began to domesticate plants and animals around 9000 B.C., or about 11,000 years ago. At that time, people began to collect seeds of wild grasses to grow them for food, and tamed certain wild animals, such as cattle, for doing work. Gradually, they started to alter these wild varieties by selecting and raising the offspring of individuals with the most desired traits. Improving plants and animals by choosing and reproducing individuals with certain desirable traits is called **selective breeding**.

For example, by selecting and growing wild wheat plants with the largest seeds, domesticated varieties that always produced larger seeds were eventually developed and thus provided greater amounts of food. See Figure 2-26. These days you must look very closely to see the similarities between wild wheat and its domesticated cousins.

Cattle are another example of selective breeding at work. Individual bulls and cows are selected for qualities, such as strength, size, and meat and milk production, and bred in a controlled manner. Calves that have the most desirable combinations of traits are selected to eventually parent the next generation. There are more than forty different cattle breeds today. In fact, most domestic animals—from cats and dogs to horses, sheep, and pigs, as well as most cultivated varieties of ornamental and food plants—are the result of thousands of years of selective breeding by humans.

Figure 2-26 This wheat (top), *as well as the variety of fruits and vegetables* (bottom), *was developed from wild strains through the process of selective breeding.*

Resistance Humans have also inadvertently altered many species of insects and disease-causing bacteria by trying to control them with chemicals. Now, many insects and bacteria are *resistant* to these chemicals. When we spray or dust insect pests with chemicals called *pesticides*, most of the insects are killed. However, due to genetic diversity, a few individuals survive the treatment because they have genes that make them resistant to the particular chemicals that were used. These individuals survive the treatment and go on to reproduce, passing this resistance on to their offspring.

In the case of disease-causing bacteria, some have become resistant to certain antibiotics. Bacteria that are resistant to antibiotics survive to produce a resistant strain of bacteria. One way to solve this problem is to alternate treatment with different drugs instead of using the same one over and over again. Resistance keeps scientists busy trying to produce new pesticides and antibiotics to control insects and bacterial diseases.

Patterns of Diversity

Have you ever noticed similarities between members of the same family? What did you notice? How would you describe the similarities? You might describe similar facial features, such as eye color and nose shape. You probably did not describe similar features, such as backbones, two arms, two legs, and the ability to walk upright. Of course these features are common among family members, but they also indicate how all people on earth are related.

Being related means sharing an *ancestor*. For example, your parents are your ancestors. You may know or have pictures of some of your other recent ancestors, such as your grandparents, great grandparents, or great great grandparents. Do you notice any similarities between these ancestors and yourself? See Figure 2-27. You also have other ancestors—ones through which you are related to other

Figure 2-27 Different generations of a family may have some similar features, but new gene combinations make every family member unique.

mammals, vertebrates, animals, and even protists. When two organisms share the same ancestor, we call this ancestor a *common ancestor*.

Determining common ancestry is the key to building modern classification systems. It also helps biologists work out the pathways of evolution. Common ancestry is determined by studying and comparing such things as homologous structures, embryo development, and biochemical molecules. Let's take a look at what these things mean.

Homologous Structures These structures are body parts that have the same basic construction. Let's consider the limbs of vertebrates as an example. Although they vary in size, shape, and purpose, they are all made in the same way. See Figure 2-28. Now, think about how the structure and shape of each of these limbs is helpful to its function. Can

Whale's flipper | Lion's leg | Human's arm | Bird's wing

Figure 2-28 Vertebrate limbs are homologous structures. The flipper of a whale, leg of a lion, arm of a human, and wing of a bird differ in size and shape, but are alike in the arrangement of the bones.

you see adaptation at work? Biologists recognize that these similar structures indicate that these animals have a common ancestor. The variations that led to these adaptations were inherited by each generation from the one before.

Embryo Development Relationships among animals can also be seen by comparing their development before birth. Most multicellular organisms, such as ourselves, begin as a single cell called an *egg*. When the egg is fertilized, it begins to divide rapidly, forming a mass of cells and then an *embryo*. Embryos change shape gradually as they develop. Eventually, they take the basic shape of their parents. Figure 2-29 compares the development of the embryos of three vertebrates. Notice that birds and mammals appear to be more similar (closely related) to one another than to the fish.

Figure 2-29 Similarities in the development of fish, chicken, and human embryos suggest an evolutionary relationship.

Evolutionarily speaking, birds and mammals have a more recent common ancestor than do fish and birds, or fish and mammals.

Biochemical Similarities Homologous structures and embryo development were used by Darwin and other early biologists to trace the evolution of life and to classify it. However, these methods are too general to establish the closest relationships. Today, these relationships are determined by looking at the biological molecules that make up organisms. The more closely related the organisms are, the more similar these molecules will be. See Figure 2-30. The most important molecule to look at is DNA, since this molecule has the instructions for making all of the others. These instructions are written in a code that scientists have learned to read. Having DNA at all is the most important similarity among the organisms on earth. The fact that every living thing on earth has DNA is the most convincing evidence that all life on earth may have a common ancestor.

Figure 2-30 This diagram compares differences between a certain DNA-produced protein obtained from different species of organisms.

PROTEIN DIFFERENCES IN ORGANISIMS

	Human	Monkey	Pig	Horse	Dog	Rabbit	Kangaroo	Duck	Turtle	Tuna fish	Moth	Yeast
Human	0											
Monkey	1	0										
Pig	10	9	0									
Horse	12	11	3	0								
Dog	11	10	3	6	0							
Rabbit	9	8	4	6	5	0						
Kangaroo	10	11	6	7	7	6	0					
Duck	11	10	8	10	8	6	10	0				
Turtle	15	14	9	11	9	9	11	7	0			
Tuna fish	21	21	17	19	18	17	18	17	18	0		
Moth	31	30	27	29	25	26	28	27	28	32	0	
Yeast	45	45	45	46	45	45	46	46	49	47	47	0

The Origin of Life

In studying how life could have evolved from the first single-celled forms, it occurred to some scientists that life itself may have developed naturally on the earth. This could not have happened if the earth was the same as it is now. But about 4 billion years ago, conditions on earth were very different. At that time, the earth's surface was covered by erupting volcanoes and shallow, hot seas. There was no oxygen in the

Figure 2-31 An apparatus like the one shown here was used to test Oparin's theory.

Figure 2-32 These microfossils of spheroid cells and filaments are among the oldest known fossils.

atmosphere, and the daily weather forecast would have included torrential rain and violent electrical storms.

In 1923 a Russian biochemist named Alexander Oparin announced his theory that life could have begun when organic molecules formed in an early atmosphere of extremely hot gases, such as hydrogen, ammonia, methane, and water vapor. These molecules, he said, collected in hot, shallow seas and gradually came together to form the major kinds of biological molecules. After millions of years, the first cells appeared. No one knows exactly how these molecules might have come together to form cells, but the possibility has been demonstrated by several experiments.

In 1953 two American scientists used an apparatus like that shown in Figure 2-31, to heat a mixture of the gases Oparin had suggested with an electrical spark. The experiment produced only two of the molecules used to make proteins. However, in other experiments where the gas mixture was changed to a more likely combination for the earth's early atmosphere (hydrogen, nitrogen, carbon dioxide, carbon monoxide, and water vapor), a variety of biological compounds were formed, including some of the compounds that form DNA.

More recently, biochemists have shown that when no oxygen is present, heating causes certain biological molecules to join together into larger ones. It has also been demonstrated that some of these molecules tend to gather together into tiny spheres that resemble some of the earliest known cells. See Figure 2-32. Many scientists today believe that these microscopic spheres could have developed the organization necessary to become living cells over many millions of years.

Summary

The history of the earth is divided into time periods called eras. The simplest forms of life lived in the earliest of these eras. The most complex organisms live in the most recent era, which includes the present. Observations of this increasing diversity led to the development of the theory of evolution by natural selection. Humans have also been able to change living things by processes such as selective breeding. Evidence of common ancestry such as homologous structures, similar stages of embryo development, and biochemical similarities are used to classify organisms by their evolutionary relationships. Experiments have shown that it is possible that life could have developed naturally on the earth about 4 billion years ago.

Unit 3

In This Unit

3.1 Solutions, Suspensions, and Colloids p. S46

3.2 Acids, Bases, and Salts p. S51

Reading *Plus*

Now that you have been introduced to solutions and non-solutions, let's take a closer look at these mixtures and some of the special properties they display. Read pages S46 to S60 and prepare a narrative of how you encounter various mixtures in your daily life. Your story should include examples related to the following questions.

1. On what are the differences in mixtures based? And how can different types of mixtures be identified?
2. How can the different properties of solutions, suspensions, and colloids be applied to the manufacture of useful products?
3. In what way are acids, bases, and salts chemically related? And what electric properties do they share?

S45

3.1 SOLUTIONS, SUSPENSIONS, AND COLLOIDS

As you know, mixtures are combinations of several substances, each of which keeps its own characteristics. The substances in a mixture can be separated by physical means. You have just learned that a solution is a mixture of two substances, one of which is dissolved in the other. In a solution, one substance "disappears" into the other, and it looks like a single substance. In other mixtures, the non-solutions, particles do not dissolve but float throughout the mixture without dissolving.

Solutions

Solutions can actually contain more than one dissolved substance. A soft drink, for example, is a mixture of water, flavorings, and carbon dioxide gas. However, before a bottle of soda is opened, the soft drink looks like a single substance. Such a mixture is called a **solution**. Every solution has two parts: the *solvent* and the *solute*. As you know, a solvent is the material in which the solute is dissolved. There is always more solvent than any other component in the solution. In a soft drink, water is the solvent. The gas and flavorings in the soft drink are the solutes. It is the gas bubbling out that gives soda its fizz. See Figure 3-1.

Solution A mixture that is formed when the particles of one substance fill the spaces between the particles of another substance, and the particles of both are evenly distributed.

Figure 3-1 The gas in a soft drink bubbles out when the bottle is opened and the pressure is released.

Although many common solutions are made from solutes added to liquids, other solutions are combinations of liquids, solids, and gases. Table 3-1 lists some examples of these types of mixtures.

Types of Solutions In order to describe a solution, you must know the solvent and solute used. For example, a salt solution describes a mixture made by dissolving salt in water.

TYPES OF SOLUTIONS		
Example	**Solute**	**Solvent**
air	gas	gas
soda water	gas	liquid
hydrogen in platinum	gas	solid
water vapor in air	liquid	gas
alcohol in water	liquid	liquid
silver amalgam	liquid	solid
sulfur vapor in air	solid	gas
sugar in water	solid	liquid
brass	solid	solid

Table 3-1

You must also know the relative amounts of the materials in the solution. If a small amount of solute is dissolved in a large amount of solvent, the solution is said to be **dilute**. A spoonful of salt in a liter of water makes a dilute salt solution. On the other hand, a large amount of solute in a liter of water makes a **concentrated** solution. A handful of salt in a glass of water makes a very concentrated solution.

You may have observed that there is a limit to how concentrated a solution can be at a certain temperature. For example, if you continue to dissolve sugar in iced tea, the solution will become so concentrated that no more sugar will dissolve. When this happens, the solution is **saturated**. A saturated solution has all the dissolved solute that it can hold at a given temperature. However, if you raise the temperature, the solution can hold more solute. For example, the amount of sugar needed to make a saturated solution of iced tea is less than the amount needed when the tea is hot.

The amount of sugar or salt that will make saturated solutions in 100 g of water at 60°C is shown in Figure 3-2. As you can see, a larger amount of sugar than salt is needed to make a saturated solution at the same temperature. Sugar

Dilute Describes a solution made by dissolving a small amount of solute in a large amount of solvent.

Concentrated Describes a solution that is made by dissolving a large amount of solute in a solvent.

Saturated Describes a solution that has all the solute that it can hold without changing the conditions.

Figure 3-2 Sugar is more soluble in water than is salt. A greater amount of sugar will dissolve in the same amount of water.

Solubility Describes the amount of solute that can be dissolved in a given solvent under given conditions.

is said to have greater **solubility** in water than salt. Solubility describes the amount of solute that can be dissolved in a given solvent under given conditions. The solubility of most solids in water increases as the temperature of the water increases. See Figure 3-3. However, the opposite is true for the solubility of gases in water. Gases are more soluble in *cold* water than in *hot* water. This is why more gas escapes when you open a warm can of soda than when you open a cold can.

Figure 3-3 This graph shows the solubility of several substances at different temperatures. Notice that most of the substances have higher solubilities at higher temperatures.

Rate of Dissolving If you want to dissolve a solid in a liquid, there are three ways you could speed up the process. First, you could raise the temperature. The heat will cause the molecules of the solute and the solvent to move faster. As a result, the two kinds of molecules mix with each other at a faster rate. Sugar, for example, will dissolve faster in hot tea than in cold tea. A second method to increase the speed of dissolving is stirring. A stirring motion helps to mix solvent molecules with the solute molecules, exposing more solute surfaces to more solvent molecules. A third way to speed up the dissolving of a solid is to crush the solid into small pieces. This exposes more surface area of the solid solute to the liquid solvent. Since the molecules of solute can escape into the solvent only from the surface of the solid, this speeds up the rate of dissolving. This is why breaking up a sugar cube helps it dissolve faster. See Figure 3-4.

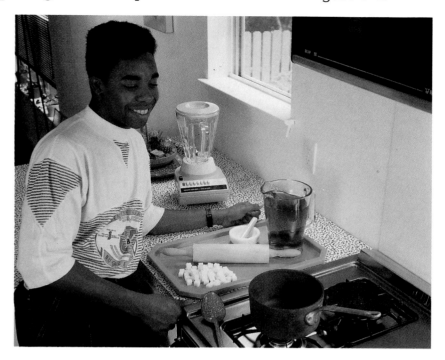

Figure 3-4 *Using the kitchen utensils shown, how would you increase the rate of dissolving for sugar in ice tea?*

Suspensions

Do you ever use Italian dressing on your salad? If so, you know that the herbs settle to the bottom while the bottle sits on the shelf. You must therefore shake the bottle before using the dressing. After you shake the bottle, the herbs settle again. Particles of clay in water act the same way. After you stir up the clay, the particles eventually sink to the bottom. These mixtures are examples of suspensions. A **suspension** is a mixture in which one of the parts is a liquid and from which visible particles settle out. See Figure 3-5.

Suspension A mixture of a solvent-like liquid and particles that slowly settle out.

Figure 3-5 *A winter-scene knickknack is a suspension of plastic snow in water. No matter how hard you shake it, the snow will settle out.*

Suspensions are used frequently in the pharmaceutical industry. Liquid medicines that require shaking before being taken are suspensions. Some paints contain solid pigments in a liquid. See Figure 3-6. These pigments must be finely ground so the paint does not separate readily. You may have noticed that paint stores have special machines that shake the cans of paint in order to mix the components.

Figure 3-6 *In order to make different colors of paint, pigments are suspended in a liquid.*

Colloids

What do cherry-flavored gelatin, cold cream, lipstick, and shaving cream have in common? They are not really solids or liquids; they are something in-between. All of these substances are colloids. A **colloid** is a type of mixture in which the particles are small enough that they do not settle out or sink to the bottom of their containers, and they are usually too small to be visible.

You do not have to look far to find a colloid. In fact, the cytoplasm in your body's cells is a colloid. Many of the foods you eat are also colloids. Jelly, for example, is one colloid that tastes great on toast!

Not all colloids are mixtures of a solid and a liquid. There are three basic types of colloids, depending on their makeup.

Colloid A mixture in which the particles are small enough that they do not settle out.

1. *Gels* are liquid particles spread out in a solid. Gels flow slowly since they are nearly solid. Some examples of gels are gelatin, jelly, and stick deodorant.

2. *Emulsions* are colloids made of two liquids. You are familiar with many oil–water emulsions, such as mayonnaise, hand cream, and milk. Many creamy salad dressings are also emulsions.

3. *Aerosols* are colloids formed when either solid or liquid particles are suspended in a gas. Fog and smoke are examples of aerosols. A paint-spray can shoots out a colloid of paint particles finely dispersed in a gas, as shown in Figure 3-7.

Figure 3-7 *There are many examples of aerosols in everyday life. The spray paint shown here is an aerosol.*

Instruments called ultracentrifuges can be used to separate the parts of a colloid. You may know of centrifuges as the machines used by hospitals to separate the components of blood for analysis. These centrifuges can spin colloids at very high speeds to separate solids from liquids. This method can be used to study the sizes of the particles in colloids. As the colloid spins, different-sized particles settle out in different layers.

The particles of a colloid scatter light. As you know, this phenomenon is known as the *Tyndall effect* and is one important way to distinguish colloids from solutions. Light scattering can be used to determine the size, shape, and concentration of the particles because all of these qualities affect the intensity of the light passing through. See Figure 3-8.

Figure 3-8 *The jar at the left contains a mixture of water and sodium chloride. The jar at the right contains a mixture of gelatin in water. Which mixture is a colloid and which is a solution?*

Summary

Mixtures contain two or more substances that can be separated by physical means. Solutions are mixtures in which one substance (the solute) dissolves into another (the solvent). Solubility can be expressed by the amount of solute in a given amount of solvent. There are procedures that affect the rate of solution. A mixture with particles large enough to settle out is called a suspension. Colloids are mixtures with characteristics falling between solutions and suspensions.

3.2 ACIDS, BASES, AND SALTS

Many compounds are soluble in water. When an ionic compound dissolves in water, the bonds that hold the ions together are broken. As a result, the positive and negative ions separate from one another and are free to move around the solution. Some molecular compounds can partially break into ions as well. Acids and bases form solutions that are characterized by the presence of certain ions when they are dissolved in water. Salts are chemical compounds that form when solutions of acids and bases are mixed.

Characteristics of Acids

When you think of an acid, you may think of a substance that "eats" through metal. However, not all acids are that strong. In fact, you might have had some acid for breakfast. Orange juice is an example of a weak acid. See Figure 3-9. Carbonated drinks also contain a weak acid.

Figure 3-9 *Citric acid was so named because it was first isolated from citrus fruits.*

S51

One characteristic of acids is that they taste sour. You may have tasted certain mild acids. Most fruits have some acid content. Oranges and lemons contain citric acid; apples contain malic acid. Green, or unripe, fruits taste sour because they have a greater amount of acid than ripe fruits have. Some of the acid in fruit is converted into sugar as the fruit ripens. Another common acid is found in vinegar. Vinegar is a 5 percent solution of acetic acid in water. All of these acids are weak acids.

Many of the weak acids come from living sources and are therefore called *organic acids*. Most organic acids contain a –COOH group as part of the molecule. This group breaks apart into –COO⁻ and H⁺ ions. The presence of hydrogen ions is what makes a compound an acid. Table 3-2 lists some organic acids with which you may be familiar.

ORGANIC ACIDS		
Name	Found in or Produced by	Chemical Formula
Acetic acid	Vinegar	CH_3COOH
Citric acid	Citrus Fruits	$C_5H_7O_5COOH$
Formic acid	Ants	$HCOOH$
Lactic acid	Sour milk	C_2H_5OCOOH
Malic acid	Apples	$C_3H_5O_3COOH$
Salicylic acid	Aspirin	$C_6H_4(OH)COOH$
Tartaric acid	Soft drinks	$C_3H_5O_4COOH$

Table 3-2

SOME COMMON ACIDS	
Name	Formula
Hydrochloric acid	HCl
Sulfuric acid	H_2SO_4
Nitric acid	HNO_3
Phosphoric acid	H_3PO_4

Table 3-3

Acid A compound that produces hydrogen ions in solution.

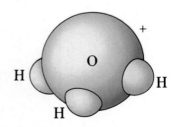

Figure 3-10 Hydronium ions contain one oxygen atom and three hydrogen atoms. The number of hydronium ions in solution determines the strength of an acid.

The Formation of Acids How can you tell if a substance will form an acidic solution when it dissolves in water? The best way is to know its chemical formula. Look at the formulas for some common acids in Table 3-3. Notice that each acid has a hydrogen at the beginning of its formula. When an acid, such as hydrochloric acid (HCl), is dissolved in water, it separates into H⁺ and Cl⁻ ions. In scientific terms, an **acid** is a compound that produces hydrogen ions (H⁺) in solution.

Since a hydrogen atom has one proton and one electron, the positive ion of hydrogen (H⁺) is just a proton. However, protons are not normally found uncombined in solution. What actually happens is that H⁺ combines with a water molecule to form H_3O^+, or a *hydronium* ion. See Figure 3-10. Any solution that contains H_3O^+ is acidic.

The Strength of Acids Why are some acids weak and others strong? The strength of an acid depends on the number of hydronium ions in solution. The more hydronium ions that the acid produces in solution, the stronger the acid. Molecules of strong acids completely break down into ions when dissolved in water, producing many hydronium ions. Therefore, their water solutions contain only ions from the acid; no molecules of the acid are left. The molecules of weak acids, however, do not all break down into ions, so there are not as many hydronium ions in solution. See Table 3-4 for some examples of strong and weak acids.

STRENGTH OF SOME ACIDS	
Strong acids	**Weak acids**
Hydrochloric acid (HCl)	Acetic acid (CH_3COOH)
Nitric acid (HNO_3)	Citric acid ($C_5H_7O_5COOH$)
Sulfuric acid (H_2SO_4)	Carbonic acid (H_2CO_3)
Hydrobromic acid (HBr)	Boric acid (H_3BO_3)

Table 3-4

Concentrated solutions of strong acids are very corrosive. They can react with (eat through) metals and must be handled with great care because they can produce severe burns. The acids listed on the left in Table 3-4 are strong acids. **Concentrated solutions of these acids should never be tasted or touched!**

In addition to reacting with metals, acids also react with substances such as limestone and marble, releasing carbon dioxide. This reaction causes the deterioration of marble buildings and statues. See Figure 3-11. It also occurs in nature, causing the formation of caves and some sinkholes. Where do you think the acid comes from to do this?

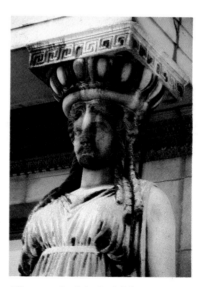

Figure 3-11 Acid has eroded the features of this marble statue.

Uses of Acids

Acids are very important to many industrial processes. The production of sulfuric acid, for example, is so important that it serves as an indicator of how well the United States' economy is doing. Sulfuric acid, a strong acid that is a colorless, oily, and dense liquid, is used to make fertilizers and automobile batteries. It is used in tanning leather for your shoes, in treating the paper on which the words of this book are printed, and even in making the vanilla flavoring for cookies, cakes, and ice cream. Sulfuric acid is also an

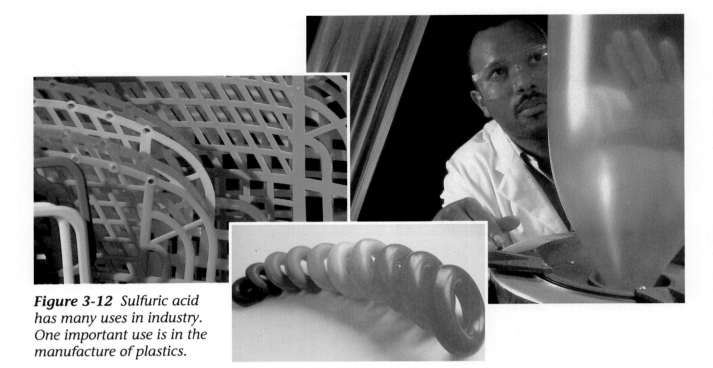

Figure 3-12 Sulfuric acid has many uses in industry. One important use is in the manufacture of plastics.

Figure 3-13 One use of nitric acid is to make dyes such as those shown here.

Base A compound that produces hydroxide ions in solution.

important component in steel production and many other manufacturing processes. See Figure 3-12.

Nitric acid, another strong acid, was first used by the alchemists of the Middle Ages to dissolve metals. They named it *aqua fortis*, which is Latin for "strong water." Today, the uses of nitric acid range from making explosives, such as nitroglycerin and dynamite, to the manufacture of ammonium nitrate, a major solid fertilizer. See Figure 3-13.

Hydrochloric acid, also called *muriatic acid*, is used to clean concrete and to balance the acid content of swimming pools. Like other strong acids, concentrated hydrochloric acid needs to be handled with extreme care.

Characteristics of Bases

In the same way that an acid produces hydronium ions, an inorganic **base** produces *hydroxide* ions (OH^-) in solution. The formulas for many common bases, along with their more common names, are given in Table 3-5. When these bases are added to water, they release negative hydroxide ions. The stronger the base, the greater the number of hydroxide ions that are released into solution. Strong bases are just as dangerous as strong acids. They can cause serious burns and must be handled with care. **Never touch or taste a strong base.** Solutions that contain bases are called "basic solutions." You may have also heard the word *alkaline* used instead of "basic" in reference to such solutions.

SOME COMMON BASES		
Chemical Name	**Formula**	**Common Name**
Sodium hydroxide	NaOH	lye, caustic soda
Calcium hydroxide	$Ca(OH)_2$	slaked lime
Magnesium hydroxide	$Mg(OH)_2$	milk of magnesia
Ammonium hydroxide	NH_4OH	ammonia water

Table 3-5

Although most bases, such as NaOH, already contain hydroxide ions, there are some bases that do not. Ammonia (NH_3), for example, produces a hydroxide ion only when it is dissolved in water.

$$NH_3 + H_2O \rightarrow NH_4^+ + OH^-$$

Since ammonia forms OH^- ions in solution, it is considered a base.

One characteristic shared by all bases is that they have a soapy, slippery feel. Soap solutions, which are mildly alkaline, demonstrate this characteristic. If you have ever had soap in your mouth, you are familiar with another characteristic of bases—their bitter taste.

Bases react with acids in a process called *neutralization*. Neutralization occurs when equal amounts of hydronium ions and hydroxide ions are in solution. These ions produce water, as is shown in the following equation.

$$H_3O^+ + OH^- \rightarrow 2H_2O$$

Once the water is formed, the solution is no longer acidic or basic; it is neutral. This characteristic reaction between acids and bases occurs under many circumstances—even in the human body. When a person has an acid stomach, or heartburn, he or she may take an antacid tablet. The antacid tablet is a weak base that dissolves in the stomach and combines with the excess acid to neutralize it.

Uses of Bases

Although bases are less commonly discussed, they are just as important as acids in many processes. Bases are used for a variety of things, from keeping you clean to unclogging sinks and even making cement. One example of a commonly used base is household ammonia, which is really an ammonium hydroxide solution. This compound is present in many cleaners for its ability to cut grease.

Sodium hydroxide, also known as caustic soda or lye, is an economic indicator, just like sulfuric acid. That is, the amount of sodium hydroxide produced indicates how the economy of the country is doing. Sodium hydroxide is used by many industries, including the paper industry. When paper is made, sodium hydroxide is used to remove pulp fibers from wood. See Figure 3-14.

Many other bases are also commonly used. More than 35 billion pounds of calcium hydroxide are produced in the United States annually. This base is used to make cement, mortar, and plaster. Another base, magnesium hydroxide, is sold as an antacid. These are only a few examples of the many bases used every day.

Figure 3-14 Sodium hydroxide is a strong base that is used in many industrial processes. The paper-making process, shown here, requires great amounts of this chemical.

The pH Scale

Since you cannot see H_3O^+ or OH^- ions in solution, how can you identify a solution as acidic or basic? Scientists use a pH scale to show how acidic or basic a dilute solution is. The **pH** of a solution is a measure of the hydronium-ion concentration of the solution. The pH of most dilute solutions ranges from 0 to 14. This scale separates acids, bases, and neutral solutions into regions. The middle point, pH = 7, is neutral. A solution with a pH of 7 is neither an acid nor a base. Acids have a pH of less than 7. The more acidic the solution, the lower the number on the pH scale. Figure 3-15 shows the pH values for some common substances. For example, the pH of lemon juice is about 2.3, while tomato juice, which is less acidic, has a pH of 4. Bases have a pH higher

pH A measure of the hydronium ion concentration of a solution.

Figure 3-15 This chart shows the pH of several common substances.

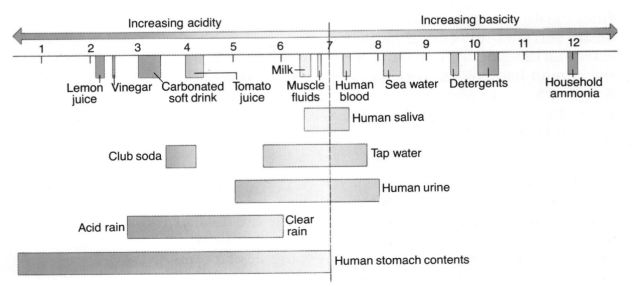

than 7. The higher the number above 7 on the pH scale, the more basic the solution.

Indicators of pH The pH scale is used to determine if a solution is acidic, basic, or neutral. Scientists can determine the pH of a solution with an instrument called a pH meter. A pH meter has electrodes that are placed in the solution. An electric current passes through the meter, and the pH is read from a dial or display. See Figure 3-16.

There are other, less expensive ways in which to determine the pH of common substances. For example, garden stores sell kits to help people determine the pH of their soil. The pH of swimming pools, which is important if the water is to be kept clean and free of bacteria, can also be determined by using a pH kit. Both of these kits work on the basis of color changes. They contain special solutions called *indicators*. An indicator is a substance that changes color in an acid or a base. Some indicators show only two color changes—one for acids and another for bases. Other indicators, however, show various color changes depending on the pH of the substance. See Figure 3-17.

Figure 3-16 *This scientist is using a pH meter to test lake water for traces of acid-rain contamination.*

Figure 3-17 *A sample's pH can be determined by adding indicator and comparing its color to those of standard solutions containing indicator.*

For an indicator to be useful, it must show a very distinct color change over a narrow range of pH. Litmus paper, which is widely used, is a strip of paper dyed with an indicator. It comes in two colors: red and blue. Red litmus paper turns blue in a base. Blue litmus paper turns red in an acid. Another common indicator is phenolphthalein, which is colorless in an acid but turns red in a base.

Figure 3-18 When pH paper touches an acidic or basic solution, it changes color. The pH of the solution may be estimated by comparing the color produced to the scale on the dispenser.

Some indicator paper is dyed with what is called a *universal indicator*. Universal indicator is a mixture of several different indicators combined so that the paper gives a different color for each pH. The advantage of this type of indicator is that it shows the exact pH of a solution rather than just showing whether the solution is an acid or a base. Paper dyed with universal indicator is called pH paper. See Figure 3-18.

Nature and pH Many plants and animals are very sensitive to changes in pH. Some fish and other aquatic life have died in lakes that have become more acidic than normal due to pollution. Lake water usually has a pH between 6 and 7. However, some lakes in northern states have a pH as low as 3. Many scientists believe that this change in pH has been caused by acid rain, which is produced by the mixing of rain and the pollution from factory smokestacks and automobile exhaust.

Scientists are experimenting with methods to bring the pH of these acidic lakes back to normal levels. In order to do this, large quantities of weak bases, such as ground limestone or sodium bicarbonate, are dumped from planes into the lakes. See Figure 3-19. Scientists hope that the addition of these bases will neutralize the excess acid. The results of these efforts, however, will only be temporary if the sources of pollution are not stopped.

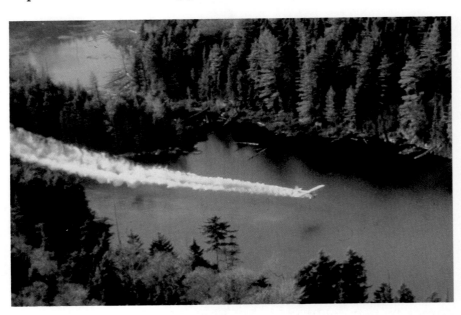

Figure 3-19 When lakes become too acidic as a result of acid rain, a weak base, such as baking soda, is often added to the water.

The human body is also very sensitive to changes in pH. The normal pH of human blood is between 7.38 and 7.42. This range is very narrow; a change in the concentration of H_3O^+ ions has a marked effect on the body. If the pH is above 7.8 or below 7, the body cannot function normally. If the pH change is sudden or not corrected rapidly, it can be fatal.

Characteristics of Salts

You are used to calling sodium chloride "salt." Scientists use the term *salt* in a more general way. A **salt** is any ionic compound formed when the negative ion from an acid combines with the positive ion from a base. You may recall that when acids and bases are mixed together, the H_3O^+ and OH^- ions form water. The other two ions that were part of the acid and base form a salt. Look at the following neutralization reaction.

Salt Any ionic compound formed when the negative ion from an acid combines with the positive ion from a base.

$$HBr + KOH \rightarrow H_2O + KBr$$

The potassium and bromide ions combine to form an ionic compound called potassium bromide. All neutralization reactions produce water and a salt.

Not all salts, however, are formed by neutralization reactions. Some are formed by the reaction of metals with acids. For instance, when pieces of magnesium metal are dropped into hydrochloric acid, hydrogen gas is released and a white solid can be collected. This white solid is magnesium chloride, a salt.

$$Mg + 2HCl \rightarrow MgCl_2 + H_2$$

Salts also come in other colors, depending on the particular ions that are combined. See Figure 3-20.

Figure 3-20 The color of a salt depends on the elements that compose it. Pictured here are four different salts, each a different color.

Uses of Salts

Salts are very common compounds. You probably use them every day without even knowing it. There are salts in most canned or processed foods. The salts either improve the flavor of the food, or they help to preserve the food.

Table salt, sodium chloride, is a common chemical compound in the earth's crust. Solid sodium chloride is found in large underground deposits in the form of gray crystals called rock salt. These deposits of rock salt can be mined. Salt can also be obtained by the evaporation of sea water. The oceans contain a huge amount of dissolved salts. Among the various salts taken from the sea, sodium chloride is the most common, although there are many others. The process of removing dissolved salts from sea water is called *desalination*.

Figure 3-21 *Salt is an important part of our diet. It is also used in many industrial processes. Lye, chlorine, hydrochloric acid, and baking soda are some of the chemicals that are made from sodium chloride.*

Approximately 40 million metric tons of sodium chloride are collected each year by desalination. See Figure 3-21.

Salts have the unusual characteristic of lowering the temperature at which water freezes. People often take advantage of this property of salts. For example, sodium chloride is spread on wet streets in the winter to prevent ice from forming at 0 °C. The salted water will need a much lower temperature to freeze. If you have ever made ice cream at home, you have seen the effect that sodium chloride has when added to the ice around the ice-cream freezer. As the ice melts, the salt dissolves in the water, lowering its freezing point. The remaining ice then cools this water to sub-zero temperatures. Since milk, sugar, and the other ingredients in ice cream freeze at a lower temperature than water does, the lower temperature of the salted water allows the ice cream to freeze. Other salts have a similar effect on water, but since sodium chloride is so abundant, it is the cheapest salt to use.

Summary

A solution containing a large number of H^+ ions is an acidic solution. A solution containing more OH^- ions than plain water is a basic solution. When an acid and a base are combined, a neutralization reaction takes place in which the H^+ and OH^- ions combine to form water. The strength of acidic and basic solutions is measured on the pH scale. Another product of a neutralization reaction is a salt. Many different kinds of salts are produced when different acids and bases react. Salts can also be made by other reactions. Salts can flavor and preserve food and also lower the temperature at which water freezes.

Unit 4

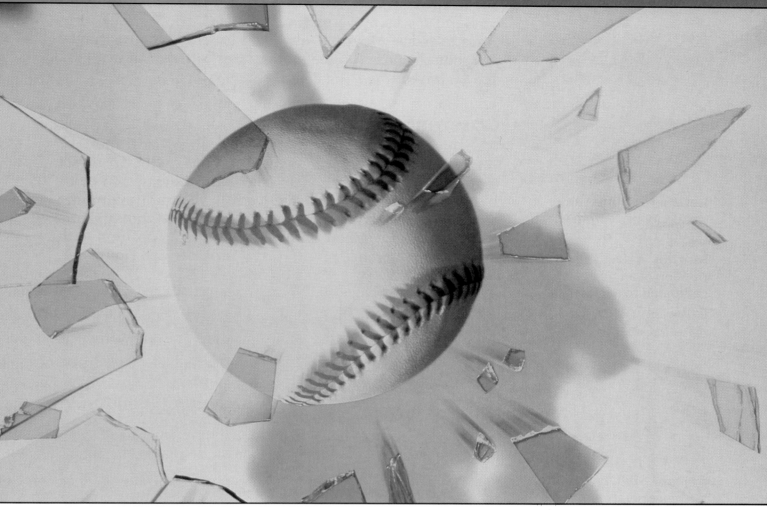

IN THIS UNIT

4.1 Motion and Force, p. 62

4.2 Laws of Motion, p. 68

4.3 Gravitation, p. 74

Reading *Plus*

Now that you have been introduced to forces and motion, let's take a closer look at how we describe motion and how motion relates to force. Read pages S62 to S76 and then prepare a demonstration that uses the principles of force and motion you learned about. Your demonstration should illustrate the concepts expressed by the following questions.

1. How do we observe and describe the motion of any object?
2. In what ways can motion be described by physical laws?
3. How are motion, force, acceleration, and gravity related?

4.1 MOTION AND FORCE

The world around us is full of motion. Leaves fall from trees, cars and trucks move along city streets, people walk from place to place. Maybe you walked to school today; maybe you rode the bus. You see motion everywhere. But what exactly is motion?

Observing Motion

What do we mean when we say something is in motion? First, we always compare a moving object with an object that appears to stay in place. The object that stays in place is called a **reference point**. On earth, we usually compare a moving object with an object on the ground that appears to be stationary. When an object changes its position with respect to a reference point, it is in **motion**. See Figure 4-1.

Reference Point Any stationary place from which motion can be observed.

Motion A change in position of an object when compared to a reference point.

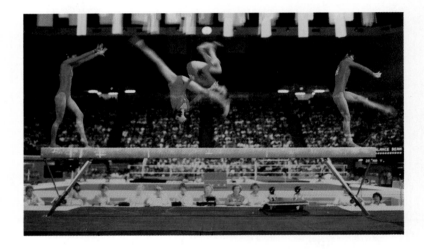

Figure 4-1 Relative to the gymnast, what are the reference points in this photograph?

For example, if you look at clouds in the sky, you can tell if they are moving by comparing them to a stationary tree within your view. If the clouds change position with respect to the tree, they are in motion. If there is no change, you know they are still.

Speed Look at Figure 4-2. What reference points do you see in these pictures? In comparing the two pictures, you can see that the balloon has moved past the trees. Now, consider that the time between the two photographs was 20 seconds. During that time, the balloon moved a distance of 100 meters. Every moving object covers a certain distance in a certain period of time. Dividing this distance by the time of travel gives the **speed** of the object. In the case of the balloon, the speed was 5 m/s.

Speed The distance covered by a moving object per unit of time.

Figure 4-2 The balloon has moved past the edge of the trees, as shown by the two photographs. The time between the two photographs was 20 seconds.

Velocity A reference point also allows us to specify direction. Objects can move either *toward* or *away* from a reference point. When we give speed a direction, we call it a **velocity**. People often use the words *speed* and *velocity* as if they have the same meaning. They do not. Velocity is a speed in a particular direction. See Figure 4-3. To completely describe a motion, you must specify two things—the speed *and* the direction of the motion. For example, 60 km/h is simply a speed, while 60 km/h west is a velocity.

Velocity The speed and direction of a moving object.

Figure 4-3 These jets are traveling at approximately the same speed. However, because they are traveling in different directions, their velocities are different.

Acceleration Most moving objects do not move at a constant speed. When you ride a bicycle, for example, your speed going uphill is usually slower than when you ride on level ground. Your speed may increase as you go downhill. This kind of motion is called **acceleration**. If the speed of a moving object changes, its motion is an accelerated motion. See Figure 4-4.

The word *acceleration* is commonly used to mean an increase in speed. However, to a scientist, acceleration is *any* change in the velocity of a moving object. Therefore, a scientist would say that when a bicycle slows down, it accelerates. We sometimes use the word *deceleration* to indicate this type of motion. However, remember that velocity includes both speed *and* direction. If the bicycle turns a corner

Acceleration A change in the speed or direction (or both) of motion.

Lines A, B, and C show the differences in acceleration for bicycle riding on level ground, going uphill, and riding downhill.

Figure 4-4 The acceleration or deceleration of a bicycle can be graphed as the relationship between velocity and time.

without changing speed, it is also accelerating. Acceleration, therefore, involves a change in speed, direction, or both. See Figure 4-5.

Frames of Reference

If you look around you, you will notice many objects that seem to be stationary because they do not move with respect to each other. These stationary surroundings are called a **frame of reference**. A frame of reference is used to specify the position and relative motions of objects. Usually, the surface of the earth is the frame of reference for the motions we observe on earth. Most of our everyday activities take place within a limited area bounded by our field of view.

Nevertheless, the earth itself is in motion. While you are reading these words, you are hurtling through space at an incredible speed. See Figure 4-6. We can only describe the motion of an object by comparing it to a reference point within a given frame of reference. For most things, stationary objects on the surface of the earth serve this purpose. We can then ignore other (cosmic) motions.

There are also other, smaller frames of reference that we can use. Imagine that you are riding in a speeding train. The walls and floor of the car you are in are stationary in relation to yourself and can be considered a frame of reference. You can toss a coin into the air and catch it, as you would while standing on the ground, without worrying that the coin is actually going forward at 135 km/h, the speed of the train. As long as the train maintains a constant velocity, all the laws of physics are the same as if you were stopped. In fact, if there were no windows in the train, you might not even notice that you were moving. See Figure 4-7.

Figure 4-5 The cars on this stretch of highway are constantly changing velocity, or accelerating, even when their speed remains constant.

Frame of Reference Any system for specifying the precise location of objects in space.

Figure 4-6 Considering the motion of the earth—as it rotates, revolves, and moves with our galaxy—just how fast are you going when you are "standing still"?

Figure 4-7 What about this photograph tells you that the train is moving?

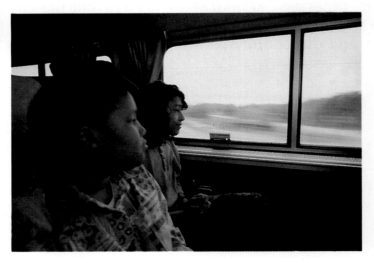

Forces and Motion

The ancient Greek philosopher Aristotle once said that a moving object would stop moving unless it was pushed or pulled. Do you think this is true? Think about a wagon. As long as you pull it, it will follow you. If you let go of the handle, it will slow down and stop. This little experiment seems to indicate that Aristotle was right. This type of push or pull is called a **force**. Aristotle thought that a force was needed to keep an object moving. However, Aristotle was not aware of a very important aspect of nature—friction.

Force Any push or pull.

Great discoveries in science are often made when someone asks the right question. Consider the question, "Why do objects move?". This question was asked by early scientists when they investigated motion. Experiments seemed to indicate that a push or pull was needed to cause motion and that an object with no force acting on it did not move. However, in the 17th century, Isaac Newton saw the problem differently. He asked the question, "Why do objects *stop* moving?".

Newton thought that moving objects are usually stopped by a force, which we call **friction**. For example, there is a force of friction between your shoes and the floor that allows you to walk. If there were no friction, you would have trouble moving forward. See Figure 4-8. Have you ever slipped on a polished floor? As you may have found out, smooth surfaces have less friction.

Friction The force that opposes motion between two surfaces that are touching.

Figure 4-8 The force of friction between the ground and your shoes helps you walk. Low friction makes it hard to walk on a slippery surface such as ice.

Wheels can also reduce the amount of friction. Have you ever fallen at a roller rink? When one object rolls over another, there is very little friction. But there is always some friction present. How do you think moving objects would behave without friction?

Newton discovered that it is the force of friction that causes a moving object to slow down and, if the force acts long enough, stop. Whenever a single force is applied to an object, the object's motion is changed. If the object is at rest, it moves; if it is moving, it changes velocity. However, if an applied force does not change the object's motion, it must have been *balanced* by another force. Only when forces are unbalanced does an object accelerate. See Figure 4-9.

Forces on a stationary object	Result	Forces on a moving object	Result
Balanced	Object remains stationary	Balanced	Object remains at constant velocity
Unbalanced	Object accelerates	Unbalanced	Object accelerates

Figure 4-9 Objects remain stationary or in motion at a constant velocity unless acted upon by an unbalanced force.

Types of Motion

Motion of an object can be described by the combination of its speed and direction. Some motion involves constant velocity (constant speed *and* direction), others involve acceleration (changes in either speed, direction, or both), and still others are the combination of two separate motions.

Free Fall A falling object is a good example of accelerated motion in one direction. A freely-falling object accelerates downward at 9.8 m/s^2 due to the pull of gravity. This means that the speed of the object increases by 9.8 meters per second each second that it falls.

Actually, a falling object will accelerate at 9.8 m/s^2 only if it falls in a vacuum. Most objects falling through air will accelerate only for a short time. The acceleration of the object is soon stopped by friction with the air. The greatest speed an object can reach while falling through air is called its *terminal speed*. For example, if you jump from an airplane, you will accelerate until you reach a terminal speed of about 190 km/h. Once you open your parachute, your terminal speed will be reduced by the added wind resistance to a safer rate. All objects falling through the air reach a certain terminal speed once the force of air resistance balances the force of gravity. See Figure 4-10.

Figure 4-10 The upward force of air resistance on the skydivers works against the downward force of gravity.

Circular Motion A race car moving around a circular track, a child on a merry-go-round, and clothes being spun in a washing machine are all examples of objects undergoing *uniform circular motion*. Uniform circular motion refers to motion at constant speed in a circular path. See Figure 4-11.

When an object goes around in a circle, its direction is constantly changing, and it is therefore accelerating. Consider a race car speeding around a curve on the track. The car must constantly change directions to follow the track. Therefore, an unbalanced force is causing the car to accelerate toward the center of the curve. The force that causes objects to move in a circular path is called **centripetal force**. *Centripetal* means "toward the center."

If you were riding in the race car mentioned above, you would feel as if you were being pushed away from the center of the turn. This is a result of your body's resistance to changing velocity. When your body is moving, it tends to continue moving at a constant velocity. See Figure 4-12. As you are moving around a curve in a car, it feels as if you are applying an outward force to the seat belt or the door of the car. However, the seat belt or door is actually applying a centripetal force to you to change your direction of travel.

As a car travels around a corner, the centripetal force is provided by friction between the road and the car's tires. If, for some reason, the tires were to lose traction as the car was moving around the curve, the car would immediately start sliding in a straight line. Without friction between the tires and the road, the car would be impossible to turn.

Figure 4-11 When this amusement-park ride spins, its frame provides the centripetal force necessary to keep the riders moving in a circular path.

Centripetal Force A force that causes objects to move in a circular path.

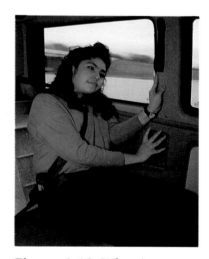

Figure 4-12 What is providing the centripetal force in this situation?

Motion in Two Directions When a ball is thrown in a horizontal direction, it moves not only outward due to the force of the throw, but also downward due to the pull of gravity. Such an object is called a *projectile*. Projectiles, once they have been thrown, are affected only by the pull of gravity. Their motion can be thought of as a combination of both a constant horizontal motion and an accelerated vertical (downward) motion. The vertical and horizontal motions are independent—that is, they have no effect on each other. The curved path a projectile takes is called its *trajectory*.

One interesting thing about trajectories is that an object thrown sideways will reach the ground at the same time as an identical object that falls straight down. The horizontal speed of the trajectory does not affect the time it takes to hit the ground. Gravity acts equally on objects regardless of their horizontal velocity. See Figure 4-13.

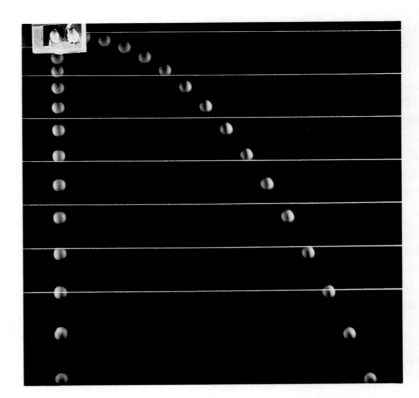

Figure 4-13 Both balls have the same vertical acceleration due to gravity, but the ball on the right also has a constant horizontal motion.

Summary

The universe is full of motion. The motion of an object can be described by comparing it to a reference point, which is an object that is considered stationary. The speed of an object is measured by the distance it travels from the reference point in a certain amount of time. Velocity is a measure of both the speed and direction of an object. Acceleration is the change of velocity of an object in either speed, direction, or both. Frames of reference are used to specify the position and relative motions of objects. The motion of any object can be described by the combinations of its speed and direction. Some motion involves constant velocity, while others involve acceleration. Still other motions are a combination of two separate motions.

4.2 LAWS OF MOTION

What is known today about the science of motion is due largely to a book published by Sir Isaac Newton in 1687. Newton's book, *Principia*, describes his findings about moving objects. The basic principles he wrote about are now known as Newton's laws of motion. These and other laws describe the motion we observe and explain the forces behind this motion.

Newton's First Law of Motion

Newton thought that a moving object with no friction or other force acting on it would continue to move in the same direction and at the same speed forever. He decided that a force had to be present whenever there was a change in speed or direction. Therefore, if an object moves at constant speed, no force is needed. To go faster, a force is applied in the direction the object is going; to slow down, a force is applied in the opposite direction. To change direction, a force is applied from the side. Every change in motion thus requires an outside force.

The property of matter that resists any change in motion is called **inertia**. A moving object, in the absence of an outside force, thus continues to move because of its inertia. Similarly, an object that is not moving needs a force to overcome its inertia and start moving. See Figure 4-14. The amount of inertia of any object depends on the **mass** of the object.

The effects of inertia were summed up in what is now called **Newton's First Law of Motion**. It states: *Every object remains at rest or moves with a constant speed in a straight line unless acted upon by some outside force.* This law is also called the law of inertia.

Consider your experience with the wagon. You should now understand that it is the force of friction that slowed down the wagon. If friction was absent, the wagon would have continued to move forever. If the wagon was at rest, it would have remained that way unless a force—provided by your push or pull—changed the situation.

In outer space there is no friction against moving objects. For example, the Pioneer-10 spacecraft, launched from earth in the early 1970s, has now left our solar system. Once it left earth's orbit, its rocket motors were no longer needed. It has since travelled billions of kilometers at a constant speed with no additional push from its engines. See Figure 4-15. Unless some outside force affects it, it will continue out into our galaxy indefinitely. This spacecraft illustrates Newton's first law of motion.

The first law of motion should remind you always to wear a seat belt when riding in a car. If the car stops suddenly, you will tend to keep moving forward. If you are not wearing a seat belt, you might hit the windshield or dashboard. A seat belt makes it less likely that you will be hurt, since it

Figure 4-14 Because of the inertia of the objects on this table, the table setting remains intact momentarily as the table is quickly slid from underneath it.

Inertia The resistance to any change in motion.

Mass The amount of matter in an object.

Figure 4-15 Why does a spacecraft remain in motion even without continuous rocket thrusts?

supplies the force needed to hold you firmly in your seat. See Figure 4-16.

Figure 4-16 *In a car crash, inertia keeps the passengers moving forward until they are stopped by a force. If seat belts are not worn, this stopping force may be provided by the car's windshield.*

Newton's Second Law of Motion

How much force would you have to apply for ten seconds to cause a bicycle to accelerate to a speed of 8 km/h? How much force would be needed to accelerate a car to this speed? You already know that a force is needed to accelerate an object. But the greater the mass of the object, the greater the amount of force needed to cause the acceleration. Therefore, the massive car would need a much larger force to accelerate than the bicycle. See Figure 4-17. This relationship between masses and the forces acting on them was given in **Newton's Second Law of Motion**: *The acceleration of an object of a certain mass is determined by the size of the force acting and the direction in which it acts.*

Figure 4-17 *The smaller mass of the bicycle means that a smaller force is needed to produce the same acceleration.*

All forces act in some direction and, of course, exist in different sizes. Scientists measure the size of a force with a unit called the *newton* (N). You would need a force of about

9.8 N to lift an object with a mass of 1 kg. You can write the second law of motion with one of the most important equations in physics:

Force = Mass × Acceleration

or

$F = ma$

Suppose you have to push a heavy box across the room. By yourself, you can supply only a small force. You might not even be able to overcome friction to get it started. If ten people pushed, however, the force would be larger and the box would move. Remember that the direction of the force is also important. If five people pushed the box one way and five pushed the other way, what would happen?

The mass of the box is also a factor. If the same ten people supplied a force to a piano case, and then to a shoe box, the acceleration of the shoe box, of course, would be greater. You can see this by rearranging the equation to:

$F/m = a$

If the force remains the same and the mass decreases, the acceleration will increase. What will happen to the acceleration if the mass increases?

Newton's Third Law of Motion

How was the Pioneer 10 initially propelled upward against the force of gravity? Hot expanding gases were produced in the rocket engines and forced out the rear of the engine. See Figure 4-18. Rocket engines exert a force on the gases to push them out. The gases also exert an equal and opposite force on the engine, and this force moves the rocket upward. This action is an example of **Newton's Third Law of Motion**, which states: *For every action there is an equal and opposite reaction.*

There are many everyday examples of Newton's third law all around you. A book lying on a table pushes down on the table with a certain force. The table, in turn, pushes up on the book with an equal but opposite force. One force acts on the table, while the second force acts on the book. See Figure 4-19. If you push on your desk with your hand, the desk pushes back. If you push harder, the desk also pushes harder. If you jump off the floor, your feet push against the floor with a certain force. The floor also pushes against your feet with a force that is equal in size, but opposite in direction.

Figure 4-18 How is Newton's third law demonstrated by a rocket engine?

Figure 4-19 The force of gravity is acting on this book, however, the book does not accelerate because it also receives an equal upward force from the table.

Sometimes it is hard to picture these opposing forces. However, there is at least one activity you can try to actually observe them. Imagine that you and a friend are standing on roller-skates on a level rink floor. If you were to push your friend away from you with both hands, you would roll backwards. The force you exerted on your friend would be equal but opposite to the force of your friend pushing back on you. See Figure 4-20. The outward force you feel in a car going around a curve is your reaction force to centripetal force. Can you think of other examples of Newton's third law of motion?

Figure 4-20 *For every action there is an equal and opposite reaction.*

Momentum

If you have ever watched American football on television, you have probably heard an announcer use the word *momentum*. The announcer may have said something like this: "Even though number 34 was hit at the 1-yard line, his momentum carried him into the end zone for a touchdown." In the world of physics, **momentum** is the inertia of an object in motion. More specifically, momentum is an object's mass multiplied by its velocity. This relationship can be shown by the following equation.

Momentum The mass of an object multiplied by its velocity.

$$\text{Momentum} = \text{Mass} \times \text{Velocity}$$

or

$$p = mv$$

Since velocity is one factor in this equation, momentum (like velocity and acceleration) has definite direction.

Since the momentum of an object depends on its mass and velocity, an object with a small mass and a high velocity can have as much momentum as an object with a large mass and a low velocity. In football, for example, a small,

S72

fast runner can be as hard to tackle as a large, slow runner. Runners who are both large and fast are the hardest to tackle due to their momentum. See Figure 4-21.

One of the most interesting things about momentum is that it stays constant unless an unbalanced force is applied. When the momentum of an object changes, it is because the object's velocity, mass, or both have been changed. Since objects usually do not lose or gain mass readily, the momentum of most objects changes as a result of changes in their velocity.

Imagine, for example, that you are back in the skating rink. Suddenly, another skater accidentally runs into you. The push from the other skater is an unbalanced force that causes you to move at a higher velocity. As a result, your momentum increases. When the skater bumps into you, he or she also receives an unbalanced force that causes him or her to move at a slower velocity, decreasing momentum. The momentum that you gained is equal to the momentum lost by the other skater. This example illustrates the **law of conservation of momentum**. This law states that *momentum can be transferred from one object to another but cannot change in total amount*. In other words, when two or more objects collide, the total momentum of the objects is the same after the collision as it was before the collision.

Figure 4-22 demonstrates the law of conservation of momentum. If the ball at one end is pulled out and allowed to swing, it falls back and hits the next one in the line. Almost immediately, the ball at the opposite end of the line flies out, as if by magic. The momentum of the first ball has been passed from one ball to the next until it reaches the end ball,

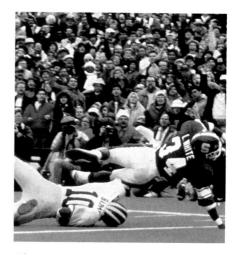

Figure 4-21 The momentum of this football player moves him forward even though he is being tackled around his legs.

Figure 4-22 This "Newton's cradle" demonstrates the law of conservation of momentum.

which is free to move. What do you think would happen if two balls at one end were pulled out and allowed to swing?

4.3 GRAVITATION

What do you think would happen if two stones of different masses were dropped from the same height? Would the larger and heavier one reach the ground first? Or would they both reach the ground at the same time? In 1590 this was the topic of an intense debate among the scientists of Europe. Most scientists believed that the heavier stone would reach the ground first.

Development of Gravitational Theory

Some 2000 years before that time, Aristotle proposed that heavier objects would fall faster. He believed that all matter was composed of four elements. From lightest to heaviest, these elements were fire, air, water, and earth. See Figure 4-23. Each element was thought to have its own natural place in the world. For example, if fire were placed below its natural position, it would tend to rise above air. Similarly, if a stone were dropped, it would pass through fire, air, and water until it came to rest on the ground, its natural place. Aristotle thought that the movement of an object depended on its mixture of these four elements. Therefore, a heavier stone was expected to drop faster than a lighter stone because it contained more of the earth element. Aristotle's physics seemed to fit everyday observations. For example, we know that if we drop a feather and a stone at the same time, the stone reaches the ground first. What did Aristotle not realize?

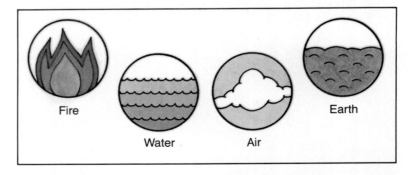

Figure 4-23 The four basic elements of the ancient Greeks: fire, water, air, and earth. Some Greeks thought that all matter was made of combinations of these elements.

By 1589 a young scientist named Galileo had confronted this view of the world. At the age of 25, he had just been appointed professor of mathematics at the university in Pisa, Italy where he began to challenge the opinions of the older professors. As it turned out, Galileo's ideas signaled the beginning of the end for Aristotle's worldview.

According to Aristotle, a 100-kg mass dropped from a height of 150 m would reach the ground by the time a 10-kg mass would fall only 15 m from the same height. Galileo, however, predicted that they would both reach the ground at the same time. We now know that Galileo was right. See Figure 4-24. Although he may never have actually dropped stones from the leaning tower of Pisa as shown in the illustration, he did do experiments that showed that the acceleration of two different masses is equal. See Figure 4-25.

Newton and Gravity

You know from studying Newton's laws of motion that for an object to accelerate, an unbalanced force must act on it. Gravity is the force that causes all things near the earth's surface to fall toward the center of the earth. Newton believed that gravitational force was an attraction between any mass, such as the earth, and the matter near it.

Through his observations and calculations, Newton showed that not only the earth, but all objects in the universe exert pulling forces on each other. For example, the earth and the moon both exert a gravitational pull on each other. Ocean tides are the most obvious effect of the moon's gravity on the earth. The earth's gravity also keeps the moon in orbit about the earth. It is the sun's gravity that keeps all of the planets orbiting in the solar system. Though it would be too small to measure, all the objects around you exert a slight gravitational pull on everything else around them.

Another aspect of gravitational force is that it decreases as the distance between two objects increases. For example, as a spacecraft moves farther from the earth, the earth's pull on it becomes less.

The relationship between gravitational force, mass, and distance is stated in **Newton's Law of Gravitation**: *The gravitational force between two objects depends on their masses and the distance between them.* See Figure 4-26. To be more

Figure 4-24 Galileo's famous experiment.

Figure 4-25 These two objects have different masses, yet they fall at the same acceleration.

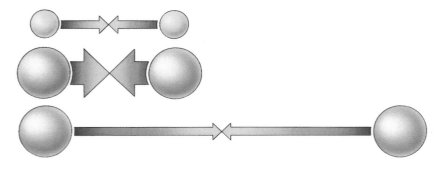

Figure 4-26 Gravitational force, represented by the arrows, varies with the mass of the objects and their distance from one another.

Figure 4-27 Astronauts on the moon can jump six times higher than they can on earth because their weight is only one-sixth what it is on earth.

Weight A measurement of the gravitational force acting on an object.

specific, a large mass exerts a stronger pull than does a smaller mass at the same distance. So, why do two objects with different masses accelerate and hit the ground at the same time? The object with the greater mass requires more force to accelerate, and because of its greater mass, the force of gravity acting on it is greater. Therefore, everything on earth would fall at the same rate if the effects of air resistance could be removed.

Like many scientific laws, the law of gravitation predicts how some parts of the natural world work. It does not explain what *causes* gravity, but it is a scientific tool that helps *explain* the effects of gravitational forces. For example, Newton's law predicts that the gravitational force on the surface of the moon is about one-sixth of that on the earth. This is because the moon's mass and size are both much smaller than that of the earth. How do you think this would affect you if you lived on the moon? See Figure 4-27.

Mass and Weight

You know that *mass* refers to the amount of matter in an object. **Weight**, on the other hand, is determined by the force of gravity pulling on an object. Since gravity is a force, it is measured in newtons. On earth, a person with a mass of 50 kg weighs 490 N. On the surface of the moon, the same person would weigh 81.7 N. This does not mean, however, that the person's mass has changed. The *weight* of an object changes as the size of the gravitational force on it changes. But the *mass*, or amount of matter in the object, does not change. As long as you stay in one place, an object with more mass will have more weight. Nevertheless, at any one place on earth, you can determine the mass of an object by measuring the gravitational force pulling on it. For this reason, mass and weight are often used as if they mean the same thing. However, scientists are always careful to distinguish between the terms *mass* and *weight*.

Summary

Isaac Newton formulated three laws that describe motion and its related forces. The first law deals with the cause of motion. The second law accounts for changes in motion. The third law describes pairs of forces that act on different objects. Momentum, the product of mass and velocity, is conserved in collisions. Gravity is described as the force that causes all objects to accelerate toward the center of a mass, such as the earth. Weight is the measure of this force.

Reading *Plus*

Now that you have learned something about the structure, design, and construction of buildings and bridges from the standpoint of western civilization, let's take a closer look at some different kinds of architecture from around the world, past and present. Read pages S78 to S88, and design a collage showing the interaction between culture and design. Your collage should include answers to at least one of the following questions.

1. How might geography and climate affect the architectural styles of a region?
2. In what ways do religion and culture affect architecture?
3. Why might different cultures solve design problems in different ways?

IN THIS UNIT

5.1 Architecture in Other Cultures, p. 78

5.1 ARCHITECTURE IN OTHER CULTURES

The development of design, as seen in the structures of Ancient Greece and Rome, Medieval and Renaissance Europe, and those of the Modern era, represents a history of Western architecture. In other areas of the world, however, different architectural styles were developed by various cultures. While the scientific principles that govern the strength and durability of structures are the same for every culture, there is a tremendous variety of designs that have been used by different peoples through the ages.

The Byzantine Empire

In A.D. 330, Constantine the Great moved the capital of the Roman Empire from Rome to the city of Byzantium and renamed it Constantinople. This new capital (now Istanbul, Turkey) became the center of the great Byzantine Empire. Although the western Roman Empire finally collapsed in A.D. 476, the empire in the East lasted until the Turkish conquest of 1453. At the height of its power, parts of Turkey, Italy, the Middle East, and Northern Africa formed the Byzantine Empire.

Architecture was the greatest form of Byzantine art. One of the architectural masterpieces of the world is the church of Hagia Sophia, built during the 530s. It is a huge building in the form of a cross, measuring 72 meters wide by 76 meters long. See Figure 5-1. Though the outside was composed of plain brick and mortar, intricate murals, mosaics, stone carvings, and metalwork once covered every interior surface. Insets of ivory, silver, and jewels were used to adorn

Figure 5-1 The Church of Hagia Sophia in Istanbul, Turkey. It was converted into a mosque by the addition of four minarets.

the pulpit as well. A huge dome, 56 meters high and 33 meters across, dominates the cathedral. See Figure 5-2. Resting on massive columns instead of walls, the dome illustrates the talent of Byzantine architects who first solved the difficult problem of placing a round dome over a rectangular building.

The Influence of Islam

Religion has influenced the development of architecture in the Islamic world, much like it has influenced western civilization. During the middle ages, the Islamic world stretched from Spain in the west to Samarkand and India in the east. The most important buildings to be designed were meeting places for prayer. These buildings are called *mosques*. Domes and *minarets* helped give the mosques a distinctive appearance. The minaret, a tall and slender tower, is used as a place from which to call people to prayer. See Figure 5-3.

Islam forbids the use of statues and paintings of humans or animals as decorations in the mosques. Instead, mosques are decorated with designs based on flowers, plants, leaves, geometric figures, and Arabic inscriptions. The result was an unparalleled richness of abstract decoration. The influence of Islamic architecture can be seen in the structures of many other cultures. See Figure 5-4.

Figure 5-2 Interior of Hagia Sophia. The Turkish government now maintains the church as a museum.

Figure 5-3 Recently completed, this mosque exhibits the typical dome and minarets that are characteristic of Islamic meeting places the world over.

Figure 5-4 Nestled among modern skyscrapers, this government building in Kuala Lumpur demonstrates the influence of Islam on the architecture of Malaysia.

India: A Contrast of Styles

The subcontinent of India was the place of origin for two of the world's major religions—Buddhism and Hinduism. Much of Indian architecture and art reflects this religious influence. For example, *stupas*, or dome-shaped shrines, were built to hold artifacts and objects associated with Buddhism. The Indian stupa is a large mound-shaped structure covered by brick or stone and surrounded by a fence with elaborate stone gates. See Figure 5-5.

Figure 5-5 A stupa in central India. What do you think the carvings represent?

Figure 5-6 Exterior of a South Indian Hindu temple showing figures of various Hindu deities.

A Hindu temple, on the other hand, is typically composed of a square building with heavy walls. The outer walls of the temple are covered with a multitude of colorful statues and carvings, and the ornamentation is more elaborate than Buddhist ornament. See Figure 5-6.

In 1506, the Mogul Empire, an Islamic empire, rose to power after years of war and conflict in India. The most famous creation of the Mogul period was the work of the emperor Shah Jahan. In the middle 1600s, he built the Taj Mahal, a magnificent tomb for himself and his wife. See Figure 5-7. It is built out of white marble and is studded with gems. Although it is an Islamic building, the Taj Mahal, as well as other Islamic structures in India, incorporated many architectural elements that were borrowed from Hindu temples. These elements include pinnacles, pavilions, and elaborate ornamentation.

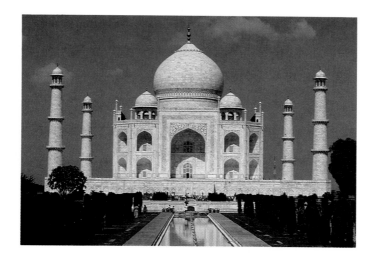

Figure 5-7 The Taj Mahal's exterior design typifies the blending of Muslim and Hindu cultures.

Traditional Chinese Design

Traditional Chinese architects seldom distinguished between religious and nonreligious structures. Most buildings—whether temples, tombs, public buildings, or private homes—followed the same basic plan and were characterized by *symmetrical* design. Huge pillars supported curved tile roofs. Therefore, walls were used primarily for privacy, not support. The roof was a major feature of Chinese architecture and generally was made of glazed tiles. Brightly-colored porcelain is also a characteristic of traditional Chinese architecture.

When Buddhism entered China from India, it brought the Indian stupa with it. The Chinese modified the stupa to develop the structure known as the *pagoda*. See Figure 5-8. Each pagoda has the same basic design. They usually have eight sides and many levels, but each higher level is smaller than the one below it. Each level of the pagoda also has its own roof with long upward-curving corners. Pagodas in China are usually constructed of wood, masonry, and highly glazed tile or porcelain.

Figure 5-8 How does this pagoda differ from Western architecture?

Chinese gateways, known as Pai-lous, were built as memorials to great men. They usually have one or three openings and were constructed of the same materials as were pagodas. See Figure 5-9. Pai-lous (or gateways) and pagodas are most typical of traditional Chinese architecture.

The last ruler of the Ch'in Dynasty (221–206 B.C.) built many public buildings and extensive fortifications, including part of the Great Wall of China. For several hundred years, roughly one third of all able-bodied Chinese men were forced to help build the wall. It is a massive structure that is 7.6 meters high and 4.5 meters wide, with towers of over 12 meters in height. It stretches east and west for 2240

Figure 5-9 A typical Chinese pai-lou.

Figure 5-10 The Great Wall of China.

kilometers across northern China and is one of the few structures made by humans that can be seen from space. See Figure 5-10.

Japan: A Natural Approach

In Japan, the architecture is strikingly different from that of Western cultures. Much of Japanese architecture has its roots in Chinese designs. Temples and shrines were constructed in a style similar to Chinese structures, with curving roofs and supporting columns. See Figure 5-11. Traditional Japanese houses are made of wood and have roofs of thatch. However, the use of tile roofs is also common. The gently-curved corners of Japanese roofs add a sense of graceful proportion

Figure 5-11 The Heian Shrine was built in 1895 to celebrate the city of Kyoto's eleven-hundredth anniversary.

that is often imitated in modern Japanese buildings. See Figure 5-12.

The entrances to shrines are often marked by a structure called a *torii*. It is a ceremonial gate that stands on columns with a curved beam across its top. Though similar to the Chinese pai-lou, the torii is typically simpler in its design. The torii in Figure 5-13 marks the entrance to the Meiji Shrine in Tokyo. This shrine is dedicated to the emperor who, in 1868, opened Japan to the West and its new ideas.

Figure 5-12 Yoyogi stadium in Tokyo retains the curving roof-lines of traditional Japanese architecture.

Figure 5-13 The torii— gateway to Meiji shrine in Tokyo, Japan.

Besides their religious shrines, Japanese designers also created beautiful gardens designed to imitate nature. These gardens are often miniature representations of the world in which rocks stand for mountains, ponds represent oceans, and sand and gravel represent lakes and rivers. See Figure 5-14.

Temples, shrines, and houses are designed to blend in with their natural surroundings and thereby emphasize harmony between the buildings and nature. See Figure 5-15. The Japanese broke from traditional Chinese architecture by using more asymmetrical designs and a less formal approach that more closely represents the unusual formations found in nature.

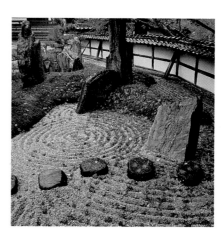

Figure 5-14 A Japanese garden.

Figure 5-15 Traditional Japanese homes complement their natural surroundings.

The Americas Before Columbus

The ancient civilizations of Central and South America developed some unique architectural styles. Although there were several distinctive cultures in these areas, three stand out as important examples of *pre-Columbian* architecture.

Aztecs The Aztecs, who dominated the area of central Mexico at the time of European exploration, built great cities, pyramid temples, marketplaces, and palaces. Altars and plazas accompanied the pyramids as the center of Aztec religious ceremony. See Figure 5-16.

Figure 5-16 A model of the Aztec temple area at Tenochtitlán.

When the Spanish conquistador Hernando Cortés discovered Tenochtitlán in 1519, it was one of the largest cities in the world. Tenochtitlán was a regularly planned city with straight streets. Built on an island in the middle of a lake, it was tied to the mainland by stone-lined causeways and bridges. Its buildings were constructed of rough stone that were plastered with polished and painted *stucco*. The original location of this grand city is present-day Mexico City.

Near Mexico City, in Teotihuacán, an earlier culture had built one of the greatest temple complexes of central Mexico. The people of Teotihuacán built the Pyramid of the Sun around A.D. 100. The base of this pyramid was larger than the Great Pyramid at Giza, Egypt. See Figure 5-17. The

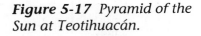

Figure 5-17 Pyramid of the Sun at Teotihuacán.

Figure 5-18 The temple of Quetzalcuatl at Teotihuacán.

Pyramid of the Sun was built with a core of clay, rock, and mud, which was faced with stone slabs that were plastered and brightly colored. Other buildings in this complex were covered with ornate carvings. See Figure 5-18.

Mayans In southern Mexico and northern Guatemala, the Mayan culture developed intricate cities that included temples and other large public buildings. At the peak of development, about A.D. 500, Mayans built great cities centered on tall pyramids of limestone that supported small temples at the top. See Figure 5-19. Besides their architectural achievements, the Mayans also devised a calendar, a form of mathematics, and an intricate writing system.

Incas In South America, the Inca Empire covered the vast area now occupied by the present-day nations of Peru, Ecuador, Bolivia, and Chile. The Incas built fortresses, irrigation systems, and paved roads. They used huge, cut stones to build the fortresses and palaces, many of which were located high in the mountains. These large stones were cut so accurately that they fit together without the need for concrete or mortar. Homes were made of stone, as well, and had thatched roofs.

One of the earliest examples of architecture in the area is the Gateway of the Sun at Tihuanaco. This archway is made of two immense upright slabs of stone that hold up a 3-meter carved stone *lintel*. The lintel by itself weighs approximately 10 tons. See Figure 5-20. Perhaps the best known example of Inca architecture is the mountain-top city of Machu Picchu. See Figure 5-21.

Figure 5-19 The great stone pyramid built by the Mayans at Tikal, Guatemala.

Figure 5-20 Gateway of the Sun, at Tiahuanaco, Bolivia.

Figure 5-21 Machu Picchu, Peru. The ruins of an Inca civilization.

North American Tribes The influence of Aztec and other Mexican cultures also spread to the North American continent. Tribes living along the Ohio and Mississippi rivers built large cities and huge ceremonial mound structures. See Figure 5-22. The largest of these mounds was 30 meters high and covered an area of about 16 acres (an area larger than 12 football fields). These mounds were used for burial as well as to support temples. Some mounds were built in the shape of birds, bears, mountain lions, and other animals.

Figure 5-22 Grass-covered mounds (left) are all that remain of an advanced North American culture whose landscape is illustrated in the drawing (right).

In other parts of North America, various cultures with different kinds of architecture flourished. Many of these differences related to the climate and the availability of materials encountered in these areas. In the hot, dry southwest, Pueblo tribes used *adobe*, a sun-dried brick, to build

Figure 5-23 How are these Pueblo buildings suited to their environment?

flat-roofed communal houses. Many of these houses were several stories high, and some were located under cliffs for protection. See Figure 5-23.

Since many North American tribes were nomadic, their architecture was developed around a need for mobility. Plains tribes, such as the Sioux, used buffalo skin to build cone-shaped *tepees*. These portable tents allowed the tribes to migrate with the buffalo herds. Tribes in the eastern end of the continent used forest materials, such as leaves, reed mats, and bark, to cover a frame of poles. The *wigwam* of the Chippewa tribe was such a structure. See Figure 5-24. The Iroquois used similar methods to build rectangular *long houses*, some of which were 30 meters in length.

Figure 5-24 Black ash bark and birch bark were used to cover this Chippewa wigwam.

The Cities of Africa

The architecture of African civilizations is reflected in the development of urban centers. Dominated by walls and passageways, early cities—like Kilwa in Kenya and Kano in Nigeria—had marketplaces, palaces, wide streets, and other structures. See Figure 5-25. The Hausa people, who lived where modern-day Nigeria is now, built cities surrounded by high walls, a common feature in the architecture of Africa. The early inhabitants probably began building walled towns when they first developed iron technology. As a town grew, its residents built a wall around it. The first outer wall

Figure 5-25 The remains of the city of Kilwa, Africa.

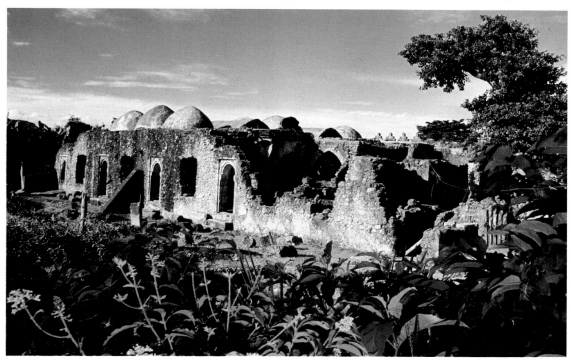

likely served as a defense against attackers. The second wall often enclosed an area for food storage.

One of the greatest fortress walls was built in Zimbabwe in the 1300s. It has walls 10 meters high made of cut stone laid in a variety of patterns. See Figure 5-26. It is over 240 meters long and consists of some 900,000 large granite blocks.

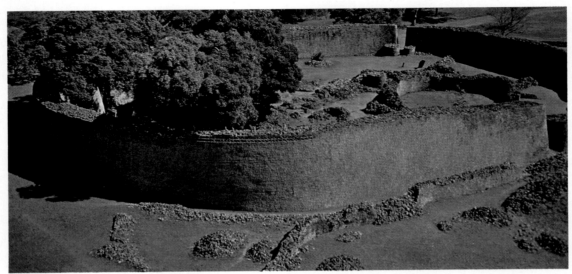

Figure 5-26 *Ancient walled cities are found in what is now Zimbabwe.*

The development of cities initiated the smaller-scale development of homesteads, such as those of the Kuria along the Kenya–Tanzania border. Here, circular houses with cone-shaped thatched roofs were clustered together and connected with a tall fence that formed a secure yard. The number of houses in the homestead depended on the size of the family.

The use of urban space was as important to Africans as their use of walls. They tried to preserve something of a rural setting, even in the largest cities. Animals were allowed to wander about, and trees shaded streets and plazas. In some towns, people even raised crops along walkways or on small plots of land.

Summary

Many kinds of architecture have developed in different countries around the world. The buildings and cities from Africa, the Orient, India, and the native cultures of North and South America represent the skill and creativity of their designers. The architecture of these civilizations differs from traditional Western architecture.

Unit 6

In This Unit

- **6.1 Rocks and the Rock Cycle, p. S90**
- **6.2 Rock-Forming Minerals, p. S99**
- **6.3 Stories in Rocks, p. S105**

Reading *Plus*

Now that you have been introduced to the movements of the earth's crust and the rocks of which it is made, let's take a closer look at the make up of rock, and the kinds of information it can give us. Read pages S90 to S110 and construct a model representing the rock cycle using as many examples and rock samples as possible. Your model should include concepts represented by the following questions.

1. In what ways do the forces at work on the earth drive the pathways of the rock cycle?
2. How are minerals related to rocks? What characteristics can be used to identify different minerals?
3. How do rocks record change over time? In what ways can we determine the history of the earth using rocks?

6.1 ROCKS AND THE ROCK CYCLE

Have you ever picked up a rock because of its interesting shape, sparkle, or color? Perhaps you own a rock collection. If so, you already know something about the variety of rocks that can be found on the earth. You have learned by now that rocks are the materials that compose the earth's solid crust. Sometimes these rocks form spectacular natural structures, such as the one pictured in Figure 6-1. When small and polished, they can be turned into gems used in making jewelry. Besides being interesting or spectacular, rocks are important to us both scientifically and economically. They not only contain the history of the earth, but are also the source of our planet's riches.

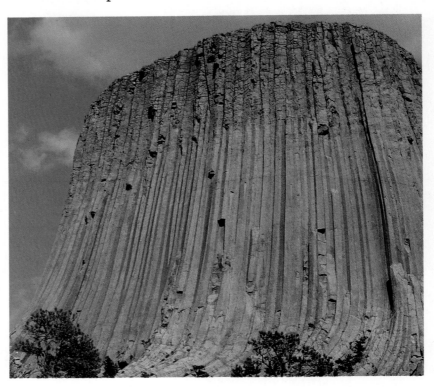

Figure 6-1 Devil's Tower in eastern Wyoming is made of igneous rock.

The Formation of Rocks

Studies of the earth indicate that the earth was probably once made mostly of molten materials. As the earth cooled, the materials at the earth's surface solidified into the first rocks. These rocks were then exposed to forces that caused them to change. Other rocks developed as a result of these changes. By studying the rocks and the forces that change them, geologists have identified three basic processes by which rocks form.

You can simulate rock-forming processes by using candle wax. Suppose you take some old candles and melt them in a pan. If you then pour the melted wax onto paper or into a mold and allow it to cool, the liquid wax will gradually harden into a solid mass. In a similar manner, rocks solidify from molten rock, or *magma*. Rocks that form when the magma cools are called *igneous rocks*. See Figure 6-2. Magma is composed of a mixture of substances called **minerals**. As the magma cools, molecules of the same mineral substances link together to form solid structures called **crystals**. Igneous rocks are solid masses of these mineral crystals.

Now, shave off some thin pieces of the solid wax. If you place a pile of these shavings under a heavy object, such as a book, the shavings will eventually stick together into a single piece. In the same way, most *sedimentary rocks* form from rock pieces that were broken up by weathering processes. See Figure 6-3. Deposits of these pieces then stick together, or *lithify*, by compaction or cementation.

Mineral A naturally occurring substance consisting of a single element or compound.

Crystal A three-dimensional structure in which each face, or surface, has a definite shape and orientation.

Figure 6-2 The dark rocks of the Pacific islands were formed of solidified magma.

Figure 6-3 The mud of this dry lake bed may eventually become sedimentary rock. The mud cracks here may also be preserved in the rock that forms.

Finally, if you warm the remaining solid wax slightly, you should be able to bend it and squeeze it by applying pressure. In a similar fashion, the heat and pressure inside the earth changes both igneous and sedimentary rocks into *metamorphic rocks*. Sometimes the minerals in the rocks are rearranged into a different pattern. In other cases, entirely new minerals are formed.

The Rock Cycle

The formation of igneous, sedimentary, and metamorphic rocks are the individual steps of an endless process that has been going on for billions of years. Old rocks are constantly being made into new rocks by a continuous process called the **rock cycle**. A simplified version of the rock cycle is shown in Figure 6-4.

Rock Cycle The continuous process of change in which new rocks are formed from old rock material.

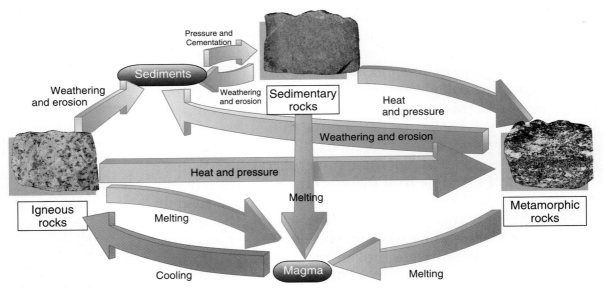

Figure 6-4 *The rock cycle is not a one-way process. Any rock may be changed into any other type of rock by going through the cycle.*

Figure 6-5 *Gneiss is a metamorphic rock formed when granite is put under high temperature and pressure.*

As you examine the diagram of the rock cycle, consider that any rock you see has been through at least one pathway in this cycle. When magma cools, for example, it may form the igneous rock, granite. Granite can then be broken by physical and chemical weathering into particles such as sand. Deposits of this sand can be cemented into sandstone, a sedimentary rock. If the sandstone is exposed to heat and pressure, it can be changed into the metamorphic rock called quartzite. Finally, quartzite may later be forced deep into the earth where it can be remelted to become magma once more. This magma, in turn, could then become more igneous rock.

But, what other pathways are there? Suppose granite is not broken down into sand, but exposed to high temperature and extreme pressure. In this case, the granite could become a metamorphic rock called gneiss. See Figure 6-5. The granite itself may also be remelted to become magma.

Scientists believe that much of the rock cycle is taking place at the edges of large pieces of the crust called *tectonic plates*. For example, magma comes to the surface both along the mid-oceanic ridges and near ocean trenches. See Figure 6-6. New igneous rocks form at ridges from dark heavy magma. In the trenches, old rock is being forced down into the mantle. There, the lighter rocks of the crust are melted and mixed with other rock material. This material may once again make its way back to the surface at another underwater ridge.

The rocks of the earth's crust have gone through the rock cycle many times. The rocks near the earth's surface are rarely older than 3 billion years. Since the earth is estimated to be 4.5 to 5 billion years old, most of the original rocks of

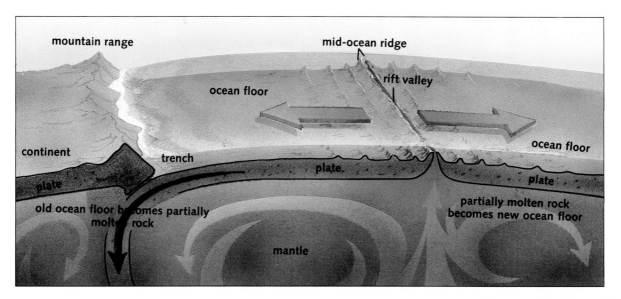

the earth's crust have been changed by the rock cycle since they were first formed.

Figure 6-6 The mid-ocean ridge circles parts of the earth much like a seam circles a baseball. New material is added to the earth's surface at the mid-ocean ridge. In other parts of the world, the plates are forced back into the mantle.

Classifying Rocks

What about you makes you a *person*? Is it your hair, your face, or perhaps the fact that you have two arms and legs? Is it your personality or IQ? Whatever it is, you have definite characteristics that make you recognizable as a person. Still, you are different from every other person on earth. You are unique!

Rocks also have definite characteristics. For example, look at the photographs in Figure 6-7. You can tell that the objects in these photographs are rocks. But, what is it about these objects that makes them rocks? Just as you share certain characteristics with all other humans, rocks share certain characteristics as well. Now, look closer. Are the two rocks the same? Again, just as you are different from other people, rocks have many differences too.

Like living organisms, rocks can be classified by their similarities and differences. One way rocks are classified is by the way they are formed. On this basis, rocks can be classified as igneous, sedimentary, and metamorphic. In addition,

Figure 6-7 Although these rocks look different, they belong to the same rock family. As with brothers and sisters of human families, members of the same rock family share many characteristics.

S93

within each of these groups, rocks can be further classified and named on the basis of their origin, their *texture* (which refers to crystal or particle size), and their mineral content.

Igneous Rocks As you know, there are two basic types of igneous rocks—*intrusive* igneous rocks and *extrusive* igneous rocks. These names indicate *where* the rocks formed. Intrusive igneous rocks form deep underground in bodies of magma that squeeze in between, or *intrude,* into layers of other rocks. Long periods of time are required for intrusive rocks to solidify, because the heat of the magma dissipates slowly underground. Extrusive igneous rock, on the other hand, come from *lava.* Lava is molten rock that exits, or *extrudes,* from inside the earth through openings such as volcanoes. Above ground, lava cools quickly and soon hardens into extrusive rock. See Figure 6-8.

Figure 6-8 Molten magma from within the earth often comes to the surface as lava. As the lava cools, igneous rock is formed.

Although intrusive and extrusive rocks are made of the same types of minerals, they are easily distinguished by their texture. These two types of rock have very different appearances due to their differing textures. For example, look at the two rocks in Figure 6-9. If you look closely, you can see

Figure 6-9 Granite (left) displays coarse texture, while rhyolite (right) displays fine texture.

that granite, an intrusive igneous rock, is composed of large, visible crystals. The slow cooling of intrusive rocks allows large crystals to grow. On the other hand, if an igneous rock cools quickly, it feels fairly smooth and may not have any visible crystals. Geologists say that such rocks, like the rhyolite shown in Figure 6-9, have *fine texture*.

Mineral content affects the color and density of igneous rocks. Basalt, for example, is dark and heavy because it contains large amounts of the elements iron and magnesium. Gabbro, which is also dark gray, is made of the same minerals as basalt. See Figure 6-10. Basalt and gabbro are both

Figure 6-10 Notice the coarse-grained texture of this sample of gabbro, an intrusive igneous rock.

made from magma that comes from deep inside the earth. Basalt and other dark extrusive rocks form the crust on the ocean bottom. The lighter-colored rocks, such as granite and rhyolite, contain elements, such as calcium and sodium, that also make them less dense. These minerals make up most of the rocks of the continental crust.

Sedimentary Rocks Most sedimentary rocks are made from sediments, which are pieces of other rocks. For this reason, sedimentary rocks are classified on the basis of the type of sediments they contain. There are three different kinds of sediments that can become sedimentary rocks—clastic, chemical, and organic sediments. Therefore, there are three major types of sedimentary rock.

Clastic sedimentary rocks are made of broken pieces of weathered rocks. These pieces can be large, rounded pebbles, sharp-edged chunks, or soil particles such as sand, silt, or clay. Clastic rocks are found in layers because of the way they are formed. When agents of erosion, such as water and wind, carry weathered rocks away from their source, the particles become *sorted* according to their size. For example, a swift-flowing mountain stream may carry large pebbles and even boulders as well as sand, silt, and clay. As the water slows down, it drops the largest particles first, then gradually smaller pieces follow. By the time the stream becomes a slow-moving river, only the smallest particles are left. This sorting produces layers of particles that are about the same size. Sorting also happens in the ocean when sediments are deposited there.

There are several types of clastic sedimentary rocks based on their particle size. Conglomerates and breccias, for example, are made of large pebbles. Although their particles may be the same size, conglomerates consist of rounded pebbles, while in breccias, the pieces are angular and have sharp edges. See Figure 6-11. Sandstones and siltstones are made of sand-sized and silt-sized soil particles respectively. Shale is composed of clay—the finest of rock particles. See Figure 6-12. In forming sedimentary rocks, larger particles, such as pebbles, sand, and silt, must be "glued" together by

Figure 6-11 Conglomerate (top), *breccia* (middle), *and sandstone* (bottom) *are all clastic sedimentary rocks.*

Figure 6-12 Shale (left) *and siltstone* (right) *are composed of nearly microscopic particles.*

a cementing agent. Shale, however, can become a rock when the excess water is squeezed out of it by compaction.

Chemical sedimentary rocks form when minerals that were dissolved by weathering processes separate from the water and crystallize. In some cases, the minerals separate by the process of evaporation. As water evaporates, the dissolved minerals form crystals and eventually a rock. Rock salt and gypsum are two examples of chemical sedimentary rocks that form by evaporation. See Figure 6-13. When conditions are right, dissolved minerals simply fall out of solution by a process called *precipitation*. Limestone is a common rock that forms in this way. Another familiar rock that precipitates from sea water is chert, which is a dense rock made of quartz. See Figure 6-14. Precipitates such as calcium carbonate and quartz are two of the materials that cement the particles of clastic rocks together. These minerals fill in the spaces around rock particles much like mortar fills the spaces between brick or stone in a wall.

Figure 6-13 *Rock salt* (top) *and rock gypsum* (bottom) *are chemical sedimentary rocks.*

Figure 6-14 *Compact limestone* (left) *and chert* (right) *are both precipitated from sea water.*

Figure 6-15 *This limestone cliff was formed from the shells of millions of dead sea animals, and is therefore a type of organic sedimentary rock.*

Organic sedimentary rocks are a type of sedimentary rock that comes from the action or remains of once-living organisms. For example, limestone (which can also be classified as a chemical sedimentary rock) forms primarily by the action of marine organisms that remove calcium carbonate from the water to build their shells. Layers of limestone can build up in oceans when these organisms die and their shells settle to the bottom. See Figure 6-15. Chalk is a type of limestone that is formed in this way. Coal is also considered to be a type of organic sedimentary rock. As you know, coal is thought to be formed from masses of dead plants that were compressed for millions of years. As this organic material was gradually compressed, it became denser and harder. Two types of coal are considered to be organic sedimentary rocks. They are lignite, a very soft, dark brown coal, and the slightly harder, black bituminous coal. See Figure 6-16.

Figure 6-16 Stages in coal formation. In the last stage, coal becomes a metamorphic rock.

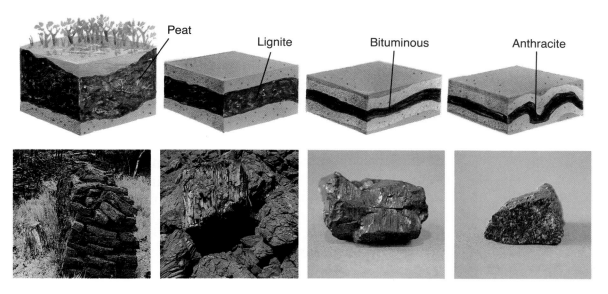

Metamorphic Rocks Just as living things go through physical changes during their lives, many rocks change physically as well. Metamorphic rocks form from other rocks when they are exposed to heat and pressure. Some of these changes are structural changes that result in the rearrangement of the crystals in a rock. In other cases, chemical changes occur, recombining the original minerals of the rock into new ones. See Figure 6-17.

Figure 6-17 Shown here are three examples of metamorphic rocks. In each case, the sample on the right is the metamorphic rock, and the sample on the left is the parent rock from which it formed.

Figure 6-18 Marble is a nonfoliated metamorphic rock. The striations in this photograph were made when the marble was cut.

Geologists divide metamorphic rocks into two groups based on how they were formed. *Foliated rocks* have obvious "bands" or "zones" of like materials that indicate they were formed by pressures coming from one direction. These bands often appear to be shiny, due to the melting and reforming of mineral crystals. *Nonfoliated rocks* have no visible alignment, like the marble in Figure 6-18. In this case, limestone was exposed to pressures from several directions.

The amount of foliation, or banding, is used to classify metamorphic rocks. For example, slate, schist, and gneiss are types of foliated rocks. Slate, with the least amount of foliation and no visible crystals, shows the least amount of alteration by the processes of metamorphism. As you can see in Figure 6-19, schist and gneiss have larger crystals and more definite bands and therefore were exposed to greater amounts of temperature and pressure.

When bituminous coal is put under intense heat and pressure, it becomes a harder, more compact form of coal called anthracite. Anthracite is very hard and has few of the impurities found in bituminous coal.

Figure 6-19 The crystal structure of gneiss (right) is much more defined than that of schist (left).

Summary

As the earth cooled, some of its molten materials solidified into the first rocks. Rocks that solidify from magma are called igneous rocks. Sedimentary rocks form from pieces of other rocks that were broken up by weathering. Heat and pressure inside the earth changes both igneous and sedimentary rocks into metamorphic rocks. The rock cycle is a continuous process in which old rocks are constantly being made into new rocks. Rocks are classified on the basis of their origin, texture, and mineral content. Igneous rocks are classified as either intrusive or extrusive. There are three different kinds of sedimentary rocks—clastic, chemical, and organic. Metamorphic rocks are divided into foliated and nonfoliated rocks.

6-2 ROCK-FORMING MINERALS

As magma and lava cool, the chemical compounds that make up the molten rock link together to form crystals of certain minerals. Most minerals are made up of at least two kinds of atoms. That is, minerals are usually made up of two or more chemical elements.

Kinds of Minerals

The earth's crust contains over 2000 different minerals, only about 20 of which are very common. These 20 minerals are the basic rock-forming minerals of the earth's crust. Whether or not they are common, all minerals can be classified into two major categories based on their chemical composition.

Silicates Silicon and oxygen are two of the most common elements in the earth's crust. See Figure 6-20. Therefore, it is not surprising that many minerals contain these two elements. Minerals that contain silicon and oxygen are called *silicates*. Silicate minerals make up about 96 percent of the earth's crust. The silicates are found in igneous, sedimentary, and metamorphic rock.

Figure 6-20 The earth's crust is made up primarily of these elements. The graph shows the percentage by mass of each element.

Figure 6-21 Pure quartz is colorless.

One common silicate mineral that you might be familiar with is quartz. See Figure 6-21. A crystal of quartz is composed of single silicon atoms, each bonded to four oxygen atoms. These atoms form a network in the shape of a *tetrahedron*, which is a three-sided pyramid. When the

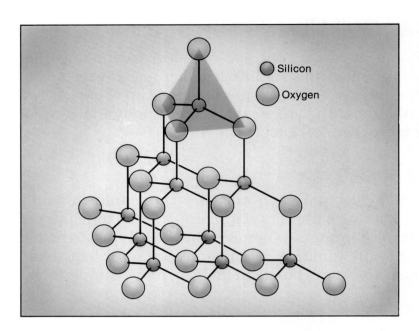

Figure 6-22 A silicon atom and oxygen atoms are joined in this pattern. How does the number of oxygen atoms and the number of silicon atoms compare?

silicon–oxygen tetrahedrons bond together to form chains or sheets, a very hard mineral is formed. See Figure 6-22.

Feldspars are the most abundant silicate minerals. In addition to silicon and oxygen they contain aluminum and calcium, potassium, or sodium atoms in their crystals. The chemical composition and the presence of impurities, such as iron or flecks of the mineral hematite, produces the variety of forms and color variations of the feldspar group. See Figure 6-23. Quartz and the feldspars together make up more than 50 percent of the earth's crust.

Figure 6-23 Three types of feldspar.

Nonsilicates Only 4 percent of the earth's crust is composed of *nonsilicate* minerals. There are several different families of these minerals based on their chemical composition. One of the most important families of nonsilicate minerals are the *carbonates*. For example, calcite is the mineral name for calcium carbonate, the substance that makes up limestone. See Figure 6-24. Dolomite is another example of a carbonate mineral. Carbonates are common in sedimentary rocks.

Other mineral families are characterized by different chemical groups. *Oxides* contain oxygen bonded to an element other than silicon, such as iron or aluminum. These

Figure 6-24 Calcite—a carbonate mineral.

minerals, for example magnetite (iron oxide), are important sources of metals. See Figure 6-25. *Sulfates*, such as gypsum, contain sulfur and oxygen. You may recall that gypsum is also the name for a chemical sedimentary rock that forms by evaporation. Sometimes the name of a rock and the mineral it is made of are the same. Other types of nonsilicate minerals are the *halides*, such as halite and fluorite, which contain a halogen, and the *sulfides*, such as galena and pyrite, which contain sulfur.

Native Elements Some nonsilicate minerals contain only one kind of element. These rare minerals are called *native elements*. Many of the minerals in this group are valuable metals. Some examples are gold, silver, and copper. See Figure 6-26. Nickel and iron may also appear as native elements, but they are very rare in the earth's crust. However, much of the interior of the earth is believed to consist of uncombined nickel and iron. Diamond and graphite are also native elements because they are made of pure carbon. They differ only in the way the carbon atoms are arranged.

Figure 6-25 Oxides, such as those shown above, are commercial sources of iron, aluminum, and other metals.

Figure 6-26 Native elements (clockwise from the top) *copper, sulfur, gold, silver, and lead.*

Identifying Properties of Minerals

Detailed analysis of minerals can be done only in a laboratory with special equipment. However, there are some simple physical properties that you can use to quickly identify a mineral sample.

Color Many minerals have bright colors. For example, cinnabar is red, azurite is a deep blue, serpentine is green, and sulfur is a bright yellow. Unfortunately, color may be one of the least dependable characteristics you can use to identify a mineral. The color of many minerals is due to the presence of small amounts of other elements called impurities.

For example, pure quartz is colorless, but with impurities it can be pink, tan, red, purple, or black. See Figure 6-27.

Luster Have you ever admired the shine of a newly waxed car? What you are admiring is the *luster* of the car's surface, or the way it reflects light. The luster of a mineral also refers to the appearance of its surface in reflected light. Each of the minerals has a characteristic luster. Some of the terms used to describe luster—metallic, glassy, waxy, pearly, greasy, or earthy—are related to familiar objects. How would you describe the luster of each of the minerals shown in Figure 6-28?

Figure 6-27 Quartz is found in many colors due to mineral impurities.

Figure 6-28 Metals, such as gold and silver, have a luster that can be easily recognized. Other types of mineral luster are named for common substances, such as pearls and earth (soil).

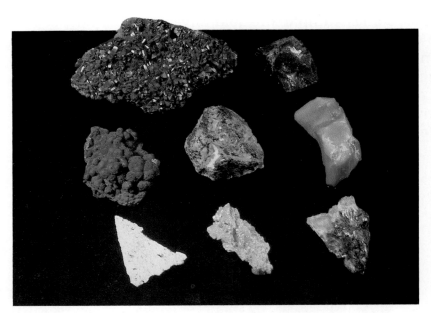

Streak If you rub a piece of chalk on a sidewalk, it leaves a white mark on the concrete. Some minerals also leave marks when rubbed on a rough surface. The color of the powder in these marks is called the mineral's *streak*. See Figure 6-29. Sometimes the color of the streak is different from the color of the mineral sample. For instance, the streak of golden-colored pyrite is greenish black, and the streak of silver-colored hematite is brick red.

Figure 6-29 Streak is a good characteristic for identifying minerals since it is not influenced by such things as tarnishing and impurities.

Hardness Another characteristic of minerals is their hardness, or resistance to being scratched. Mohs' Scale of Mineral Hardness is a set of ten standard minerals that is used to compare the hardness of all minerals. See Table 6-1. You probably do not have a sample of each of the minerals on Mohs' Scale, but you can still determine the hardness of an unknown mineral by using some common objects. For example, if a certain mineral can scratch a penny, its

MOHS' SCALE OF MINERAL HARDNESS		
Mineral	Hardness	Common Test
Talc	1	Easily scratched by fingernail
Gypsum	2	Can be scratched by fingernail
Calcite	3	Barely scratched by copper penny
Fluorite	4	Easily scratched with steel knife blade
Apatite	5	Can be scratched by steel knife blade
Feldspar	6	Easily scratches glass
Quartz	7	Scratches both glass and steel
Topaz	8	Scratches quartz
Corundum	9	No simple test
Diamond	10	No simple test

Table 6-1

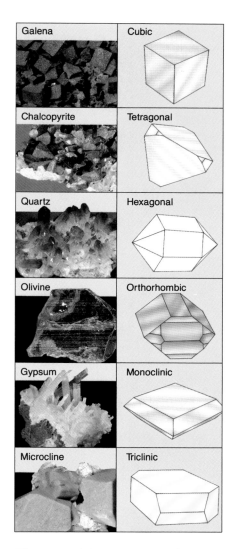

Figure 6-30 Representative minerals and their crystal shapes.

hardness is greater than 3. If a steel nail can scratch it, you know its hardness is less than 5.

Crystal Shape Each mineral has a characteristic crystal shape that can be useful in identifying the mineral. This shape is due to the orderly arrangement of the atoms that make up the mineral. See Figure 6-30.

Cleavage and Fracture These terms describe how minerals break. Minerals that always break along flat surfaces have a property called *cleavage*. The cleavage of a particular mineral always occurs in the same way and produces crystal faces that have characteristic angles. Halite, for example, breaks into cubes, while calcite pieces tilt diagonally, and mica peels off in thin sheets. See Figure 6-31.

Figure 6-31 The diagram on the left demonstrates cleavage in three directions, as would be found in galena. The diagram on the right demonstrates cleavage in one direction, which is characteristic of mica.

Some minerals do not have cleavage but instead break into irregular pieces. This pattern of breaking is called *fracture*. Minerals can show fracture in several different ways. Some fractures have a curved surface that resembles the inside of a clam shell. For example, quartz and obsidian break

Figure 6-32 Obsidian fractures when it breaks.

in this way. See Figure 6-32. Fracturing may also produce jagged and uneven surfaces.

Special Properties Minerals show a variety of other interesting properties, some of which are unique to a certain mineral. Magnetite, for example, is a type of iron oxide that is *magnetic*. Rocks that contain magnetite attract iron objects just as a magnet does. See Figure 6-33. Other minerals are *radioactive*, for example, pitchblende, which contains the element uranium. Still other minerals, such as calcite, give off light when exposed to ultraviolet light or X rays. This property is called *fluorescence*. See Figure 6-34. Some

Figure 6-33 The loadstone shown here is magnetic and will attract a variety of iron objects.

Figure 6-34 Notice the change in color of the various fluorescent minerals in this rock as they go from ordinary light (left) to ultraviolet light (right).

minerals will continue to glow after the ultraviolet light is cut off. These minerals display the property of *phosphorescence*. Calcite has two other interesting properties as well. When a weak solution of hydrochloric acid is placed on a sample of limestone (which is made of calcite) the acid begins to bubble. Furthermore, a very clear calcite crystal shows the property of *double refraction*, which occurs because of the way light is transmitted through the crystal. See Figure 6-35.

Summary

Minerals can be classified into two major categories based on their chemical composition. Minerals that contain

Figure 6-35 The mineral calcite exhibits double refraction.

silicon and oxygen are called silicates; those that do not are called nonsilicates. Some nonsilicate minerals are native elements, many of which are valuable metals. The physical properties of color, luster, streak, hardness, crystal shape, and cleavage or fracture can be used to identify a mineral. Other interesting mineral properties include magnetism, radioactivity, fluorescence, phosphorescence, and double refraction.

6-3 STORIES IN ROCKS

Imagine that you find a family diary from 100 years ago. As you lift it, the binding breaks and the pages fall out in bunches. Several bunches fall together, landing in a pile on the floor. Other bunches scatter away from the pile. How are you going to put the diary together again? How will you decide the order in which the pages belong? Is it possible that some pages may be missing? See Figure 6-36.

In a similar way, the earth's history is written in layers of rock. Trying to piece this history together is like putting the diary mentioned above back together in order. However, piecing together earth's geologic history is more difficult because many of the "pages" have been widely scattered and even destroyed. For example, individual rock layers have been separated by processes such as continental drift and mountain-building. Others are lost due to erosion or remelting. Still others are out of order because they were overturned by movements of the crust. Although the story is incomplete, certain clues in the rocks can be used to piece together a fairly complete picture of the earth's past.

Figure 6-36 What clues could you use to put this diary back together?

Uniformity

Have you ever noticed how uniform or consistent some things are? For instance, next time you open a loaf of bread, notice how uniform the texture of the loaf is. A slice of bread from a different loaf of the same brand of bread still shows the same uniformity of texture. The bread-making process insures this uniformity. When the same process is used, even at different times, the results are similar. See Figure 6-37.

Uniformity is an underlying principle of the basic geological processes. In the same way that a bread-making process makes uniform loaves of bread, the geological processes that form rocks and mountains today formed similar rocks and mountains in the past. This idea was first presented as the *principle of uniformitarianism* in 1785 by James Hutton, a Scottish physician and geologist. Simply put, this principle

Figure 6-37 These loaves of bread are uniform in color and texture because they were made by the same process.

states that similar processes, in different times, produce similar results. Observations support Hutton's theory. For example, sediments may be carried away by water or wind and then deposited in new places, but they are always deposited in similar ways regardless of where or when. Over long periods of time, these sediments are compressed and cemented to form sedimentary rocks.

Records of Environmental Change

Did you ever think to look for seashells on top of a mountain? Probably not. But nevertheless, fossil seashells are found in the rocks of mountains throughout the world. See Figure 6-38. For this to be true, either the rocks have moved or the environments in which they formed have changed.

Like a mystery story, certain rocks contain clues about what happened in the past. Geologists can "read" these clues to figure out the mystery of their history! With these clues, they can figure out what the environment was like when a rock was formed and how those conditions have changed. For example, what does finding fossil seashells in the rocks of a mountain tell you? A geologist would say that the rocks first formed in an ocean, and then were uplifted to become a mountain.

The kinds of fossils contained in rocks can tell us where the rocks were formed—in an ocean, by a stream, or on land. This is possible since the organisms that live in each place are different from one another. Sedimentary structures, such as ripple marks, mud cracks, and sand dunes, also indicate what the environment was like when and where the rocks were formed.

The shape of mineral grains may also help geologists determine what sort of environment existed when rocks were formed. For instance, the rounded quartz grains of sandstone probably formed on a beach. Geologists know this because the action of waves rounds the sand grains, and sand-sized sediments are often deposited near shore.

The size of the fragments in sedimentary rocks is another clue to the conditions under which the rocks formed. Large fragments are deposited first as a stream slows. In a similar way, the movement of waves also separates fragments by size. Large fragments are deposited close to shore to form sandstone, while smaller particles are carried out into deeper water. See Figure 6-39. Ocean currents may carry clay particles far from land. In quiet water, away from the beach, clay settles to the bottom to form shale. Eventually, all the particles settle to the bottom. In the deepest, quietest parts

Figure 6-38 Fossils, such as brachiopods, indicate that the origin of this rock was in an underwater environment.

Figure 6-39 The lighter sediments are carried much further out into the sea than the larger, heavier sediments.

of the ocean, limestone forms as dissolved calcite precipitates from the water.

Relative Age One clue to the order of the pages in the earth's diary is the *law of superposition*, which simply states: younger rock layers lie on top of older rock layers. In other words, the earth's diary reads from back to front. As the layers of sedimentary and extrusive igneous rocks formed, the oldest layers were deposited first. Successive layers formed one at a time, one on top of the other. Thus, when you look at the layers of rock in a cliff, the law of superposition tells you that the top layer was deposited last; and as you go down the cliff, the layers get older. See Figure 6-40.

Another geologic principle states that layers of sedimentary and extrusive igneous rock are laid down in flat, level beds. Once this happens, however, these level beds may become tilted or offset by the processes of folding and faulting.

Look at Figure 6-41. Can you list the rock layers from oldest to youngest? If you said that **H** is the oldest layer and **A** is the youngest layer, you are correct. However, this is not the only information that can be determined from this diagram. Notice that the lower layers lie at an angle and are broken and offset, while the top layers remain level and unbroken. The boundary between layers **E** and **D** indicates that pages are missing from the earth's diary in this area. Such a boundary is called an **unconformity**. There is also a body marked "magma" that cuts across the lower layers. Do you notice anything else? How can these patterns be interpreted?

Layer **H** was deposited first, followed by **G**, **F**, and then **E**. Next, these layers were broken by a fault and tilted by mountain-building, and then invaded by a magma intrusion. A period of erosion then wore down the mountains to a flat, level plain. Finally, the process of deposition began again, laying down layers **D**, and then **C**, **B**, and finally **A**.

Figure 6-40 Geologists can study the exposed rocks of the Grand Canyon to learn about the geologic history of the region.

Unconformity A boundary between rock layers that indicates that layers are missing due to erosion.

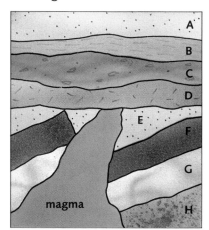

Figure 6-41 List the relative ages of these rocks from oldest to youngest.

S107

Relative Age The age of a rock (older or younger) in comparison to other rocks.

Figure 6-42 This sequence of events leads to the kind of unconformity in which horizontal layers overlie tilted layers. This process requires millions of years.

Using the law of superposition, geologists can determine the **relative age** of rocks by studying the order in which they were laid down. It is important to remember that the relative age of a rock cannot be given in years; it can only state that one rock is older or younger than another rock. You might think that this information is not very useful; however, it is very important to a geologist who is trying to put together the geologic history of an area. By determining the relative age of rock layers and studying the way they are tilted or broken, a sequence of geologic events can be developed. See Figure 6-42.

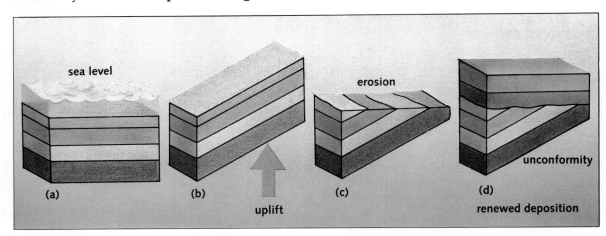

Half-life The time it takes for half of the atoms of a radioactive isotope to decay.

Figure 6-43 After 4.5 billion years, half of all the uranium atoms will have changed into lead.

Numerical Dating You may recall that some elements tend to be unstable and therefore undergo *radioactive decay*. By giving off atomic particles, these radioactive elements change into more stable elements. For example, uranium decays to eventually form lead—a very stable, nonradioactive element.

Some radioactive isotopes of elements are very unstable and decay rapidly; others decay more slowly. The time it takes for half of the atoms of a radioactive isotope to decay is called its **half-life**. Each successive half-life reduces the remaining number of atoms by one-half, turning them into atoms of a more stable element. See Figure 6-43. Some isotopes have a half-life shorter than one second; others have a half-life of billions of years.

Since radioactive isotopes decay at known rates, they act like an internal clock to mark the passage of time. Thus they can be used to find the approximate age of a material. *Numerical dating* (also know as absolute dating) is the process of estimating the age of a sample using the radioactive isotopes in the sample. Rocks, minerals, fossils, water, and ice are dated using this method. The isotopes most commonly used in numerical dating are listed in Table 6-2.

ISOTOPES COMMONLY USED IN NUMERICAL DATING		
Isotope	**Half-life (yrs)**	**Material to which applied**
Tritium	12	Ground water, sea water, ice
Carbon-14	5,730	Wood, bones, shells
Potassium-40	1,300,000,000	Rocks, minerals
Rubidium-87	48,000,000,000	Rocks, minerals
Thorium-232	14,000,000,000	Rocks, minerals
Uranium-235	704,000,000	Rocks, minerals
Uranium-238	4,500,000,000	Rocks, minerals

Table 6-2

One problem with dating rocks is that scientists do not know how much radioactive isotope was originally in a rock when it was formed. Therefore, the age of a rock cannot be found by determining the amount of isotope remaining. Scientists can, however, measure the amounts of both the parent isotope (such as uranium) and the decay product (lead). The ratio of decay product to parent isotope gives the approximate age of the sample. The older the mineral or rock, the more decay product there will be.

Numerical dating, however, cannot be used to date sedimentary or metamorphic rocks, which are made of pieces of other, much older rocks. In sedimentary rock, measuring radioactive elements and their decay products gives the age of these older rock pieces, not the age of the younger sedimentary rock itself. Furthermore, as a metamorphic rock forms, heat and pressure alters the rock's original minerals and "resets" the numerical clock, affecting the amounts of both the original isotopes and the decay products.

Sedimentary and metamorphic rocks must be dated by relative dating. A nearby igneous formation can be dated numerically, and then the age of the neighboring sedimentary or metamorphic formations can be estimated by their position relative to the igneous rock. See Figure 6-44.

Figure 6-44 This igneous rock intrusion may be helpful in determining the relative ages of the surrounding rocks.

Fossil Correlation Sometimes rocks are moved far away from where they were first formed—for example, when a continent breaks apart and then becomes separated by an ocean. When this happens, it becomes difficult to compare layers of rock between the two sections. However, by using the fossil remains of plants and animals, the age of rock layers can be determined.

On a geologic scale, plant and animal species exist for only a limited time. Because life forms change over time, fossils can be used as time indicators. Such fossils are called *index fossils*. Different index fossils appear in sedimentary rock of different ages. By numerically dating the fossils, or by dating similar fossils that occur elsewhere, the approximate age of sedimentary rock can be determined.

Fossil correlation is the use of index fossils to determine the relative age of widely separated rock layers. If similar fossils are found in rock layers separated by an ocean, the two rock layers would still be the same age. By comparing index fossils, and the rock layers in which they are found, geologists have been able to combine their observations to make a **geologic column**. See Figure 6-45. In this way, fossils and the rock layers they are found in can be used to sort out the pages of the earth's diary.

Geologic Column An arrangement showing rock layers in the order in which they were formed.

Figure 6-45 Layers having the same color and fossils (left) are the same geologic age no matter where they are found. A combined geologic column (right) can be constructed with each layer in order by its age.

Summary

Earth's history is written in layers of rock. Geologists piece this history together using the underlying principle of uniformity. Certain rocks contain clues that geologists can use to figure out what the environment was like when a rock was formed and how the conditions have changed. Using the law of superposition, geologists can determine the relative age of rocks. Radioactive isotopes can be used to find the approximate numerical age of geologic material. By using index fossils, the relative age of widely separated rock layers can be correlated.

Unit 7

In This Unit

- 7.1 Upward and Outward, p. S112
- 7.2 Stars: A Universe of Suns, p. S118
- 7.3 One Small Step, p. S126

Reading *Plus*

Now that you have been introduced to astronomy and human exploration of the universe, let's take a closer look at the members of our Solar System and beyond, and how humans further their investigation of space. Read pages S112 to S132 and write a short science-fiction story about one aspect of space or space exploration. Your story can be based on ideas taken from the following questions.

1. What are the differences and similarities among the various bodies of the Solar System?
2. How do astronomers get information from the stars?
3. In what way is our knowledge of the universe increasing?

7.1 UPWARD AND OUTWARD

The Earth, our home, is only one member among the multitudes of objects in space. Looking at the Earth from space, we see it as an entire planet. See Figure 7-1. From this vantage point, we have been able to get first-hand information about our own planet in space. But for most of history, people have looked up and out into the blackness of space and could only wonder at the marvels they saw.

Recently, however, we have been able to get a better look at some of our closest cosmic companions. The new knowledge we have gained, and are continuing to learn, is sometimes stranger than fiction. Let's look at some of this information our ancestors only dreamed about.

Figure 7-1 View of Earth from the first Apollo mission to the Moon in 1969.

The Moon, Our Nearest Neighbor

In some ways, the Moon resembles the Earth since both have a crust and mantle. The Moon may also have a hot core like the Earth, although it is probably much smaller. The astronauts who visited the Moon, however, were more impressed by the differences between the surfaces of the Moon and the Earth. First, there is no atmosphere on the Moon to shield it from the Sun's heat or help hold the heat during the night. During the day, the temperature on the Moon's surface rises to above 100°C; during the night it drops below −100°C. There is also no water, nor is there any evidence of living things on the Moon.

The surface features of the Moon, therefore, have not been changed by wind or water. Many of the features that can be seen on the Moon today have been there since its early history. The Moon is not free from all change, however. The lack of an atmosphere allows its surface to be hit constantly by small particles from space. This breaks up some of the rocks on the surface. Since the small pieces of rock are not washed or blown away, a fine dust covers almost all the Moon's surface. See Figure 7-2.

If it were not already the Earth's only natural satellite, the Moon could qualify as a small planet. In fact, some astronomers think that at one time the Moon actually was a small planet that was captured by the pull of Earth's gravity and pulled into orbit around the planet. Others say that the Earth and the Moon were formed at the same time from the same cloud of dust and gas. Still other astronomers claim that a third body hit the Earth and knocked material out into orbit around the planet. This material later came together to form the Moon.

Figure 7-2 Moon dust is stirred up by this astronaut on the Moon.

It is unlikely that the Moon was once part of the Earth, however, because there are too many differences in the chemical makeup of the two bodies. Furthermore, if the Moon was captured by the Earth, the principles of gravity should have given the Moon a different orbit than it has today. The theory that the Earth and Moon were both formed together is the most likely theory. However, the information so far gained from exploring the Moon has not settled the question about its origin.

The Sun: Source of Life

The Sun is the source of most of the energy used on the Earth. Not only does it heat our atmosphere, drive our climates, and warm our oceans, the Sun's energy is used by plants to make food—food that eventually ends up on our tables. Without the Sun, there would be no life on Earth as we know it. But what is this burning sphere in the sky? And why does it shine?

Composition of the Sun No one has ever been able to see into the interior of the Sun. No space probes from Earth have ever penetrated the Sun's surface. Yet by mathematical models, the inner structure of the Sun can be deduced.

The Sun is thought to be made up of several layers that blend into one another. Because all the layers are gases, there are no sharp dividing lines between them. Because the Sun is so hot, the gases in the Sun may well be a mixture of bare atomic nuclei and free electrons—a form of matter called *plasma*. On Earth, plasmas are found only where the temperatures are very high, as in the flame of a welding torch.

The Sun can be divided into six layers. See Figure 7-3. The changes that produce the Sun's huge supply of energy take place in its center, or *core*. The temperature in the core has never been measured directly, but scientists have

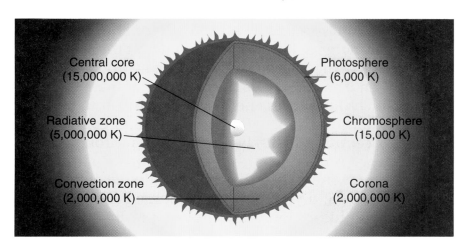

Figure 7-3 The three layers of the Sun's interior and its three atmospheric layers.

estimated the temperature to be at least 15,000,000 K. Energy coming from this central power plant is carried outward through the *radiative* and *convection* zones until it reaches the Sun's surface.

Energy from the Sun's interior reaches the surface where it causes a thin layer of gases to give off a brilliant light. This glowing layer, called the *photosphere*, produces the visible light that comes from the Sun. The photosphere, the lowest part of the Sun's atmosphere and the part of the Sun that we see, has a temperature of about 6000 K.

Just above the photosphere are two more layers that make up the Sun's atmosphere. The first layer, the *chromosphere*, gives off a faint red light that cannot be seen against the bright background of the photosphere. The chromosphere blends into a much less dense layer, called the *corona*, that surrounds the Sun like a halo. There is no outer boundary for the corona and some of the Sun's material escapes completely. Streams of particles from the Sun's corona make up the *solar wind*. This "wind" moves past the Earth at speeds of 300 to 700 km/s, much faster than a speeding bullet.

Fortunately for living organisms on Earth, the Earth's magnetic field pushes the solar wind to either side as it passes and causes the wind to flow around the planet like a river flowing around an island. Some of the charged particles composing the solar wind, however, pass through the magnetic field and become trapped. When these particles strike the gases in the Earth's atmosphere, light is given off. These greenish-white and blue lights are known as *auroras*, or northern lights and southern lights. See Figure 7-4.

Figure 7-4 The northern lights—aurora borealis—as seen from Alaska.

Energy Production by the Sun Each second, the Sun gives off an amount of energy equal to 200 billion hydrogen bombs. We are fortunate that this solar furnace is 150,000,000 km away!

How the Sun works, however, remains an intriguing question for astronomers. It was first thought that the Sun burned fuel, much like the fuel we know on Earth—wood, coal, and oil. But even if the Sun were burning one of these fuels, the rate at which it produces energy would cause it to burn out in a few thousand years. Therefore, the Sun could not be using any common fuel. Though scientists throughout the 19th century proposed other explanations for the Sun's energy output, one by one they had to be ruled out.

Then in 1905, a young physicist named Albert Einstein came up with his famous formula, $E = mc^2$. By establishing the fact that energy and mass are equivalent, Einstein paved the way for the discovery of a new source of energy—fusion. The process of fusion, in which hydrogen is converted to helium, can account for the Sun's tremendous output of energy. Even though it would take 4 million tons of hydrogen each second to produce the energy output of the Sun, the Sun contains so much hydrogen that it should continue to burn another 10 billion years before the hydrogen runs out.

This seems to be the answer to how the Sun and other stars produce energy. But this is not the only way that energy is produced on a large scale in the universe. In fact, objects have been discovered that seem to produce energy on such a scale that not even fusion can account for it.

The Solar System: Our Planetary Family

The Solar System is made up of the Sun, the planets, and billions of smaller bodies. Each of these bodies follows an orbit around the Sun. These orbits are not perfectly circular. Each orbit is in the shape of an ellipse. Many orbits, such as the Earth's, are only slightly elliptical.

The bodies of the Solar System are held in orbit by the force of gravity exerted by the huge mass of the Sun. In fact, the Sun makes up more than 99 percent of the Solar System's mass. Thus, the Sun is the controlling body of the Solar System. The nine planets are the best known and, in most cases, the largest bodies of the Solar System. See Figure 7-5.

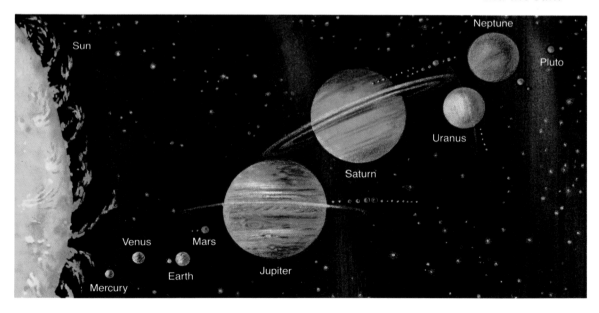

Figure 7-5 The planets compared with each other and the Sun.

The Planets The planets can be separated into two major groups—the inner planets and the outer planets. The inner planets include Mercury, Venus, Earth, and Mars. These planets are the smaller ones and consist of solid rock with metal cores. The four large outer planets include Jupiter, Saturn, Uranus, and Neptune. These giant planets are made of ice and gases and have small solid cores. Pluto is usually the farthest planet from the Sun. It is very small and seems to be different from all the other planets. Table 7-1 compares some of the major characteristics of the planets.

CHARACTERISTICS OF THE PLANETS								
Planet	Distance from Sun (AU)	Revolutions (years)	Radius at Equator (km)	Rotation Period (days)	Mass Compared to Earth	Density (g/cm^3)	Surface Gravity (Earth = 1)	Surface Temperature (°C)
Mercury	0.39	0.24	2,439	58.6	0.055	5.43	0.38	–193 to 342
Venus	0.72	0.62	6,052	–243.0	0.82	5.24	0.91	455
Earth	1.00	1.00	6,378	0.997	1.00	5.52	1.00	–88 to 58
Mars	1.52	1.88	3,397	1.026	0.11	3.9	0.38	–124 to –31
Jupiter	5.20	11.86	71,398	0.41	317.8	1.3	2.53	–160 to –149
Saturn	9.54	29.46	60,000	0.43	94.3	0.7	1.07	–176
Uranus	19.19	84.07	26,200	–0.65	14.6	1.3	0.92	–216
Neptune	30.06	164.82	25,225	0.72	17.2	1.5	1.18	–240
Pluto	39.53	248.6	1,100	6.38	0.0025	2.0	0.09	–223 to –208

Table 7-1

Scientists believe that the way the Solar System was formed explains the differences between the inner and outer planets. The inner planets are made up mostly of solid rock and metals. They were formed close to the Sun's heat and had their lighter materials driven off. The outer planets were formed farthest from the Sun's heat and held on to the lighter materials. The strong gravity of the massive outer planets enable them to hold on to the lighter materials. These materials, in the form of ices and gases, make up most of the mass of the outer planets.

Smaller Members of the Solar System

There are other bodies in the Solar System, some of which are closely associated with the planets. These bodies, though much smaller than the planets, far outnumber the nine planets. While you have already learned about meteors and comets, there are two other important members of the Solar System—satellites and asteroids.

Satellites The Earth has a very familiar *satellite*—the Moon. Other planets have satellites, or moons, of their own. Mars has two small satellites, Phobos and Deimos, while the gas giants have many satellites. Jupiter has 16, Saturn has 19, while Uranus and Neptune have 15 and 8 respectively. See Figure 7-6. Mercury and Venus, however, do not have any natural satellites.

Figure 7-6 The moons of Saturn in a mosaic made of several photographs.

Jupiter's four large satellites—Io, Europa, Ganymede, and Callisto—were discovered by Galileo in 1610, using his primitive telescope. More recently, active volcanoes have been observed on Io. One of the moons of Saturn, named Titan, is the only satellite in our Solar System known to have a substantial atmosphere. Nine of the moons of Uranus were discovered by *Voyager 2* in 1986, while all but two of Neptune's moons remained undiscovered until the visit by *Voyager 2* in 1989. Neptune's largest moon, Triton, rotates around its planet in the opposite direction from Neptune's rotation. This has lead astronomers to believe Triton was captured by Neptune's gravitational pull. Pluto has one moon, named Charon, that is about half the diameter of Pluto. Thus they make almost a double-planet system, each one orbiting the other in a period of about six days. See Figure 7-7.

Figure 7-7 An artist's rendering of Pluto and its moon Charon.

Asteroids Between the orbits of Mars and Jupiter is a wide region that does not contain a planet. In this zone, a huge number of rocks and boulders, called *asteroids,* are found. See Figure 7-8. The total number of asteroids is not known, but at least 4000 have been discovered. Ceres, the largest asteroid, makes up about 30 percent of the total mass of all the known asteroids. At one time, scientists thought that the asteroids might be the wreckage of a small planet that was torn apart by Jupiter's tremendous gravity. However,

Figure 7-8 Asteroids are fragments of rock that orbit the Sun.

Figure 7-9 This meteorite was found in Antarctica and is estimated to be 1.3 billion years old.

scientists now think that asteroids are the parts of a planet that never formed. Some asteroids follow very circular orbits around the Sun, while others have very elliptical orbits that might cross the paths of the inner planets including Earth. While most meteors seem to be associated with the dust trails of comets, most meteorites are fragments of asteroids that have wandered out of their usual place in the Solar System. See Figure 7-9.

Summary

The Solar System is dominated by the Sun and contains nine planets, which are of two major types. The inner planets are composed of solid rock and metals, while the outer planets are made of ices and gases. Most of the planets of our Solar System have satellites, or moons. The most numerous of the smaller bodies in the Solar System are asteroids, some of which may fall to Earth as meteorites.

7.2 STARS: A UNIVERSE OF SUNS

There are more stars in space than all the drops of water in the oceans of the world or all the grains of sand on the beaches of the Earth. Each star you see in the night sky is similar to the Sun. See Figure 7-10. Whether or not those stars have planets of their own is not yet known. However, some scientists feel that chances are high that other planetary systems will be found. What has been found, though, is quite intriguing in its own right.

Except for the Sun, the stars are so distant from the Earth that even the largest and most powerful telescopes show them only as pinpoints of light. Since stars are so far away, it might seem difficult to learn about them. But information about stars is carried in their light. By interpreting this information, scientists have learned what stars are made of, how large they are, and how they are moving in space.

How We Measure Stars

How far away is a star? When looking at the night sky, you can see that not all stars have the same brightness. Actually, some of the brightest objects seen are in fact planets, not stars. Venus, Mars, Jupiter, and Saturn can easily be seen without a telescope. Planets, as you know, can be sorted out from stars by observing their positions from night to night. Since stars are much farther away from the Earth than the

Figure 7-10 The stars visible from Earth are just a small fraction of those that actually exist.

planets, individual stars appear to keep the same position among other stars.

When we look at the sky, we see the stars in constellations as if they were located on a kind of ceiling over our planet. For example, the stars in the Big Dipper all seem to be at the same distance from the Earth. This is because of their great distance. But the stars are actually widely scattered and at different distances from us. See Figure 7-11.

Figure 7-11 The stars that make up the Big Dipper are at different distances from the Earth.

Some stars *do* seem to move very slightly when observed over a six-month span. As the Earth moves in its orbit around the Sun, these stars seem to shift their positions in relation to more distant stars. This is similar to the way a nearby tree appears to move against the distant background when viewed from a moving car. Trees that are farther away appear to move less against a distant background. This apparent motion, called **parallax**, can be used to give an idea of how far away the trees are. By using parallax, the distances to nearby stars can also be measured. See Figure 7-12 on the next page.

Scientists use other methods to measure the distances to stars that are very far away. One method makes use of the brightness of the star. Have you ever judged the distance to a street light by how bright it looked? If you have, you were estimating distance using brightness. The actual amount of light a star sends out is called its **luminosity**. Stars have different luminosities; some are 10,000 times brighter than the Sun, while others are only a tiny fraction as bright. Once the luminosity of a star is known, its distance can be found by comparing its luminosity to its brightness as actually seen from the Earth. You might judge

Parallax The apparent shift of position of an object (such as a star) when seen from different locations or angles.

Luminosity A measure of the actual brightness of a star.

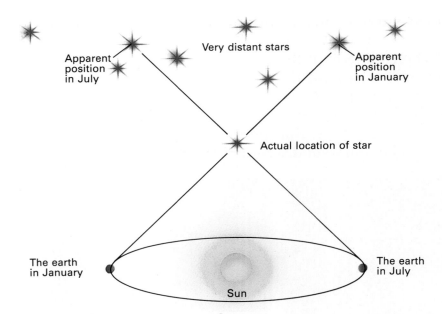

Figure 7-12 The movement of the Earth causes nearby stars to appear to move relative to those in the distant background.

the distance to a street light by making the same kind of comparison.

Astronomers have different ways of finding the luminosity of stars. One of the best ways makes use of a *variable star*, a kind of star whose brightness changes during periods of one day to many days. The length of time it takes for these stars to change their brightness is related to their luminosities. Once the luminosity of these stars is found, the distance to them (and to other stars close to them) can be found. This is done by comparing their perceived brightness with their luminosities. For example, if a star that has a high luminosity appears dim, it means that the star is probably very far away.

Types of Stars

The messages carried by the light from stars have shown scientists that there are many kinds of stars. Some stars are larger than the Earth's orbit around the Sun, while others are smaller than the Earth itself. There are stars made of gas thinner than air and others that are harder than diamond. One kind of star repeatedly blows up like a balloon, then shrinks and blows up again, in an endless cycle.

Red and Yellow, Black and White One difference among stars is their color. You already know that the color of starlight is related to the temperature on the star's surface. The coolest red stars have surface temperatures of about 3000 °C, while the hottest blue stars have surface temperatures over 25,000 °C. See Table 7-2.

| CLASSIFICATION OF STARS |||
Color	Surface Temperature (°C)	Example
Violet	Above 28,000	10 Lacertae
Blue	10,000–28,000	Rigel, Algol
Blue	7500–10,000	Vega, Sirius
Blue–White	6000 to 7500	Canopus, Procyon
White–Yellow	5000 to 6000	Sol (Sun), Capella
Orange–Red	3500 to 5000	Arcturus, Aldebaran
Red	Less than 3500	Betelgeuse, Antares

Table 7-2

Astronomers have discovered that there is also a relationship between the star's surface temperature and its brightness. Generally, the higher the temperature of the star, the brighter it is. When stars are plotted on a chart according to their surface temperature and brightness, they fall into three main groups. See Figure 7-13. This chart is known as the Hertzsprung–Russell diagram, named after the Danish and American astronomers who first designed it. Most stars fall into a narrow band on the chart. The band begins with hot stars at the upper left and ends with cool stars at the lower right. Since most known stars fall into this group, they are called the *main sequence* stars. A second group of stars appears at the upper right of the chart. These are bright stars, even though their red color indicates low surface temperatures. The stars in this group are called *red giants*. One red giant, Betelgeuse (BET ul jooz), is so large that if it were placed

Figure 7-13 Stars are grouped in this chart according to their surface temperature and luminosity.

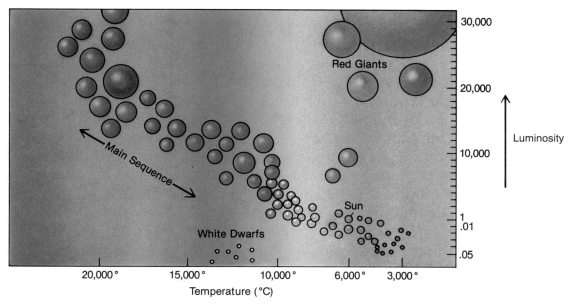

at the center of our Solar System, its surface would reach well past the orbit of Mars. A third group of stars appears at the bottom of the chart. These are called *white dwarfs*. They are very hot and produce white light, but they are very small and therefore not very bright. Some white dwarfs are only the size of the Earth.

Binary Stars Many of the stars you see in the night sky are not single stars, as is our own Sun. In fact, about 60 percent of all stars are actually double-star systems, or *binary stars*. In binary star systems, two stars revolve around each other. See Figure 7-14. Some binary systems are positioned so that the stars periodically eclipse one another from the point of view of an observer on Earth. Our closest neighboring star, Alpha Centauri, is actually a multiple-star system composed of the stars Alpha, Beta, and Proxima Centauri. Since it is located nearer the south celestial pole, it is not visible in the skies of North America. The nearest star visible from most of the United States is Sirius, which you already know is a binary star.

Figure 7-14 Revolution of a double star. These three photographs (covering a period of about 12 years) show the mutual revolution of the components of the nearby star, Kruger 60.

Variable Stars Variable stars, as you have seen, can be used to judge distance since their luminosities change. One type of variable star is the *pulsating variable*. Energy from below the surface of such a star heats the gases of the visible, shining surface. As the surface becomes hotter and brighter, it expands. Once the surface expands, however, the gases cool and the star again becomes dimmer. The cooler gases then contract, causing them to once again get hotter, and a new cycle begins.

The first pulsating variable discovered was Mira, whose luminosity changes every 331 days. The time from bright to dim and back to bright again is the star's period. Most known pulsating variables have a period of 100 days or less. Mira has an unusually long period. The North Star, in contrast, is a pulsating variable that has a period of only 4 days.

The magnitude of other stars varies for different reasons. One type of variable, called *eclipsing binaries*, are actually pairs of stars that move around each other. Their brightness appears to vary as one star periodically blocks the light from the other star, causing an eclipse. If, for example, a pair of stars happens to rotate around each other in line with the Earth, one will periodically block our view of the light from the other, varying their total brightness.

Exploding variables, such as the *nova* in Figure 7-15, are stars that have bursts of energy that make them appear

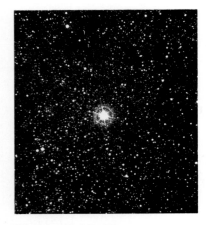

Figure 7-15 Novas occur when stars explode.

many thousands of times brighter for days or even years. The term *nova* comes from the Latin for "new." They were given this name because in ancient times they were thought to be new stars. A nova eventually returns to its original brightness, but a *supernova* is such a huge explosion that the star blows itself apart. See Figure 7-16.

Galaxies: Collections of Stars

Galaxies consist of billions of stars. There are many galaxies in the universe, including the Milky Way, of which our sun is a part. The Milky Way is a spiral galaxy of over 100 billion stars, stretching 100,000 light-years across. See Figure 7-17. Have you ever noticed, on a very clear night, a band of stars so dense that it looks like a starlit cloud? See Figure 7-18. This band is actually a portion of our galaxy, viewed from our position in one of its spiral arms.

Figure 7-16 This supernova, photographed in 1987, appeared in the sky of the Southern Hemisphere. It was the first supernova visible to the naked eye in nearly four centuries.

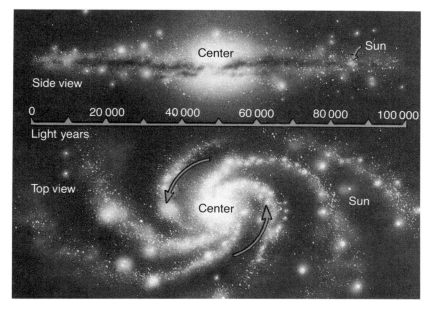

Figure 7-17 Earth is part of the Milky Way galaxy, shown here in relation to the Sun.

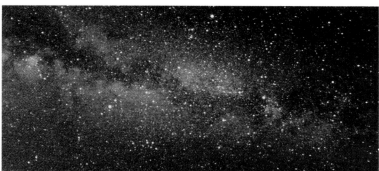

Figure 7-18 If you look up at the sky on a very dark, clear night, you can see the Milky Way.

Stars are held in a galaxy by gravity, which tends to pull the stars toward one another. The Milky Way is shaped like a disk with a bulge at its center. Our Solar System is located

Figure 7-19 The spiral galaxy M31 in the constellation Andromeda looks very much like our own galaxy.

closer to the edge than to the center of the galaxy. Thus, by looking in the direction of the galaxy's center, we see many more stars than when looking elsewhere. If you look at the constellation of Andromeda, it is possible to see a small, fuzzy patch of light among the stars. A telescope shows that this is a galaxy about 2.2 million light-years away. See Figure 7-19. Our own Milky Way would look similar if we could see it from far out in space.

Types of Galaxies Telescopes that allow us to look deep into space beyond our own galaxy show that the universe contains billions of other galaxies. Just as stars are seen to differ from each other, not all galaxies are alike. In addition to the spiral shape of the Milky Way and Andromeda galaxies, some galaxies are shaped like footballs, or like spheres. Other galaxies have no general shape. The major types of galaxies are spiral, barred spiral, elliptical, and irregular. See Figure 7-20.

Figure 7-20 Types of galaxies: *barred spiral* (left), *elliptical* (middle), *irregular* (right).

Except for the galaxies that share our "local group," all galaxies that can be observed have one thing in common: They are all moving away from each other at a very high speed. Scientists can tell that the galaxies are moving apart because of the information that can be read from light. The colors in the spectrum of the stars in these galaxies are shifted toward the redder colors. This is called the *red shift* of the starlight. The red shift in the light from the galaxies is caused when the light waves are spread out as the light source moves away. You may have heard the same effect in the pitch of sound waves that come from a moving source.

Strange Objects in Space

As if red and blue stars, novas and supernovas, spiral and elliptical galaxies were not enough, even stranger objects are among those to be found in the universe. Scientists have only begun to examine such objects, some of which are discussed below.

Pulsars At first, scientists thought certain stars were the source of alien radio messages. These stars, called *pulsars*, give off rapid pulses or bursts of radio waves, light, and X rays. As a pulsar spins on its axis, it sends out a stream of radiation in a pattern similar to that of the rotating light in a lighthouse. See Figure 7-21. When the Earth is in the path of one of these streams of radiation, radio telescopes can pick up the signals as rapid pulses. A pulsar at the center of the Crab Nebula rotates thirty times a second.

Figure 7-21 An artist's conception of a rotating pulsar.

Quasars One of the most puzzling of all objects detected by astronomers is the quasi-stellar radio source, or *quasar* for short. Quasars emit vast amounts of radio waves and light. They are several billion light-years away and may actually be the brightest objects in the universe, producing more energy than that produced by hundreds of galaxies combined. See Figure 7-22. Since they seem to be so far away, quasars may be among the earliest objects in the universe. Scientists still do not fully understand the nature of quasars.

Figure 7-22 Quasar 3C 275.1, the first to be found at the center of a cluster of galaxies, appears as the brightest object near the center of this image. This quasar is 7 billion light-years away. The light we now observe left the quasar more than 2 billion years before our Solar System was formed.

Black Holes Astronomers have been searching for ways to explain the huge amounts of energy produced by quasars and other objects. Some have theorized that very large stars may collapse with such force that they pass beyond the stage of being a neutron star to become a "black hole." As you know, a black hole is an object so dense that its gravity will allow nothing—not even light—to escape from it. Because no energy of any kind can come from a black hole, there is no way to observe it directly. However, some astronomers hypothesize that they may be able to locate black holes by the effect they have on nearby objects. For example, a black hole close to a star would strip matter from its neighbor. The material lost from the star would then give off energy such as X rays as it disappeared into the black hole. This process could explain some of the powerful energy sources now detected in space. Yet, black holes remain unobserved theoretical objects. None have actually been discovered.

Summary

There are billions upon billions of stars in the universe. Astronomers use parallax and absolute magnitude to estimate the distance to stars. Astronomers estimate the surface temperature of a star by its color. Most stars are part of a binary system, and some stars are variables. Galaxies are huge collections of stars that come in a variety of shapes and sizes. Some of the more unusual objects in space are pulsars, quasars, and the elusive black holes.

7.3 ONE SMALL STEP

Stonehenge is the ruins of an ancient monument in southern England. See Figure 7-23. It consists of two rings of rectangular stone columns. Other stone rectangles lie on top of some of these columns. The tallest column is 7.3 m high. Each stone weighs 30 to 60 tons. Through the center of the ring is a broad avenue. At the end of the avenue is a stone 5.4 m high called the Head Stone. Why did people, beginning in 2,600 B.C., build this structure? Why did they carry these heavy stones from as far away as 380 km?

How We See the Universe

For centuries, astronomers watched the evening skies. They noticed that the stars kept their same relative positions. However, five points of light, the visible planets, slowly moved with respect to the stars. Ancient people could only wonder at their nature. They thought of these planets as gods.

In the 1600s, the telescope was developed. The planets could then be seen as spheres that are illuminated by the Sun. Some surface features were visible through the new telescopes, but space probes gave us our first truly clear images of the planets. By 1989, eight of the nine known planets had been visited by probes sent from Earth.

Figure 7-23 *What was the purpose of Stonehenge?*

The Telescope is Born The invention of the telescope may have occurred many times in different places. By the 1600s, it was being used as a field glass for viewing distant objects on Earth. In 1609, however, Galileo made his own telescope and pointed it toward the sky. See Figure 7-24. Although his instrument was primitive, Galileo made many discoveries, such as the craters and mountains on the Moon, the phases of Venus, and the four large satellites of Jupiter.

Isaac Newton constructed the first reflecting telescope. See Figure 7-25. He realized that concave mirrors have three advantages over lenses. First, they reflect light instead of refracting it and therefore produce no distortion. Second, a good mirror absorbs less light than even the most transparent glass, so the image will be brighter. Third, a mirror is not as heavy as a lens of the same size. Therefore, a mirror is easier to support, and much larger telescopes can be constructed. The largest optical telescopes used today are reflecting telescopes. Composed of 36 interlocking mirrors with a total diameter of 10 m, the new Keck telescope, located on top of Mt. Mauna Kea in Hawaii, is the largest reflector in the world—twice the size of the famous Hale telescope at Mt. Palomar. The images this giant telescope can produce rival those of the orbiting Hubble Space Telescope.

Figure 7-24 Galileo's telescope used lenses to focus light from distant objects.

Other Types of Telescopes Visible light is not the only type of radiation emitted by stars and galaxies. Radio waves, X rays, and ultraviolet waves are also produced by these objects. For example, stars, large planets, and the cores of many galaxies emit radio waves. Quasars emit many radio waves. Since radio waves are not absorbed by matter, as are visible light waves, radio waves from outer space may cross the entire universe with little loss through absorption.

Astronomers can capture radio waves with a special instrument called a radio telescope. Though some radio telescopes are like large television antennas, most are shaped like metallic dishes. Radio waves are reflected by metallic surfaces just as light is reflected by a mirror. At the focus of a radio telescope is a radio receiver. A radio astronomer uses a tuner to select the radio frequency he or she wishes to study.

A single radio telescope provides an image of the sky that is less precise than the image provided by an optical telescope. Radio telescopes can be linked together, however, so that the image received by one can be compared and mixed with the image received by another. Combining images from

Figure 7-25 A replica of the reflecting telescope made by Isaac Newton.

Figure 7-26 *A radio image (left) of quasar 2300-189 at a distance of 1.2 billion light-years. This array of 27 radio telescopes (right) provides a much clearer image of the universe than a single dish does.*

several radio telescopes can produce a much clearer radio image of the sky. See Figure 7-26.

Other telescopes have been sent into orbit around the Earth. Ultraviolet and infrared telescopes, as well as X-ray and gamma-ray collectors now give astronomers a much broader look at the universe around us. See Figure 7-27.

Figure 7-27 *Optical images (left) and infrared images (right) of M51—the Whirlpool Galaxy.*

The Exploration Begins

Suppose that observers from outer space have been watching the Earth since it was formed. For more than 4 billion years they would not have seen anything leave the Earth. Then suddenly, starting about 40 years ago, they would have noticed that the people of the Earth began to send off tiny spacecraft. At first, the spacecraft only flew around the Earth. Then they landed on the Moon. Later, the observers would have seen other spacecraft fly to the Earth's neighboring planets. The observers might conclude that the people of the Earth had begun a new age of exploration.

Rockets and Early Satellites Even though Jules Verne had written about a fictitious trip to the Moon in 1865, the first person to scientifically study the use of rockets for space travel was Konstantin Tsiolkovsky, a Russian inventor. In the 1920s, he designed a spaceship with a bullet-shaped passenger cabin in the nose and fuel tanks filled with liquid hydrogen and liquid oxygen in the tail. Although Tsiolkovsky never launched a rocket himself, his thinking was fundamental to the development of rocket flight.

At about the same time, an American physicist, Robert Goddard, was actually experimenting with rockets. In 1926 Goddard launched the world's first liquid-fueled rocket. See Figure 7-28. During World War II, the Germans constructed many rockets for delivering bombs. At the end of the war, scientists in the United States used captured German rockets to scientifically explore the upper atmosphere. An American two-stage rocket, with a captured German rocket as its first stage, reached a record altitude of 393 km in 1949.

Figure 7-28 Robert Goddard launched a gasoline-fueled rocket in 1926. The rocket was 1.2 m long, and it rose to a height of 56 m at an average speed of 103 km/h.

The development and use of rockets proceeded slowly until October 4, 1957, when the Soviet Union surprised the world by placing into orbit an artificial satellite called *Sputnik 1*. This 83.6-kg satellite carried instruments to measure the density and temperature of the upper atmosphere. See Figure 7-29. A month later, on November 3, 1957, the Soviet Union put another satellite into orbit. This second satellite carried a passenger—a dog named Laika. The purpose for orbiting the animal was to study the effects of space travel on a living organism.

The first American satellite, *Explorer 1*, was launched on January 31, 1958. The rocket was 2.05 m long and carried a scientific package for measuring temperature, cosmic rays, and meteors. *Explorer 1* discovered the Van Allen belts of radiation surrounding the Earth.

Figure 7-29 Sputnik 1 *was the first artificial satellite to orbit Earth.*

Human Exploration On April 12, 1961, Yuri Gagarin, a 27-year old cosmonaut, was launched into space. He orbited the Earth once, reaching an altitude of 327 km. American astronaut John Glenn was launched into space on February 20, 1962, and made three orbits around the Earth. The first woman to go into space was Valentina Tereshkova, who was launched in June of 1963. In 1965, Alexei Leonov became the first human to "walk" in space. Dressed in a protective spacesuit, he left his pressurized cabin and floated for 10 minutes in space. All these space firsts had been accomplished in less than 10 years after the launch of *Sputnik 1*; the next 10 years would be even more dramatic.

Figure 7-30 Neil Armstrong and "Buzz" Aldrin spent 2½ hours walking on the Moon.

Figure 7-31 By landing on a runway, the Space Shuttle can be used again and again.

Figure 7-32 An artist's conception of the U.S. Space Plane.

As a challenge to the Soviet Union's leadership in space, the United States determined to place an astronaut on the Moon before the end of the 1960s. In preparation for this journey to the Moon, many additional Earth-orbiting missions were launched. During the *Mercury* and *Gemini* missions, American astronauts learned to live and work in space. On July 16, 1969, *Apollo 11* was launched from Cape Canaveral, Florida toward the Moon. The *Apollo 11* lunar module landed on the surface of the Moon on July 20 at 8:17 P.M. See Figure 7-30. Neil Armstrong and Edwin "Buzz" Aldrin became the first humans to set foot on another world.

The Space Shuttle, a vehicle that can lift off like a rocket and return to Earth like an airplane, is now the United States' major vehicle for space travel. See Figure 7-31. It was designed to be used in the construction of a space station. Many missions involving the delivery of military, scientific, and telecommunications satellites as well as orbiting telescopes have been launched. A space plane that will be able to enter orbit after taking off horizontally from a runway is now under development by the United States. See Figure 7-32.

Interplanetary Probes You have already been introduced to several interplanetary probes—*Pioneer 10*, *Voyagers 1* and *2*, *Magellan*—that have sent back much more information about our neighbors in the Solar System than could ever have been gained by looking through telescopes. Another space probe currently in operation is the *Galileo*. Having passed Mars and the asteroid belt, it will reach Jupiter in 1995 where a smaller probe will be sent into the Jovian atmosphere. See Figure 7-33. This probe will send valuable information back to *Galileo* for only 75 seconds before it is crushed by Jupiter's atmospheric pressure.

Another probe, called the *Ulysses*, is on its way to an historic rendezvous with the Sun. It will be the first probe to ever journey outside the ecliptic plane. Rather than orbiting the Sun like the planets, it will travel at right angles to their paths and thus cross over the Sun's south pole. Once the

Ulysses reaches the Sun in 1994, it will complete a polar orbit, giving us our first view of the top and bottom of our own star, the Sun.

Theories of Cosmology

Since light travels at a certain speed, the farther we look out into space, the further back in time we can see. This ability may help us answer some very fundamental questions about our universe. **Cosmology** is the study of how the universe began, and how it may eventually end. There is no way to witness the actual events that occurred billions of years ago, but scientists have proposed several theories to account for the universe as we see it today.

The Big Bang The idea that the universe is expanding has led scientists to develop the *big-bang* theory. This theory states that, at one time, all the matter of the entire universe was concentrated in an infinitesimally small volume. Then, about 15 to 20 billion years ago, there was a tremendous explosion that caused the universe to begin expanding. About one billion years after the big bang, huge clouds of hydrogen and helium began to form. These clouds were the beginnings of galaxies. In time, stars started to develop in the galaxies. Within about 10 billion years after the big bang, all of the galaxies that can now be seen had formed. Today, we see the results of the big bang as a rapidly expanding universe with galaxies moving away from each other. See Figure 7-34.

As you know, scientists have found that all space is filled with weak radio waves. One explanation of this cosmic background radiation is that it is evidence left by the great explosion that was the big bang. But one thing the big-bang theory cannot explain, however, is the uniformity of this background radiation. This radiation is the same no matter

Figure 7-33 Artist's impression of Galileo's entry probe descending through Jupiter's atmosphere.

Cosmology The study of the structure, origin, and evolution of the universe.

Figure 7-34 Before the big bang, all matter and energy in the universe was concentrated in a very tiny space. When the big bang occurred, matter was sent outwards in all directions.

what direction we look. Because of the size of the universe, it is not likely that the radiation coming from one side has ever been in contact with the radiation from the opposite side. How then is this radiation the same temperature everywhere?

Pulsating Universe If there is enough matter in the universe, gravity should eventually stop the outward expansion of the galaxies caused by the big-bang explosion. A variation of the big-bang theory, called the *pulsating* or *oscillating* universe theory, states that the universe will eventually contract, bringing the galaxies back together into another hot mass, until another big bang sends them flying out again. This is repeated over and over, so the universe will continue to expand and shrink forever. See Figure 7-35. However, the estimated density of the universe is such that there is a balance between the two possibilities. It cannot, at this time, be determined whether expansion will stop, or if the universe will expand forever.

Figure 7-35 Most astronomers think that the universe will continue expanding (A–E) and eventually die. Others think there is enough matter in the universe to make it stop expanding and collapse back on itself, leading to another big bang, endlessly repeating this cycle.

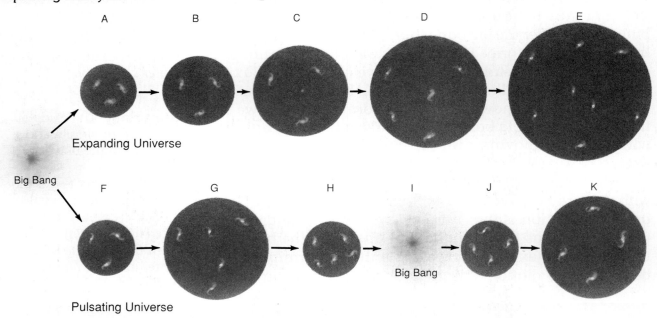

Summary

A variety of telescopes has enabled humans to explore the universe from their home on Earth. With the development of the rocket, space probes have been sent on close-up missions to the members of our Solar System. The universe contains billions of other galaxies that are apparently moving apart as the universe expands. The way the universe may have begun and how it might end is a question that scientists still are trying to answer.

Unit 8

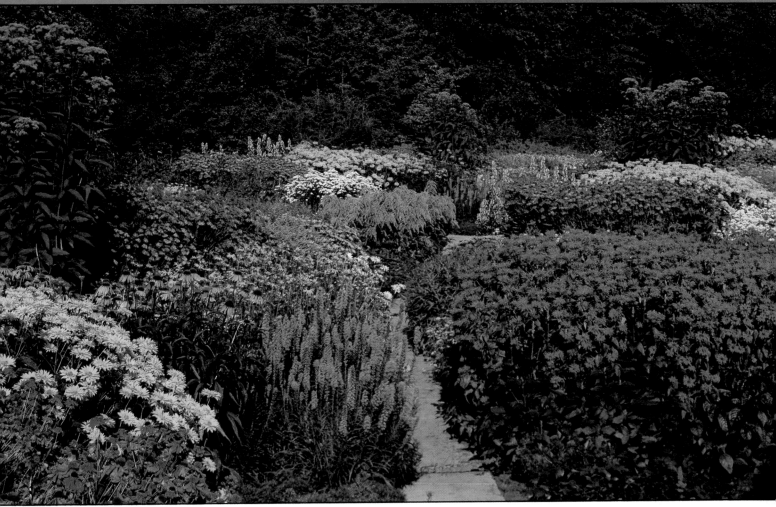

In This Unit

8.1 Plant Structure, p. S134

8.2 Uses of Plants, p. S141

Reading *Plus*

Now that you have learned how to grow plants, let's take a closer look at the parts of a plant, how plants grow, and how we use plants. Read pages S134 to S144 and prepare a presentation that explains how plants are adapted to their environments and why plants must be protected and studied. Your presentation should answer the following questions.

1. How do the leaves, stems, and roots work together in a healthy plant?
2. Why might plants be more affected by the environmental changes than are humans?
3. In what ways do animals and plants depend on each other?

8-1 PLANT STRUCTURE

Have you ever wondered why a plant's body is so different from yours? These differences are related to the way plants and animals live. Being an animal, your way of life involves walking around, searching for and capturing other organisms for food, then eating that food and digesting it inside your body. In this way, you obtain the energy and nutrients you need for growth and the functioning of your cells. Having arms, legs, eyes, and a mouth, not to mention all your other body parts, makes you able to accomplish these necessary tasks. A plant's way of life, however, is very different from yours. Plants gather the *inorganic* nutrients carbon dioxide and water from their surroundings. Then, by capturing and using energy from the sun, plants convert those inorganic nutrients into *organic* substances such as sugar, which they use for food. Their bodies are adapted to the way they obtain energy and interact with the environment.

Figure 8-1 Land plants need vascular tissues in order to grow large.

Most of the plants you see and use every day are *vascular plants*. These plants have become adapted to living on land by developing a body that consists of leaves, stems, and roots. Inside the leaves, stems, and roots of a vascular plant are special tubes that transport water, minerals, and food throughout the plant. See Figure 8-1. These tubes form the vascular system that links the parts of the plant together into a functional organism. To understand how plants function, let's examine the basic parts of a plant's body.

Leaves

Leaves are the food-making organs of plants. Light, carbon dioxide, and water obtained by leaves are used during *photosynthesis*, the process by which plants use energy from the sun to convert carbon dioxide and water into sugar and oxygen. In order to carry out photosynthesis, leaves also exchange carbon dioxide and oxygen, and help water to move through the plant. As you know, leaf shapes and sizes vary greatly. These variations are often adaptations that allow leaves to carry out their activities.

Some leaves have very special shapes, such as the jaw-like leaves of the Venus' flytrap or the vase-shaped leaves of the pitcher plant. See Figure 8-2. These carnivorous plants actually trap and digest insects to obtain some of their nutrients. Lucky for us, none of these plants are large enough to "eat" people!

Figure 8-2 The vase-shaped leaves of a pitcher plant (top) and the jaw-like leaves of a Venus' flytrap (bottom) are adaptations for trapping insects.

Structure of Leaves The cells of a typical leaf are arranged into a flattened structure, called the *blade*, which has many raised *veins*. Leaves are usually connected to a stem by a stalk called a *petiole*. See Figure 8-3. Most leaves are thin and flat, which exposes as many cells as possible to sunlight. This shape also provides good contact with the air, from which leaves take in carbon dioxide. The veins of a leaf are made of vascular tissues that bring water into the leaf.

A leaf's veins also carry the sugar it has made to other parts of the plant. Each vein is a bundle of long, tubular cells that resemble small water pipes. The main vein into a leaf divides many times into smaller and smaller veins. As a result, water and food molecules do not have far to travel between a vein and any cell.

The cells of a typical leaf blade are organized into several specialized layers. See Figure 8-4. A layer of cells, called the *epidermis*, covers both the upper and lower surfaces of the leaf and prevents the loss of water from inner cells. The epidermis is tough, giving strength to a leaf, and its cells are clear, allowing light to pass through.

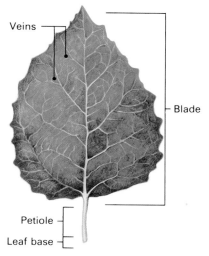

Figure 8-3 *A typical leaf has a blade with many veins and a petiole.*

Figure 8-4 *The inner structure of a leaf.*

Chloroplasts Organelles of plant cells that contain chlorophyll and are the sites of photosynthesis.

Stomata Small pores in leaves that allow gases to enter and leave plants.

Inside the leaf are two kinds of food-making cells that contain special organelles called **chloroplasts**. The chloroplasts contain chlorophyll, which absorbs light for photosynthesis. The tightly packed upper layer of food-making cells performs most of the photosynthesis in the leaf. The food-making cells on the lower side of the leaf are loosely packed and have many air spaces between them. These air spaces are connected to the outside by small openings called **stomata**. Each stoma is surrounded by two *guard cells* that can change shape to open and close the stomata. The top surface of a leaf has very few stomata and is coated with a waxy material that prevents water from evaporating. But the lower surface of a leaf may have thousands of these

Figure 8-5 Stomata occur primarily on the lower epidermis of leaves.

openings that allow carbon dioxide, oxygen, and water vapor to go into and out of the leaf. See Figure 8-5.

Leaf Functions During a typical day, all leaf functions occur simultaneously and are linked to each other. The food-producing activities of a leaf begin when light first strikes it. Light causes the guard cells to open the leaf's stomata, allowing gases to move into and out of the leaf. Some of the sunlight is captured by chlorophyll and used to begin the process of making sugar. Once made, these sugars move into the leaf veins and are carried to other parts of the plant. Since plant cells also need energy for cell processes, plants use some of the sugar they make to supply this energy. Unused sugar is transported out of the leaf and is stored elsewhere as *starch*. Some of the oxygen made during photosynthesis is used by the plant to "burn" sugar during cellular respiration. Excess oxygen moves out of the leaf into the surrounding air. At the same time, more carbon dioxide is drawn back into the leaf through the stomata. This cycle of activity, shown in Figure 8-6, continues throughout the daylight hours. At night, the stomata close and photosynthesis stops. But the leaf can continue to make sugar by using energy stored during the daytime.

Figure 8-6 The activities of a leaf include making sugar from water and carbon dioxide and exchanging gases.

Water that is used during photosynthesis or lost through open pores is replaced by water from the veins. Leaf veins are connected to the plant's vascular system, which extends all the way down into its roots. When water is used or lost by the leaves, the roots of the plant take up more water to replace it. See Figure 8-7. As long as there is enough water available, it continues to move through a plant during the day. But if the soil around the roots becomes too dry, the leaves will not receive enough water. What do you think this does to the food-making activities of the leaf? Just as a factory must slow down or stop its production when there are not enough raw materials, photosynthesis also slows down or stops. To conserve water, the stomata close, which keeps water vapor from escaping. If it still does not get water, the leaf will eventually wilt and may even dry up.

Stems

What do a cactus pad, a potato, a stalk of asparagus, and a tree trunk have in common? They are all stems of plants. The stem is the midsection of a plant's body that in most plants grows above the ground. For many plants, stems have specialized functions. The potato plant, for example, stores food in specialized stems (the potatoes). In the cactus, the stem stores water. However, all plant stems have two basic functions—they support the leaves and transport food and water between the leaves and the roots.

Structure of Stems You may have noticed that some plant stems are soft and flexible while others are stiff and hard. Plants that have stems that are soft and flexible, such as the stems of beans, tomatoes, grasses, and petunias, are called *herbaceous* plants. Plants whose stems are stiff and hard, such as those of trees and shrubs, are called *woody* plants. Although the stems of both types are able to support their leaves, there is a big difference in their sizes. Woody plants are able to grow much taller and larger because the woody stem provides more support.

Both herbaceous and woody stems contain two kinds of vascular tissues. One kind carries water and minerals upward from the roots to the leaves, while the other generally carries food downward from the leaves to the stems, roots, and any developing fruits and seeds. Both kinds of tissue are bundled together in long strands, called **vascular bundles**. These bundles run through the entire length of the stem. In some stems, they are scattered throughout a matrix of unspecialized cells, and in others, they are arranged in a ring. See Figure 8-8 on the next page.

Notice that in each bundle, the cells of the inner side are larger and have thicker walls than the cells on the outer side of the bundle. The larger, thick-walled cells are the water-carrying cells. When these cells are filled with water they help to stiffen the stems of an herbaceous plant. Wood is also made of these cells. The small, thin-walled cells are the food-carrying cells. The cells located between the two types of vascular cells are special dividing cells. These cells are able

Figure 8-7 *When water is lost from the leaves, water from the roots is pulled upward through the plant's vascular system.*

Vascular Bundles Groups of water-carrying and food-carrying cells.

Figure 8-8 *Two ways that vascular bundles are arranged in stems.*

to produce new cells that mature into each of the other types of stem-tissue cells.

Stem Growth While you are growing, most of your body cells are able to divide to make more cells. However, in plant stems, only certain cells are able to divide to make more cells. Groups of these cells are found in two places in a stem—between the two types of vascular tissues in the vascular bundles and inside the *buds*.

The dividing cells inside the vascular bundles cause stems to grow in width. As layers of new cells are added, the stem becomes thicker. Herbaceous stems thicken only a limited amount and usually die at the end of a growing season. Woody stems, on the other hand, can become very thick since they do not die each year.

In woody stems, the vascular bundles are arranged in a solid ring that produces new vascular tissue and a new layer of wood each year. See Figure 8-9. The wood consists mostly of old water-carrying cells. As new water-carrying cells pro-

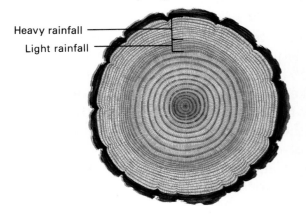

Figure 8-9 *You can determine the age of a tree by counting its rings.*

duced on the inside of the ring mature, the food-carrying and dividing cells of the ring are pushed outward. As a result, the outermost cells of a woody stem are crushed and die. These dead cells become part of the protective covering of a woody stem—its bark. Because the actively growing and transporting cells are located just under the bark, woody plants can be killed if their bark is severely damaged or removed.

Inside the buds of a stem are groups of dividing cells that cause stems to grow in length. See Figure 8-10. In woody plants, buds are usually formed at the end of the previous year's growing season. When conditions are favorable for stem growth, a plant produces chemicals that stimulate dividing cells to produce new cells rapidly. The stem becomes longer by adding new cells at its tip. Some of the new cells become new stem tissue, while others become the tissues that make up leaves. The new stem cells grow in length just after they form, but once they mature, their size does not change. Normally, only the bud at the end of a stem grows. If this bud dies or is broken off, buds on the side of the stem will begin to grow.

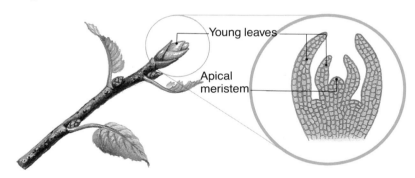

Figure 8-10 Stems grow only at their tips. The end of a stem where the dividing cells are located is called the apical meristem.

Roots

The roots are the part of a plant's body that, in most cases, grows underground. When you look at a plant, consider that this underground portion is about the same size as the aboveground portion. Just like stems and leaves, roots also have several functions. One function of the roots is to anchor the plant in the ground. Thus the plant is less likely to be blown away by wind or washed away by rain. Some roots, such as carrots, sweet potatoes, and turnips, also store food. However, the most important function of roots is to absorb water and minerals for the plant.

Types of Root Systems If you have ever had to pull weeds in a garden or a lawn, you have probably noticed that some weeds come up easily, while others are held so firmly that the top of the plant usually breaks off before the roots come out of the ground. One reason for this is that there are two basic types of root systems in plants.

Taproots are the large, central roots that grow almost straight down. See Figure 8-11. They can be as long as a plant is tall, anchoring it firmly into the ground. In some trees, the main taproot may reach down hundreds of feet into the ground. Taproots can be long and slender, thick like carrots, or bulb-shaped like radishes. However, the smooth, unbranched structure of carrots and radishes is not typical of most taproots. These shapes were developed through years of selective breeding. Most taproots branch, but the branches are much smaller than the main root.

Figure 8-11 Taproot root systems grow deep into the soil, while fibrous root systems spread out over a large area under the surface of the soil.

Fibrous root systems, on the other hand, consist of many thin roots that are all about the same size. Fibrous roots often form a dense network near the surface of the soil. This network holds soil particles together and helps to keep them from being carried away by water. Grasses, which play an important role in preventing soil erosion, have fibrous roots.

Root Growth Roots grow in length by making new cells at the ends of the roots, or *root tips*. See Figure 8-12. The dividing cells are located just inside the very end of the root tip, which is protected by the *root cap*. After they are formed, new root cells lengthen and become one of the specialized types of root tissue. Cells that are one centimeter or more from the end of a root tip cannot lengthen further. Thus for a root to continue to grow in length, new cells must be produced at the root tip.

Vascular tissues develop in the center of the root, while support and storage tissues develop around the vascular tissues to form the outer portion of the root. Some of the new cells that become part of the center of a root retain their ability to divide. These cells cause a root to thicken by producing more vascular and support cells. The outermost cells become the epidermis, through which water and minerals

Figure 8-12 The three parts of a root tip are the regions of division, elongation, and maturation (x150).

are absorbed. As the root thickens, the epidermal cells are crushed against surrounding soil particles forming an outer layer of dead cells. This part of the root can no longer absorb water.

Water and minerals are taken up by the actively growing tips of new root branches. These branch roots grow outward from the central cylinder. Most of the water and minerals taken in by the roots are absorbed through *root hairs*, which are extensions of the epidermal cells on root tips. The root system of a single plant may have billions of these tiny, threadlike structures, which are only a single cell wide. See Figure 8-13. As long as there is enough water in the soil, water is drawn into the root hairs by *osmosis*. This water continues to move through or between the root cells until it reaches the water-conducting cells. There, it can travel up to the leaves.

The addition and lengthening of new cells exerts enough pressure to drive a root tip through the soil and even into tiny cracks in rocks. If a root grows into a crack, the pressure from the thickening of a root enlarges the crack by forcing its sides apart. Roots are an important factor in weathering rocks and in forming soil.

Figure 8-13 The fine, white fuzz on the root of this germinating corn seed consists of root hairs.

Summary

The body of a vascular plant consists of roots, stems, and leaves. These structures all contain and are connected by two types of vascular tissue. One carries the water and minerals up to the leaves, while the other carries the food produced in the leaves down to the stems and roots or into the fruits and seeds. Leaves are adapted to produce food by photosynthesis. Plants use some of this food, while the rest is stored. Stems hold up the leaves and transport materials through continuous strands of vascular tissue called vascular bundles. Roots anchor a plant in the ground and absorb water and minerals through root hairs. Both stems and roots grow in length by producing new cells at their tips.

8-2 USES OF PLANTS

Plants are used in a multitude of ways, from providing food and covering for our bodies, to keeping us well. Plants are more important than you may think. Modern technology is largely based on the use of plant materials.

Figure 8-14 How many different plant parts are represented in this photograph?

Plants for Food

Have you ever eaten a flower? If you have eaten broccoli or cauliflower, you have. In fact, many different plant parts make excellent food. See Figure 8-14. For example, an onion bulb is actually a bud, and celery is the petiole of a leaf. When you eat lettuce, spinach, or cabbage, you are eating whole, mature leaves. The tea in a tea bag and seasonings, such as parsley, chives, basil, rosemary, mint, and sage, are also leaves. Some herbal teas contain flower parts as well.

Our most important sources of food, however, are stems, roots, fruits, and seeds. Of course, you already know that potatoes and asparagus are stems. Potatoes, however, grow underground, while asparagus grows above ground, which is why asparagus is green and potatoes are not. Also, about half of the sugar you eat comes from the stems of a large grass called *sugar cane*. The rest of the sugar you eat comes from the roots of sugar beets. Carrots, turnips, radishes, and sweet potatoes are also plant roots. Tapioca, which we make into a pudding, comes from the roots of the cassava plant. In the tropics, the cassava root is the main food for millions of people.

Bananas, apples, oranges, and grapes are probably the fruits most familiar to you. But some of the foods you know as "vegetables" are really fruits, as well, because they develop from an ovary and contain seeds. Tomatoes, eggplant, squash, peppers, cucumbers, and green beans are examples of these fruits. However, it may surprise you to know that the cereal grains, such as wheat, rice, corn, oats, and barley, are also fruits.

The nutritious part of a cereal grain, such as rice, is the single, large seed contained in each fruit. See Figure 8-15. The cereal grains are the most important source of food for most of the people of the world. They were among the first plants to be cultivated and improved by humans.

Figure 8-15 Rice is the principal food in many parts of Asia. The rice grain is actually a fruit.

Other nutritious seeds, such as beans, peas, and peanuts, are found in the fruits of a group of plants called *legumes*. These plants produce seeds that are rich in protein. In fact, the soybean is often used in place of meat. Peas and beans were also among the first types of plants to be cultivated, over 11,000 years ago.

Plant Fibers

You have probably heard the term "fiber" used in commercials for breakfast cereals and food supplements. But what is this fiber anyway? The fiber you eat is the material with which plants make their cell walls. This material, called **cellulose**, cannot be digested by your body. However, it helps to move food through your digestive system. Some plants are able to make very long strands of cellulose fibers. These fibers are used to make things such as cloth, rope, and paper. See Figure 8-16.

Cotton is the world's most important plant fiber. It has been grown by humans for over 5000 years. The thread used to make cotton cloth is spun from long fibers that grow from the seeds of the cotton plant. Cotton cloth is used to make clothing, bed sheets, towels, rugs, and diapers. Cotton cloth is very comfortable because it holds in body heat to keep you warm and absorbs perspiration, while allowing air flow to your skin, to keep you cool. Cotton fibers are also used to make bandages, stuffing for cushions, paper, and even explosives.

Linen is another important plant fiber that is used to make cloth. Linen fibers come from the stems of the flax plant. See Figure 8-17. Like cotton, linen is also very strong and absorbant. The ancient Egyptians used linen cloth to wrap mummies. Now it is used to make cool, summer clothing as well as table cloths, napkins, handkerchiefs, and paper. In the United States, paper money is 25 percent linen and 75 percent cotton fiber.

Hemp and sisal are two very long and coarse plant fibers that are used to make rope and twine. At one time hemp fibers were used to make paper. Another plant fiber used to make paper comes from the inner bark of a mulberry tree. See Figure 8-18. Most paper is made of ground-up wood.

Medicine from Plants

Before there were drug stores and hospitals, people obtained the medicines they needed from plants. Plants make a variety of chemicals, in addition to sugar and starch. Many of these chemicals have special properties that can help cure diseases or relieve certain kinds of discomfort. Although

Figure 8-16 Cotton cloth is made from fibers produced in the fruits of cotton plants.

Cellulose The material that makes up plant cell walls and plant fibers.

Figure 8-17 The stem of the flax plant is the source of a strong, durable fiber that is used to make linen cloth.

Figure 8-18 Kozo, a fiber that comes from mulberry trees, is the chief ingredient in a fine, very light-weight paper that is popular in Japan.

S143

Figure 8-19 *The leaves of the foxglove contain digitalis, a powerful heart stimulant.*

many modern medicines are usually made synthetically, most of these medicines were first discovered in plants, and some are still extracted from those plants.

Aspirin, one medicine you have probably taken, contains *salicylic acid*, which originally came from the bark of willow trees. It is an effective pain reliever and also controls fever. *Ephedrine*, a drug you may have taken for a stuffy nose, comes from a rare type of gymnosperm called *ephedra*. Foxgloves and periwinkles, plants you may have in your garden or neighborhood, are another source of important medicines. See Figure 8-19. *Digitalis*, which is extracted from the leaves of the foxglove, is a stimulant that is used to treat heart disease. The drugs made from periwinkles help to relieve the symptoms of two types of cancer, Hodgkin's disease and leukemia. *Quinine*, the drug used to treat malaria, comes from the bark of the tropical chinchona tree and makes life in the tropics bearable for millions of people who suffer from this disease.

Taxol is a new drug that shows promise as a treatment for cancer. This drug comes from the bark of the Pacific Yew, a tree that grows in northern California. Unfortunately, the amount of extract required for treatment requires that many yew trees be cut down, and it takes many years for a yew tree to mature. Even though cancer is a dreaded human disease, collecting this useful medicine poses a threat to the existence of other species.

World-wide research continues to identify plants with medicinal value. Much of this research is concentrated on the tropical rain forests, where the richest variety of plant species on earth is found. Unfortunately, these forests are being cut down at an alarming rate. Untold numbers of plant species are becoming extinct as a result. The cures for many human diseases may have been found in some of these plant species. Perhaps the cure for AIDS is hidden within an undiscovered plant in the rain forest. But once these plants are gone, their secrets will be lost forever.

Summary

Plants are an important source of food for humans. Edible plant parts include roots, stems, leaves, fruits, and seeds. Plants also contain fibers that are used to produce cloth and other valuable products. Many plants produce chemicals that can be used as medicines. Many modern medicines were originally derived from plants. The destruction of the rain forests, however, may preclude the discovery of new medicines.

GLOSSARY

A

Abiotic nonliving, and having no living origin (8)

Acid a chemical that has a sour taste, turns litmus paper red, reacts with a base to form a salt, and has a pH value below 7 (53)

Acidic having a pH value of less than 7 (53)

Action the act of doing something, or the state of being in motion (252)

Adaptations features of organisms that arise over time and enable the organisms to survive in a given environment (74)

Aerated having oxygen added (49)

Alkaline having a pH value greater than 7 (53)

Appendage something that is attached to something else, such as a tail or a finger (84)

Arch a curved structure, often constructed of stone, that supports the weight of the material above an open space (285)

Asteroids rocks or boulders that revolve around the Sun, usually between the orbits of Mars and Jupiter (408)

Astronomy the science that studies the stars, planets, and other celestial objects (379)

Atmosphere the air around the earth (179)

Axis a straight line around which an object rotates (399)

B

Bacteria a large class of single-celled organisms, considered neither plant nor animal, belonging to Kingdom Monera. Their actions cause a number of diseases, as well as certain processes including decay, fermentation, and soil enrichment. (18)

Basalt a dark, fine-grained volcanic rock (339)

Base a chemical that has a bitter taste, turns litmus paper blue, reacts with an acid to form a salt, and has a pH value greater than 7 (53)

Beam a long, rigid piece of stone, wood, or metal used for horizontal support in a ceiling, roof, etc. (282)

Big bang a theorized explosion, believed to have occurred 15 to 20 billion years ago, resulting in the formation of the universe (436)

Biosphere an area or region where conditions are suitable for living things to survive (448)

Biotic living or having a living origin (8)

Black hole an object with a gravitational field so strong that neither matter nor light can escape once it has been drawn in (427)

Boiling the process in which liquid rapidly changes into a gas at its boiling point (the maximum temperature at which the liquid can exist as a liquid) (152)

Buoyant force the directional force exerted upward by a liquid on a submerged body (203)

Buttress a structure built against a wall to reinforce or support it (286)

C

Camouflage the method by which an organism disguises itself or blends with its surroundings to conceal itself from its enemies (76)

Cantilever a beam or structure that is fixed at one end only (284)

Carnivore a consumer of meat; a meat eater (18)

Cementation the process in which the spaces between loose particles are filled with a hardening or bonding agent (354)

Classify to sort into groups (94)

Comets small frozen masses of dust and gases that travel definite paths through the solar system. As they pass near the sun, they are vaporized, leaving trails resembling hair. (408)

Commensalism a relationship between organisms in which one organism benefits from the relationship, while the other neither benefits from nor is harmed by the relationship (12)

Compaction the process in which overlying particles compress the spaces between underlying particles (354)

Compressive force the force that causes material to become more compact and pressed together (273)

Concentrated solution a strong solution, having a relatively large amount of solute dissolved in the solvent (170)

Concentration the strength of a solution, determined by the amount of solute dissolved in the solvent (170)

Consumers organisms that depend upon other organisms as food sources (16)

Glossary **S145**

Counteract to undo the effect of something; to neutralize **(53)**

Crystallization a process in which the chemicals in a solution form solids. This is one form of lithification. **(354)**

D

Decomposers consumers that break down dead food sources into substances that enrich the soil **(18)**

Deflection the act of bending **(276)**

Desalination the removal of salt from a substance, especially sea water **(156)**

Dikes intrusions that form in vertical cracks in existing rocks **(355)**

Dilute solution a weak solution, having a relatively small amount of solute dissolved in the solvent **(170)**

Dissolving the spreading of the particles of a solute evenly throughout a solvent **(140)**

Distillation a two-step process involving (1) the heating of a solution to change one part of it into a gas or vapor and (2) cooling the vapor back into liquid form and collecting it **(156)**

Diversity differences or variety among living things **(66)**

Dome a hemispherical roof **(285)**

E

Elastic force the force that is exerted by a material when it is stretched **(203)**

Elastic rebound theory H. F. Reid's theory of the cause of earthquakes whereby energy is stored along faults and then released suddenly as the faults slip **(327)**

Elastic response the tendency of a material to return to its original shape or size after it has been stretched **(275)**

Electrical force the force that causes unlike charged materials to attract each other and like charged materials to repel each other **(203)**

Endangered species those species that are near extinction and would not survive in the wild without human protection **(92)**

Environment the physical surroundings of an organism, including all the conditions and circumstances that affect its development **(5)**

Epicenter the point on the surface of the earth directly above the focus or hypocenter of an earthquake **(330)**

Epoch a sub-division of a geologic period on the geologic time scale **(366)**

Era the largest division of the geologic time scale, denoting an interval of time when certain rocks were formed **(366)**

Evaporation the gradual process by which a liquid changes into a gas or vapor **(152)**

Extrusive rocks rocks produced by molten lava that flows onto the surface of the earth; also called volcanic rocks **(349)**

F

Fault the boundary between two rock sections that have been displaced relative to each other **(324)**

Focus the point where a rupture starts an earthquake; also called a hypocenter **(328)**

Food chain a chain-like linked diagram that shows the relationship between various organisms and their sources of food energy **(21)**

Food web a web-like linked diagram that shows as many food relationships as possible between various organisms and their sources of food energy **(22)**

Friction the rubbing of objects against one another. This produces heat and causes surfaces to wear away. **(232)**

Frictional force the force that resists the motion of objects rubbing against each other **(203)**

G

Geologic time scale the sequence of events of the earth's geologic story told in sedimentary rock **(366)**

Germinate to sprout; to begin to grow from a seed **(450)**

Gravitational force the mutual force of attraction exerted by particles of matter **(203)**

H

Habitat the place where an organism lives **(10)**

Hard water water that contains certain dissolved chemicals **(146)**

Herbivore a consumer of plants; a plant eater **(18)**

Humus a brown or black component of soil, resulting from the decay of material that was once living **(454)**

Hydroponics the process of growing plants without soil using solutions to provide the necessary nutrients for growth **(471)**

Hypocenter the point where a rupture starts an earthquake; also called a focus **(328)**

Hypothesis a possible explanation for why things happen the way they do. In an experiment, a hypothesis is tested. **(160)**

I

Igneous rock rock produced by the cooling and stratification of magma, either at or below the earth's surface **(344)**

Inertia the property of matter to remain at rest or resist change in its state of motion **(246)**

Inferences conclusions, drawn from observations, that attempt to explain or to make sense of the observations **(160)**

Insoluble unable to dissolve or pass into solution **(148)**

Intrusion the pushing up of magma into cracks in existing rock **(355)**

Intrusive rocks rocks produced by the cooling of magma that pushes up into the earth's crust; also called plutonic rocks **(350)**

Invertebrates animals without backbones **(100)**

L

Lava magma that emerges on the surface of the earth **(339)**

Lichens plant-like organisms that consist of a fungus and an alga existing together to the mutual benefit of each other **(13)**

Life cycle the series of developmental stages in an organism's life, from birth to reproduction **(116)**

Light-year the distance light travels through space in one year **(431)**

Lithification the process by which sediments become rock **(354)**

Lithosphere the outer shell of the earth containing the crust, the plates, and the continents **(321)**

Loam a mixture of sand, clay, and organic matter **(456)**

Local Group a family of galaxies, including the Milky Way, that are held together by mutual gravitational forces **(434)**

Magma melted rock deep within the earth **(339)**

Magnetic force the attractive or repelling forces between a magnet and iron, between two magnets, etc. **(203)**

Magnitude the strength, or measured size, of an earthquake **(329)**

Mass the measure of the amount of matter in an object **(207)**

Metamorphic rock rock that has undergone change as a result of intense heat and pressure **(344)**

Metamorphosis change; a transformation of the nature of a thing **(120)**

Meteorites meteors that reach the ground without burning up completely in the earth's atmosphere **(406)**

Meteors lumps of rock and/or metals that enter the earth's atmosphere and burn **(406)**

Milky Way galaxy a spiral-shaped system, of which our Solar System is a part, containing billions of stars, huge clouds of dust particles, and gases **(433)**

Mimicry the resemblance in behavior, look, sound, or smell of an organism to another organism or object in its surroundings. Mimicry helps to protect organisms from their natural enemies **(79)**

Mutualism a relationship between organisms in which all benefit from the relationship **(12)**

N

Natural selection the selecting or favoring of organisms that are best able to survive and reproduce in their environment. Charles Darwin used it to explain how the features of a species can change over many generations. **(86)**

Neutron star a collapsed star composed mostly of neutrons, compressed to the size of an asteroid; contains the mass of many Suns **(426)**

Newton the international metric unit (SI) used to measure force. Written as N, it is approximately equal to the weight of a 100-g mass. **(213)**

Niche an organism's role in its community, including its behavior and its place in the food chain **(10)**

Noncontact forces forces exerted by agents that do not touch the receivers (200)

Nutrients chemicals needed for the functioning and growth of living things (456)

O

Omnivore an eater of both plants and animals (18)

Organism living thing with organs and/or parts that work together (66)

Osmosis the movement of water across a semipermeable membrane from areas where water particles are more concentrated to areas where they are less concentrated (464)

P

Paleontology the study of fossils, ancient life forms, and their evolution (364)

Parasitism the relationship between two organisms in which one organism, called a parasite, lives to the detriment or harm of the other, called the host (13)

Percolation rate the time it takes for a certain volume of water to pass through a sample of soil (458)

Period a sub-division of a geologic era on the geologic time scale (366)

pH scale a scale, ranging from 0 to 14, used to measure the alkalinity or acidity of a solution (53)

Photosynthesis the process by which green plants and plant-like organisms use energy from sunlight to convert water and carbon dioxide to sugars that the organisms can use for food (16)

Plane of the ecliptic the plane of Earth's orbit around the Sun. All of the other planets' orbits lie roughly in this plane. (395)

Plate tectonics the theory that large crustal plates of rock move on the earth's surface (321)

Plutonic rocks rocks produced by the cooling of bodies of magma in the earth; also called intrusive rocks (350)

Pollination the transfer of dust-like pollen from the stamen of one plant to the pistil of another plant of the same species in order to produce seeds (12)

Predator a consumer that hunts or captures a live food source (18)

Prey an organism that is hunted or captured and eaten by another consumer (18)

Producers Organisms that produce food for themselves and others (16)

Pumice a dark, porous, light-weight volcanic rock (339)

Q

Quasars bright energy sources, about the size of our Solar System, that are billions of light years from Earth (436)

R

Reaction in physics, the action that opposes a given action (252)

Recycle to reuse the remains of things (40)

Red giant a star of great brightness, volume, and a relatively large diameter; emits red light (426)

Relative time scale an organization of events based only on their sequence. No dates are included. (365)

Retrograde motion as seen from Earth, the apparent backward motion of a planet before continuing its forward motion (389)

Revolution the motion of one body around another (395)

Richter Scale of Magnitude a scale of measurement in which the number expressing the magnitude of an earthquake is calculated by means of a formula (329)

Rotation the spinning motion of a planet or other object around its axis (395)

S

Saturation that point at which no more solute can dissolve in a solvent (176)

Scavenger a consumer of dead food sources (18)

Sedimentary rock rock produced by cemented mineral particles deposited by wind, water, ice, or chemical reactions (344)

Seismic waves intense vibrations in the earth during an earthquake; also called shock waves (328)

Seismograph the instrument used to record the vibrations of the earth during an earthquake (329)

Seismologists people who study earthquakes (329)

Shear force a force that tends to cause a material to tear or be cut apart **(273)**

Shock waves the release of stored energy in the form of intense vibrations; also called seismic waves **(328)**

Sills intrusions that form in horizontal cracks of existing rocks **(355)**

Soft water water that is mostly free of dissolved matter **(146)**

Solubility the amount of substance able to pass into solution under certain conditions; the measure of the amount of solute needed to saturate a solution **(177)**

Soluble able to dissolve or pass into solution **(148)**

Solute a substance that is dissolved in a solution **(141)**

Solution substances in which the solute particles are spread evenly throughout the mixture **(135)**

Solvent a substance that dissolves a solute to make a solution **(141)**

Stomata microscopic openings on the lower surface of leaves that regulate the passage of substances into and out of a plant **(468)**

Stratovolcano a cone-shaped volcano made up of layers of lava, ash, and cinders; also called a composite cone volcano **(336)**

Succession the process in which a new group of organisms, which are better suited to the changes that have taken place in a particular area, replaces an old group of organisms **(36)**

Supergiant a star of very great brightness and enormous size, at least 100 times larger and brighter than the sun **(426)**

Supernova an exploding star **(426)**

T

Tensile force the force that causes a material to stretch **(273)**

Texture a characteristic of soil based on particle size, roughness, and tactile properties **(455)**

Tolerance the ability to endure a set of conditions **(26)**

Transpiration the movement of water out of plants through the leaves **(466)**

Truss a framework of connected beams or bars built to add strength and support to a bridge, roof, or other structure **(282)**

Tyndall effect a test for determining true solutions. True solutions are transparent, but in a nonsolution, the particles are large enough to scatter or reflect light. **(135)**

V

Vertebrates animals with backbones **(100)**

Volcanic rock rock produced by molten lava that flows onto the surface of the earth; also called extrusive rock **(349)**

Volcano an opening in the earth's surface through which lava and other materials erupt **(332)**

W

Water cycle the process, involving evaporation and condensation, by which the earth's water is constantly purified **(158)**

Weight the measure of the gravitational force acting on an object **(205)**

Z

Zodiac in astronomy, twelve different constellations, named mostly for animals, that lie close to the ecliptic (NOTE: These constellations do not correspond to the "birth signs" you may have read about.) **(393)**

GLOSARIO

A

Abiotic/Abiótico que no es vivo y no tiene origen vivo **(8)**

Acid/Ácido sustancia química de sabor agrio, tiñe de rojo el papel tornasol, reacciona con una base, formando una sal y tiene un valor de pH menor de 7 **(53)**

Acidic/Acídico que tiene un valor de pH menor de 7 **(53)**

Action/Acción el acto de hacer algo, o el estado de hallarse en movimiento **(252)**

Adaptations/Adaptaciones características de los organismos, que aparecen con el tiempo y les permiten sobrevivr en un medio ambiente dado **(74)**

Aerated/Aireado que se le ha agregado oxígeno **(49)**

Alkaline/Alcalina la sustancia que tiene un valor de pH mayor de 7 **(53)**

Appendage/Apéndice algo que está añadido a otra cosa, como una cola o un dedo **(84)**

Arch/Arco una estructura curva, muchas veces construida de piedra, que sostiene el peso del material sobre un espacio abierto **(285)**

Asteroids/Asteroides rocas o peñascos que giran alrededor del Sol, en general entre las órbitas de Marte y Júpiter **(408)**

Astronomy/Astronomía la ciencia que estudia las estrellas, los planetas y otros cuerpos celestes **(379)**

Atmosphere/Atmósfera el aire que rodea a la Tierra **(179)**

Axis/Eje una línea recta, alrededor de la cual gira un objeto **(399)**

B

Bacteria/Bacterias una clase muy grande de organismos unicelulares que no se consideran ni plantas ni animales, y que pertenecen al reino de las Moneras. Sus acciones son causa de numerosas enfermedades, así como ciertos procesos, entre ellos la descomposición, la fermentación y el enriquecimiento del suelo. **(18)**

Basalt/Basalto una roca volcánica oscura, de grano fino **(339)**

Base/Base sustancia química que tiene un sabor amargo, tiñe de azul el papel tornasol, reacciona con un ácido, formando una sal y tiene un valor de pH mayor de 7 **(53)**

Beam/Viga una pieza larga y rígida de piedra, madera o metal, usada como un sostén horizontal en un cielo raso, techo, etc. **(282)**

Big Bang/Gran explosión una explosión teórica que se cree que se produjo hace unos 15 a 20 mil millones de años y dio como resultado la formación del universo **(436)**

Biosphere/Biosfera un área o región donde las condiciones son aptas para que sobrevivan los seres vivos **(448)**

Biotic/Biótico que es vivo o que tiene origen vivo **(8)**

Black hole/Agujero negro objeto con un campo gravitacional tan fuerte que ni la materia ni la luz pueden escapar una vez que han sido atraídas a él **(427)**

Boiling/Ebullición proceso en que el líquido cambia rápidamente a gas en su punto de ebullición (temperatura máxima a la que un líquido puede existir como líquido) **(152)**

Buoyant force/Fuerza de flotación la fuerza direccional ejercida hacia arriba por un líquido sobre un cuerpo sumergido **(203)**

Buttress/Contrafuerza una estructura construida contra una pared para reforzarla o sostenerla **(286)**

C

Camouflage/Camuflaje método por el que un organismo se disfraza o se confunde con el medio que lo rodea, para esconderse de sus enemigos **(76)**

Cantilever/Viga voladiza una viga o estructura que está fija sólo en un extremo **(284)**

Carnivore/Carnívoro consumidor de carne; que come carne **(18)**

Cementation/Cementación proceso en que se llenan con un agente endurecedor o aglutinante los espacios entre partículas sueltas **(354)**

Classify/Clasificar poner en grupos **(94)**

Comets/Cometas pequeñas masas congeladas de polvo y gases que recorren rutas definidas a través del sistema solar. Al pasar cerca del Sol, por lo común se vaporizan dejando estelas que parecen una cabellera. **(408)**

Commensalism/Comensalismo relación entre organismos en la que uno de ellos se beneficia de la relación mientras el otro ni se beneficia ni se perjudica con ella **(12)**

Compaction/Compactación proceso en el que partículas sobrepuestas comprimen los espacios entre las partículas que están abajo de ellas **(354)**

Compressive force/Fuerza de compresión la fuerza que hace que el material se vuelva más compacto y comprimido **(273)**

Concentrated solution/Solución concentrada una solución fuerte que tiene una cantidad relativamente grande de soluto disuelto en el disolvente **(170)**

Concentration/Concentración lo fuerte que es una solución, determinada por la cantidad de soluto disuelto en el disolvente **(170)**

Consumers/Consumidores organismos que dependen de otros como fuentes de alimentos **(16)**

Counteract/Contrarrestar deshacer el efecto de algo; neutralizarlo **(53)**

Crystallization/Cristalización proceso en el que las sustancias químicas en una solución forman sólidos. Es una forma de petrificación. **(354)**

Decomposers/Descomponedores consumidores que descomponen fuentes muertas de alimentos en sustancias que enriquecen el suelo **(18)**

Deflection/Deflección el acto de doblarse **(276)**

Desalination/Desalinización la remoción o removimiento de la sal de una sustancia, en especial del agua de mar **(156)**

Dikes/Diques intrusiones que se forman en hendiduras verticales en las rocas ya existentes **(355)**

Dilute solution/Solución diluida una solución débil que tiene una cantidad relativamente pequeña de soluto disuelto en el disolvente **(170)**

Dissolving/Disolver esparcir por igual las partículas de un soluto a través de un disolvente **(140)**

Distillation/Destilación proceso de dos etapas que supone (1) calentar una solución para convertir una parte de ella en un gas o vapor y (2) enfriar el vapor para convertirlo de nuevo en líquido y recogerlo **(156)**

Diversity/Diversidad diferencias o variedad entre los seres vivos **(66)**

Dome/Cúpula un techo hemisférico **(285)**

Elastic force/Fuerza elástica la fuerza que ejerce un material cuando se le estira **(203)**

Elastic rebound theory/Teoría del rebote elástico teoría de H. F. Reid sobre la causa de los terremotos, según la cual la energía se almacena a lo largo de fallas y luego se libera repentinamente cuando hay un deslizamiento en la falla **(327)**

Elastic response/Respuesta elástica la tendencia de un material a volver a su forma o tamaño original después de haber sido estirado **(275)**

Electrical force/Fuerza eléctrica la fuerza que hace que cuerpos con cargas diferentes se atraigan mutuamente y que los que tienen cargas iguals se repelan entre sí **(203)**

Endangered species/Especies en peligro especies que están próximas a extinguirse y que no sobrevivirían en su estado natural sin la protección humana **(92)**

Environment/Medio ambiente el medio físico en el que está un organismo y que incluye todas las condiciones y circunstancias que afectan su desarrollo; también se le llama ambiente **(5)**

Epicenter/Epicentro el punto de la superficie de la Tierra que está directamente sobre el foco o hipocentro de un terremoto **(330)**

Epoch/Época la subdivisión de un período geológico, en la escala del tiempo geológico **(366)**

Era/Era la división básica de la escala del tiempo geológico, que denota un intervalo de tiempo en el que se formaron ciertas rocas **(366)**

Evaporation/Evaporación el proceso gradual por el que un líquido se convierte en gas o vapor **(152)**

Extrusive rocks/Rocas de extrusión (o extrusivas) rocas producidas por lava fundida que fluye a la superficie de la Tierra; se llaman también rocas volcánicas **(349)**

F

Fault/Falla el límite entre dos secciones rocosas que han sido desplazadas entre sí (324)

Focus/Foco el punto en que una ruptura inicia un terremoto; se llama también hipocentro (328)

Food chain/Cadena alimenticia diagrama enlazado en forma de cadena que muestra la relación entre diversos organismos y sus fuentes de energía alimenticia (21)

Food web/Red alimenticia diagrama enlazado en forma de red que muestra tantas relaciones alimenticias como es posible, entre diversos organismos y sus fuentes de energía alimenticia (22)

Friction/Fricción el frotamiento de un objeto contra otro. Esto produce calor y hace que las superficies se desgasten. (232)

Frictional force/Fuerza de fricción la fuerza que opone resistencia al movimiento de los objetos que se frotan uno contra otro (203)

G

Geologic time scale/Escala del tiempo geológico la sucesión de eventos de la historia geológica de la Tierra, tal como la revelan las rocas sedimentarias (366)

Germinate/Germinar brotar; empezar a crecer a partir de una semilla (450)

Gravitational force/Fuerza de gravedad la fuerza de atracción mutua que ejercen las partículas de materia (203)

H

Habitat/Hábitat el lugar en el que vive un organismo (10)

Hard water/Agua dura agua que contiene ciertas sustancias químicas disueltas (146)

Herbivore/Herbívoro consumidor de plantas; que come plantas (18)

Humus/Humus un componente del suelo, de color café o negro, resultante de la descomposición de materia que una vez fue viva (454)

Hydroponics/Hidropónica el proceso de cultivar plantas sin suelo, usando soluciones para darles las sustancias nutritivas necesarias para su crecimiento (471)

Hypocenter/Hipocentro el punto en que una ruptura inicia un terremoto; se llama también foco (328)

Hypothesis/Hipótesis una posible explicación de por qué las cosas suceden como suceden. En un experimento se comprueba una hipótesis. (160)

I

Igneous rock/Roca ígnea la roca producida por el enfriamiento y la estratificación del magma, en la superficie o bajo la superficie de la Tierra (344)

Inertia/Inercia la propiedad de la materia de mantenerse en reposo o de resistir al cambio en su estado de movimiento (246)

Inferences/Inferencias conclusiones sacadas de observaciones que tratan de explicar o dar sentido a las observaciones (160)

Insoluble/Insoluble incapaz de disolverse o convertirse en solución (148)

Intrusion/Intrusión el empuje hacia arriba del magma, a través de hendiduras de rocas ya existentes (355)

Intrusive rocks/Rocas intrusivas las rocas producidas por el enfriamiento del magma que empuja hacia arriba a través de la corteza terrestre; se llaman también rocas plutónicas (350)

Invertebrates/Invertebrados animales sin columna vertebral (100)

L

Lava/Lava magma que emerge sobre la superficie de la Tierra (339)

Lichens/Líquenes organismos que parecen plantas que constan de un hongo y un alga, que existen juntos para su beneficio mutuo (12)

Life cycle/Ciclo vital la serie de etapas de desarrollo en la vida de un organismo, desde el nacimiento hasta la reproducción (116)

Light-year/Año luz la distancia que recorre la luz a través del espacio en un año (431)

Lithification/Petrificación proceso por el que los sedimentos se convierten en piedra (354)

Lithosphere/Litosfera la capa externa de la Tierra que incluye la corteza, las placas y los continentes (321)

Loam/Greda (tierra negra) una mezcla de arena, arcilla y materia orgánica **(456)**

Local group/Grupo local una familia de galaxias que incluye la Vía Láctea y que se mantienen juntas gracias a fuerzas de gravedad mutuas **(434)**

M

Magma/Magma roca fundida en las profundidades de la Tierra **(339)**

Magnetic force/Fuerza magnética la fuerza de atracción o repulsión entre un imán y un pedazo de hierro, entre dos imanes, etc. **(203)**

Magnitude/Magnitud la fuerza o tamaño medido de un terremoto **(329)**

Mass/Masa la medida de la cantidad de materia de un objeto **(207)**

Metamorphic rock/Roca metamórfica la roca que ha pasado por un cambio, como resultado de intenso calor y presión **(344)**

Metamorphosis/Metamórfosis cambio; una transformación en la naturaleza de una cosa **(120)**

Meteorites/Meteoritos meteoros que llegan al suelo sin quemarse por completo en la atmósfera de la Tierra **(406)**

Meteors/Meteoros masas hechas de mezclas parecidas al carbón de roca o de hierro que entran en la atmósfera de la Tierra y se queman **(406)**

Milky Way galaxy/Galaxia de la Vía Láctea sistema en forma de espiral, del que forma parte nuestro sistema solar y que contiene miles de millones de estrellas, enormes nubes de partículas de polvo y gases **(433)**

Mimicry/Mimetismo capacidad que tiene un organismo para confundirse con su medio ambiente. Este puede adoptar el modo de actuar, la apariencia, el sonido o el olor de otro organismo u objeto del medio que lo rodea, para protegerse de sus enemigos naturales. **(79)**

Mutualism/Mutualismo relación entre organismos en la que todos se benefician **(12)**

N

Natural selection/Selección natural selección de organimos mejor equipados para sobrevivir y reproducirse en su ambiente. Teoría de Charles Darwin, establecida para explicar cómo pueden cambiar las características de una especie a lo largo de muchas generaciones. Se conoce también como la supervivencia del más fuerte. **(86)**

Neutron star/Estrella de neutrones una estrella desintegrada, compuesta en su mayor parte de neutrones y comprimida al tamaño de un asteroide; contiene la masa de muchos soles **(426)**

Newton/Newton la unidad métrica internacional (SI) usada para medir la fuerza. Se escribe como N, y es aproximadamente igual al peso de una masa de 100 g. **(213)**

Niche/Nicho la función de un organismo en su comunidad, incluyendo su conducta y su lugar en la cadena alimenticia **(10)**

Noncontact forces/Fuerzas sin contacto las fuerzas ejercidas por agentes que no tocan a los que las reciben **(200)**

Nutrients/Sustancias nutritivas sustancias químicas necesarias para el funcionamiento y crecimiento de los seres vivos **(456)**

O

Omnivore/Omnívoro que come tanto plantas como animales **(18)**

Organism/Organismo ser vivo con órganos y/o partes que funcionan en conjunto **(66)**

Osmosis/Ósmosis movimiento del agua a través de una membrana semipermeable, de áreas donde las partículas están más concentradas a áreas donde hay menos concentración de ellas **(464)**

P

Paleontology/Paleontología el estudio de los fósiles, formas antiguas de vida, y su evolución **(364)**

Parasitism/Parasitismo relación entre dos organismos en la que uno de los organismos, llamado *parásito,* vive, causándole detrimento o daño al otro, llamado *huésped* **(13)**

Percolation rate/Índice de percolación el tiempo que tarda cierto volumen de agua en pasar por una muestra de suelo **(458)**

Period/Período subdivisión de una era geológica en la escala del tiempo geológico **(366)**

pH scale/Escala de pH la escala que va de 0 a 14 y que se usa para medir la alcalinidad o acidez de una solución **(53)**

Photosynthesis/Fotosíntesis el proceso en el que las plantas y los organismos similares a ellas, de color verde, usan energía de la luz del sol para convertir el agua y el bióxido de carbono en azúcares que los organismos pueden usar como alimento **(16)**

Plane of the ecliptic/Plano de la eclíptica el plano de la órbita de la Tierra alrededor del Sol. Las órbitas de todos los demás planetas están más o menos en este plano. **(395)**

Plate tectonics/Tectónica de placas la teoría de que en la superficie de la Tierra se mueven grandes placas de corteza de roca **(321)**

Plutonic rocks/Rocas plutónicas rocas producidas por el enfriamineto de masas de magma en la Tierra; se llaman también rocas intrusivas **(350)**

Pollination/Polinización la transferencia del polen del estambre al pistilo de plantas de la misma especie, para producir semillas **(12)**

Predator/Depredador consumidor que caza o captura una fuente alimenticia viva **(18)**

Prey/Presa organismo que es cazado o capturado y comido por otro consumidor **(18)**

Producers/Productores organismos que producen alimento para sí mismos y para otros **(16)**

Pumice/Piedra pómez una roca volcánica oscura, porosa y de poco peso **(339)**

Q

Quasars/Quásares fuentes de energía brillante, más o menos del tamaño de nuestro sistema solar que están a distancias de miles de millones de años luz de la Tierra **(436)**

R

Reaction/Reacción en la física, la acción que se opone a otra acción dada **(252)**

Recycle/Reciclar volver a usar los restos de las cosas **(40)**

Red giant/Gigante rojo una estrella de inmenso brillo y volumen y de un diámetro relativamente grande; emite luz roja **(426)**

Relative time scale/Escala del tiempo relativo organización de hechos basada sólo en su sucesión. No incluye fechas. **(365)**

Retrograde motion/Movimiento retrógrado tal como se ve desde la Tierra, es el movimiento aparente hacia atrás de un planeta, antes de continuar su movimiento hacia adelante **(389)**

Revolution/Revolución movimiento de un cuerpo alrededor de otro **(395)**

Richter Scale of Magnitude/Escala de magnitud Richter escala para medir en la que el número que expresa la magnitud de un terremoto se calcula mediante una fórmula **(329)**

Rotation/Rotación movimiento giratorio de un planeta o algún otro objeto alrededor de su eje **(395)**

S

Saturation/Saturación el punto en el que no puede disolverse más soluto en el disolvente **(176)**

Scavenger/Animal de carroña consumidor de fuentes alimenticias muertas **(18)**

Sedimentary rock/Roca sedimentaria la roca producida por partículas minerales cementadas, depositadas por el viento, el agua, el hielo o reacciones químicas **(344)**

Seismic waves/Ondas sísmicas intensas vibraciones en la Tierra durante un terremoto; se llaman también ondas de choque **(328)**

Seismograph/Sismógrafo instrumento usado para registrar las vibraciones de la Tierra durante un terremoto **(329)**

Seismologists/Sismólogos personas que estudian los terremotos **(329)**

Shear force/Fuerza de corte la fuerza que tiende a hacer que un material se desgarre o se rompa **(273)**

Shock waves/Ondas de choque liberación de energía almacenada en forma de intensas vibraciones en la Tierra durante un terremoto; se llaman también ondas sísmicas **(328)**

Sills/Repisas intrusiones que se forman en hendiduras horizontales de rocas ya existentes **(355)**

Soft water/Agua blanda agua que en su mayor parte está libre de materia disuelta **(146)**

Solubility/Solubilidad la cantidad de sustancia capaz de convertirse en solución bajo ciertas condiciones. La medida de la cantidad de soluto necesario para saturar una solución **(177)**

Soluble/Soluble capaz de disolverse o convertirse en solución **(148)**

Solute/Soluto sustancia que se disuelve en una solución **(141)**

Solution/Solución sustancias en las que las partículas disueltas se esparcen por igual por toda la mezcla **(135)**

Solvent/Disolvente sustancia que disuelve a un soluto para formar una solución **(141)**

Stomata/Estomas aberturas microscópicas en la superficie inferior de las hojas que regulan el paso de las sustancias que entran y salen de una planta **(468)**

Stratovolcano/Estratovolcán un volcán en forma de cono, formado por capas de lava, cenizas y escoria volcánica; se llama también volcán de cono compuesto **(336)**

Succession/Sucesión proceso en el que un nuevo grupo de organismos mejor provistos para los cambios que se han producido en una región determinada, reemplaza a un viejo grupo de organismos **(36)**

Supergiant/Supergigante una estrella de un brillo muy grande y de enorme tamaño, por lo menos 100 veces más grande y más brillante que una estrella gigante **(426)**

Supernova/Supernova una estrella en explosión **(426)**

T

Tensile force/Fuerza de tensión la fuerza que hace que un material se estire **(273)**

Texture/Textura característica del suelo basada en el tamaño de las partículas, su aspereza y sus propiedades tactiles **(455)**

Tolerance/Tolerancia la capacidad de soportar un conjunto de condiciones **(26)**

Transpiration/Transpiración la salida de agua de las plantas a través de las hojas **(466)**

Truss/Armazón una estructura de vigas o barras conectadas que se construye para añadir fuerza y sostén a un puente, un techo o a otra estructura **(282)**

Tyndall effect/Efecto de Tyndall prueba para determinar verdaderas soluciones. Las soluciones verdaderas son transparentes; en cambio, en una no solución, las partículas son suficientemente grandes para dispersar o reflejar la luz. **(135)**

V

Vertebrates/Vertebrados animales con columna vertebral **(100)**

Volcanic rock/Roca volcánica la roca producida por lava derretida que fluye a la superficie de la Tierra; se llama también roca de extrusión o extrusiva **(349)**

Volcano/Volcán abertura en la superficie de la Tierra, por la que brotan lava y otros materiales **(332)**

W

Water cycle/Ciclo del agua proceso que incluye evaporación y condensación, por las que el agua de la Tierra se purifica constantemente **(158)**

Weight/Peso la medida de la fuerza de gravedad que actúa sobre un objeto **(205)**

Z

Zodiac/Zodíaco en astronomía, doce constelaciones diferentes, con nombres en su mayor parte de animales, que están cerca de la eclíptica (NOTA: Estas constelaciones no corresponden a los "signos de nacimiento" sobre los que tal vez has leído algo.) **(393)**

INDEX

Boldface numbers refer to an illustration on that page.

A

Abiotic factors, 8, 26–34, **34**, S2, S6
Acceleration
  description of, S63–S64
  force and, S70–S71
  due to gravity, S66
Acids
  in the atmosphere, 179
  characteristics of, S51–S52
  defined, S52
  formation of, S52
  strength of, S53
  uses of, S53–S54
Acid rain
  air pollution and, 53–54, S18
  in the United States, **53**
  effects of, 55, S19
  pH of, 53
Adaptation, S39–S40
Adaptations
  of animals, 76–79, 84, **76–79**
  defined, 74
  diversity and, 74
  of plants, 80, **80–81**
  of seeds, 82, **82–83**
African architecture, S87–S88
Agricola, 320, **320**
Agricultural engineering, 496–497
Aldrin, Edwin "Buzz," S130
Algae
  description of, S25–S26
  lichens and, 13, S10
Amphibians, S33
Ancient Greece
  architecture of, 296, **296**
  belief about mountains, 319
Ancient Rome
  architecture of, 297, **297**
  belief about mountains, 319
Andromeda, 434, **434**, S124
Angiosperms, S29
Animal Kingdom, 98–109
  classification of, 106, S30–S34
Ants, 15, **15**
Anther, 475, **475**, 476, **476**
Aphids, 15, **15**
Apollo 11 spacecraft, S130
Appendages, 84

Arch, 284–288, **285**
  making a model of an, 287
Arched bridge, 288, **288**
Architecture
  careers in, 312–313
  of dome of Florence, 314–315, **314**
  Gropius and, 301–302
  through history, 293–300
  in other cultures, S78–S88
  Lambert and, 306
  Wright and, 301–303
Aristarchus, 389
Aristotle
  gravitation and, S74–S75
  model of the universe, 388
  motion and, S65
Armstrong, Neil, S130
Arthropods, S32
Asteroids, 408, S117–S118
Astronomer, 381
Astronomical unit (AU), 393, 430
Astronomy
  careers in, 442–443
  defined, 379, 382
  history of, 382, 388–390
Atmosphere, 179
Auk, *See Great Auk*
Auroras, S114
Autumnal equinox, 401
Axis, 399
Aztec architecture, S84

B

Bacteria
  blue-green, S25
  as decomposers, 18, **19**
  description of, S25
  in human digestive tract, 15, **15**
  nitrogen-fixing, S7
  resistance to chemicals of, S41
Bark, S139
Barnacles, 15, **15**
Basalt, 339, **348**, 349, S94–S95
Bases
  described, 53, S54
  characteristics of, S54–S55
  uses of, S55–S56
Beam
  compared to arch, 286–287
  as a cantilever, 284, **284**
  defined, 282

I-, 283, **283**
  compared to truss, 283
Bear, 70, **70**
Big-bang theory, 436–437, S131–S132
Binary stars, S122
Biochemical similarities of organisms, S43
Biochemistry, 188
Biomes
  defined, S12
  types of, S12–S17
Biosphere
  in Arizona, **416**
  defined, 448, **448**, S2
  in space, 416
Biotic factors, 8–9, **8**, S2
Birds, S34
Black holes, 378–379, 427, 444, S125–S126
Blue-green bacteria, S25
Bridges
  cantilever, 284, **284**
  classification of, 288–289, **288–289**
  famous, 289, **290**
  Tacoma Narrows, 267, **267**, 291, **291**
Brown pelican, 92, **92**
Brunelleschi, Filippo, 314–315
Buoyant force, 203
Burdock seeds, 15, **15**
Byzantine architecture, S78–S79

C

Camouflage, 76–77, **76–77**
Cantilever, 284, **284**
Cantilever bridge, **288**
Carbon dioxide–oxygen cycle, S7
Careers
  in agricultural engineering, 496–497
  in architecture, 312–313
  in astronomy, 442–443
  in chemistry, 188–189
  in electrical engineering, 258–259
  in genetic engineering, 498–499
  in geology, 372–373
  in medicine, 126–127
  as seed analyst, 451
  in veterinary science, 60–61
Carnivore
  defined, 18, S6

top, 21, S6
Cartier, Jacques, 43
Cathedral of Florence, 314–315, **314**
Cavendish, Lord, 207
Cellulose, S143
Cementation, 354
Cenozoic Era, 366, 368, S36
Chameleon, 76, **76**
Chemistry, 188–189
Chinese architecture, S81–S82
Chloroplasts, S135
Classification
  of animals, 98–100, 106, S30–S34
  biological system of, 102
  of fungi, S26–S27
  of invertebrates, 100–101, S30–S32
  of living things, 98, S24–S35
  of monerans, S24–S25
  of plants, 110–111, S27–S29
  of protists, S25–S26
  scientific names and, 112
  system of Linnaeus, 112
  of vertebrates, 100, S32–S34
Clay, 456
Cleavage, S103
Coal
  anthracite, S98
  bituminous, S97
  formation of, S96–S97
  lignite, S97
Coelenterates, S30
Colloids, S50–S51
Color
  of minerals, S101–S102
  of stars, S120–S121
Comets, 408–410, **408–410**
  Halley's observation of, 409–410
Commensalism, 12, **12**, 14, **15**
Community
  climax, S11
  defined, S5
  forest floor, 46
  habitat and niche in, 10
  of living things, 10–15
  movement of energy in, 16, **16**
  pond, 48, **48–49**
  roles in, 16–18
  rotting-log, 38–39
  interactions in, 12
  woodland, 44

Compaction, 354
Compost pile, 19
Compressive force, 273–274
Concentration, of a solution, 170, 171–173
Conglomerate, 352, **352**, S95
Coniferous forest biome, S13
Coniferous tree, 44–45, **44–45**
Conservation
    defined, S18
    methods of, S20–S22
    of water, S22
    of wildlife, S22
Constellations, *see Stars*
Consumer
    in a community, 16
    primary, 21
    secondary, 21
Convergent boundary, 321, **321**
Copernicus
    Galileo and, 392
    model of planets, 393
Coquina, **352**
Cortés, Hernando, S84
Cosmology, S131
Cows, 14, **14**
Cross-breeding, 478
Cross-pollination, 477, **477**
    as plant breeding technique, 478
Crystallization, 354
Crystals
    defined, S91
    growing, 183–184, **183–184**
    shape of, S103
Cullinan Diamond, 374–375
Cycles
    carbon dioxide–oxygen, S7
    nitrogen, S7–S8
    rock, 357–359, S91–S93
    water, 158–159, S6–S7

D

Darwin, Charles, 88–89, **88**, S35–S36
    theory of natural selection, S36–S38
Deciduous forest biome, S14
Deciduous tree, 44–45, **44–45**
Decomposers
    bacteria and fungi as, 19, **19**
    as consumers, 18
    in a rotting-log community, 40
Deflection, 276
Desalination, 156, S59–S60
Desert biome, S14–S15
Design
    Bauhaus School of, 301, **301**
    in the environment, 304–308, **304**, **305**
    of structures, 293–308
    through history, 293–300
    zoning laws and, 305
Diamonds, 374–375, **374–375**
Digitalis, S144
Dikes, 355
Dinosaurs, 363
Dissolving, 140
Distillation, 156–158
Distillery, 158, **158–159**
Divergent boundary, 321, **321**
Diversity
    defined, 66
    evolution and, S35
    extinction and, 90, S19–S20
    inherited characteristics and, 114–115
    reasons for, 74
    within a species, 114–115
Dodder plant, 14, **14**
Dome, 284–285, **285**
    cathedral of Florence, 314–315, **314**
Dragonfly, 71, **71**
Drying, 354

E

Early civilization, 295, **295**
Earth
    Aristarchus' theory of motion of, 389
    characteristics of, 399–401, S112, S115–S116
    crust of, S99–S101
    early atmosphere of, S43–S44
    as frame of reference, S64
    geologic history of, S105–S110
    history of life on, S35–S36
    journey from, 429–435
    orbit of, S115
    Ptolemy's theory of motion of, 389
Earthquakes, 326–331
    cause of (elastic rebound theory), 327–328, **327–328**
    seismograph measuring, 329–330
Echinoderms, S32
Ecliptic
    model of, 399–401, **399**
    plane of, 395
Ecosystem
    defined, S2
    parts of, S2–S3
Einstein, Albert, S114
Elastic force, 203, 275–277
Elastic rebound theory, 327–328, **327–328**
Electrical engineering, 258–259
Electrical force, 203, 205
Elements
    of the Greeks, S74
    native, S101
Embryo development, S42–S43
Endangered species, 92, S19
Endoskeleton, S32
Energy
    in black holes, S125–S126
    diagram, **16**
    in leaf functions, S136
    movement in community, 16
    pathways in ecosystems, S5–S6
    production by sun, S114–S115
    in quasars, S125
    sun as source of, S113, S134
    in variable stars, S122
Engineering
    agricultural, 496–497
    electrical, 258–259
    genetic, 498–499
Environment
    abiotic and biotic factors of, 9, **9**
    defined, 5
    design in the, 304–308, **304–307**
    interactions in, 7
    preventing problems in, 51, S20–S22
Environmentalists, S20
Ephedrine, S144
Epicenter, 330, **330**
Epidermis, S135
Epochs, 366, **368**
Eras, 366, **368**
Estuary, S15
Evolution, 86, S36–S40
Exoskeleton, S32
Experiments
    controlling variables, 148
    hypothesis in an, 160
    inference in an, 160
Extinction
    brown pelican and, 92, **92**
    caused by DDT, 92
    defined, S17
    of endangered species, 92
    game of, 90–91
    as loss of diversity, 90, S19–S20

F

Fault
    causing earthquakes, 326
    formation of, 324
    in elastic rebound theory, 327–328, **327–328**
Fertilizer
    N-P-K rating of, 469–470
    pH rating of, 470
Fiber, S143
Finch, 88–89, **88**
Fish, 70, **70**, S33
Flatworms, S30–S31
Flowers
    parts of, 473, 475, **475**
    in plant reproductive cycle, 476, **476**
    pollination of, 473–478
Focus, 328, **328**
Fold, 324
Food chain
    diagram of, 21
    food webs and, 22
    primary consumer in, 21
    secondary consumer in, 21
    top carnivore in, 21
Food pyramid
    defined, S5
    energy and, S5–S6
Food web, 22–23
Force
    agent, receiver, and effect of, 198, **198–199**
    balanced, 240, **240**, S66
    buoyant, 203
    centripetal, S67
    compressive, 273–274, **273**
    defined, S65
    describing, 197
    elastic, 203, 275–277
    electrical, 203
    estimating, 216–219
    frictional, 203, 225–235
    gravitational, 203, 205
    of inertia, 246, **246–247**
    magnetic, 203
    as mass times acceleration, S71
    measuring, 210–215, **210–213**
    motion and, 236–241, S65–S66
    in pairs, 251–254, **252–253**
    in sailing, 196
    shear, 273–274, **274**
    tensile, 273, **273**, 274
    unbalanced, 240

Index **S157**

Force meter, 210–213, **210–213**
Forest-floor community, 46
Fossil correlation, S109–S110
Fossils
   defined, 361, **361**
   dinosaur, 363
   evolution and, S35
   formation of, 361, **361**
   index, S110
   study of (paleontology), 364
   types of, 361
Fracture, S103–S104
Frame of reference, S64
Fresh water biome, S16–S17
Friction, S65
Frictional force
   case studies, 225–226, **225–226**
   cause of, 231
   characteristics of, 229–230
   harm of, 232
   introduced, 203
   measuring, 227–228
   effect on motion, 241
   reduction of, 231–233
Frog, 70, **70**, 116–117, **116**
Fruit, **476**
Fungi
   classification of, S26–S27
   club, S27
   in compost, 19
   kinds of, 19, **19**
   lichens and, 13, **13**
   sac, S27
   thread-like, S26

G

Gabbro, **348**, 350, S94
Gagarin, Yuri, S129
Galapagos Islands, 88–89
Galaxy
   Andromeda, 434, **434**, S124
   defined, 433
   description of, S123
   formation of, 436–437
   Local Group, 434, **434**
   Milky Way, 433, **433**, S123–S124
   motion of, S124
   NGC 1275, 444, **444**
   quasars forming, 436
   types of, S124
Galileo
   Copernicus' model and, 393
   gravitation and, S74–S75
   Jupiter's moons and, S117
   telescopes and, S127
   thought experiments, 237
   trial of, 392

Galileo spacecraft, S130
Games
   Extinction, 90–91
   interaction string, 7–9
   Puffball, 242–245
Gateway of the Sun, S85
Genetic engineering, 498–499
Geocentric, 390, 440
Geologic column, S110
Geologic time scale
   dates on, 367
   described, 366
   diagram of, **368**
   subdivisions of, 366, S36
   table of, 364
Geology, 372–373
Germination
   defined, 450
   in plant reproductive cycle, **476**
Girder beam bridges, 289, **289**
Glenn, John, S129
Goddard, Robert, S129
Gold, 182, S101
Granite, **348**, 350, S92, S94–S95
Grasshopper, 77, **77**
Grassland biome, S14
Gravitational force
   defined, 205
   galaxies and, S123
   introduced, 203
   Lord Cavendish's proof of, 207
   measured, 214, **214**
   on the moon, 206
   Newton's Law of, S75
   Newton's theory of, 206–207
   related to weight, 205, S76
   Solar System and, S115
   theories of, S74–S75
Great Auk, 43, **43**
Great Smoky Mountains National Park, 24
Great Wall of China, S81–S82
Gropius, Walter, 301–302
Gymnosperms, S29
Gypsum, 352, **352**, S96, S101
Gypsy moth caterpillar, 43

H

Habitat
   in community, 10
   defined, 10
Hagia Sophia, S78–S79
Half-life, S108
Halley, Edmund, 409
Halley's comet, 409–410, **409–410**
Hardness, of minerals, S102–S103

Heliocentric, 390, 440
Herbivore, 18, S6
Hertzsprung-Russell diagram, S121
Homologous structures, S42
Hornet fly, 79
Host, in parasitism, 13
Humus, 454
Hutton, James, 355–356, S105
Hydronium ion, S52
Hydroponics, 471–472
Hydroxide ion, S54
Hypocenter, 328
Hypothesis, 160

I

Ichneumon flies, 14, **14**
Igneous rock
   classification of, 348, S94–S95
   described, 344
   extrusive (volcanic), 349, **349**, S94
   formation of, 349–350, **350**, S94
   identification of, 345, **345**
   intrusive (plutonic), 350, **350**, S94
   simulated formation of, 346–347, **346–347**, S90–S91
Impatiens plant, 465, **465**
Incan architecture, S85
Indian architecture, S80
Indicators, pH, S57–S58
Inertia
   defined, 246, S69
   examples of, 246–247, **246–247**
   experiencing, **246–250**
   as momentum, S72
Inference, 160
Inks, 161
Insects
   diversity of, 122
   life cycles of, 118–120
   resistance to chemicals, S41
Insolubility, 148
Interactions
   examples of, 4–5, **4–5**
   in the environment, 7
   string game, 6–8, **6–8**
Intrusions, 355–356
   dikes, 355, **355**
   sills, 355, **355**
Invertebrates
   classification of, 102–103
   defined, 100, S30
   examples of, **100–101**
   subgroups of, 102–103, S30–S32
Islamic architecture, S79

J

Japanese architecture, S82–S83
Jewelry, 374–375
Jupiter, 412, **412**, 416, S116–S117
   red spot of, 413, 413

K

Katydid, 76, **76**
Kepler, Johannes
   discovery of elliptical orbits, 402
Krakatoa, 336, **336**

L

Lambert, Phyllis, 306, **306**
Lamination, 280
Landfills, S21
Landscape, 488
Lava, 339, **339**, 349, **349**, S94
Laws
   of conservation of momentum, S73
   of Gravitation, Newton's, S75–S76
   of Motion, Newton's, S68–S72
   of superposition, S107
Leaves
   function of, S136–S137
   stomata of, 468, S135–S136
   structure of, 467–468, **468**, S134–S136
   water and, 468
Lemmings, 43, **43**
Leonov, Alexei, S129
Lichens
   in a forest-floor community, 46–47
   as fungus and algae, 13, **13**
   as mutualism, 12
   as pioneer species, S10
Life cycle
   of frogs, 116–117, **116**
   of insects, 118–120
   metamorphosis and, 120
Light
   levels for plant growth, 487
   speed of, 429
Light tolerance, 32, **32**
Light-year, 431
Limestone, 352, **352**, 356, **356**, S96
Limiting factors, S4
Linnaeus, Carolus, 112, **112**

Linnaeus' system of classification, 112
Lithification, 354
Lithosphere, 321
Litmus paper, 470, S57
Lizard, **66**, 76, **76**
Loam, 456
Luminosity, S119
Lungworm, 15, **15**
Luster, S102
Lyell, Charles, 355–356, **356**

M

Machu Picchu, S85
Magellan spacecraft, 415, 442–443, S130
Magma, 339, S91
Magnetic force, 203
Malthus, Thomas, S37–S38
Mammals, S34
Marble, 356, **356**, S98
Marine biome, S15–S16
Mars
    characteristics of, S115–S116
    colonizing, 416, **419**
    Ptolemy's theory about, 389–390
    retrograde motion of, 389, **389**, 396–397, **396**
Mass
    acceleration and, S71
    defined, 207
    inertia and, S69
    measuring, 223, **223**
    weight compared to, 208, S76
Mayan architecture, S85
Medical research, 126–127
Medieval Europe, 298, **298**
Meiji Shrine, S83
Mercury, **415**, S115–S116
Mesozoic Era, 366, **368**, S36
Metamorphic rock
    classification of, S98
    formation of, 355–356, **355**, S91, S97–S98
    Hutton and, 355–356
    identification of, 345
    introduced, 344, **344**
    Lyell and, 355, 356
Metamorphosis
    in rock, 355
    life cycles and, 120
Meteorites, 406–407, S118
Meteors
    defined, 406
    compared to meteorites, 407, S118
    as a shooting star, 407

Mexican milk snake, 79, **79**
Milky Way, 433, S123
Mimicry
    eyespots as, 79
    of the hawkmoth, 79, **79**
    of the hornet fly, 79
Minarets, S79
Minerals
    cleavage of, S103
    color of, S101–S102
    crystal shape of, S103
    defined, S91
    fracture of, S103–S104
    hardness of, S102–S103
    kinds of, S99–S101
    luster of, S102
    nonsilicates, S100–S101
    precipitation of, S96
    properties of, S101–S104
    silicates, S99–S100
    special properties of, S104
    streak of, S102
Mistletoe, 15, **15**
Mixtures, S46
Modern civilization, 300, **300**
Moisture tolerance, 29–30
Molecules
    biological, S44
    rate of dissolving and, S48
Mollusks, S31–S32
Momentum, S72–S74
Monerans, classification of, S24–S25
Mont Pelée, 337, **337**
Moon
    composition of, S112
    observations about, 383, **383**, 386, **386**, 405, **405**
    origin of, S112–S113
    surface of, 386, 406, **406**
Moons, *see Satellite*
Mosques, S79
Mosses, 71, **71**
Motion
    circular, S66–S67
    continual, 237
    defined, S62
    of earth, S64
    effect of force on, 240, S65
    effect of friction on, 241
    free fall, S66
    Galileo's experiments on, 237, **237**
    overcoming inertia, 246
    after initial force, 237
    laws of, S68–S72
    projectile, S67–S68
    thought experiments on, 237–241
    types of, S66–S68
    uniform circular, S67
Mound builders, S86

Mountains
    formation of, 324
    formed by erosion, 320, **320**
    plate tectonics and, 321
    rise and fall of, 319–321
Mount Everest, 318, **318**, 319
Mount Pinatubo, 335
Mount Saint Helens, 332–334, **332–334**
    location of, 338
    magma movement in, 339
    as a stratovolcano, 336
Mutation, S39
Mutualism, 12, **12**, 14, 15

N

Native American architecture, S86–S87
Native elements, S101
Natural selection
    adaptation and, S39–S40
    Charles Darwin and, 86–89, S35–S39
    defined, 86, S38
    diversity and, 86
    finches and, 88–89, **88**
    peppered moths and, 87, **87**
Neptune, 411, 413, **413**, **414**, S116–S117
Neutralization, S55
Newton, Isaac
    friction and, S65
    gravity and, 206, **206**, S75–S76
    laws of force and motion, 207, S68–S72
    telescopes and, S127
    theory of force, 252
Newtons, 213
Newton's First Law of Motion, S69–S70
Newton's Law of Gravitation, S75
Newton's Second Law of Motion, S70–S71
Newton's Third Law of Motion, S71–S72
Niche
    in a community, 10, **10**
    defined, 10
    table, 11
Nitrogen, 469–470
Nitrogen cycle, S7–S8
Nomad, 294, **294**
Nonsilicate minerals, S100–S101
Nova, S123
N-P-K rating, 469–470
Numerical dating, S108–S109

Nutrients
    defined, 456
    in hydroponic solutions, 471
    in plants, 456

O

Obsidian, **348**
Ocean, 180
Oceanography, 62–63
Oil spill, 145, **145**
Omnivore, 18
Onion, 70, **70**
Oparin, Alexander, S44
Orchid, **14**
Organism
    defined, 66
    life cycles of, 116
Origin of life, S43–S44
Osmosis
    defined, 464, **464**
    in plants, 464–465, S141

P

Pagoda, S81
Paleontology, 364
Paleozoic Era, 366, **368**, S36
Parallax, S119
Parasite, 13, 18
Parasitism, 13, **14**, 15
Paricutin, 337
Pelicans, 71, **71**
    brown, 92, **92**
Peppered moth, 87, S39
Percolation rate, 458
Pesticides, 484–485
    alternatives to, 485
    resistance to, S41
Petiole, S135
pH
    defined, S56
    of fertilizer, 470
    of human body, S58–S59
    indicators of, S57–S58
    nature and, S58–S59
Phosphorous, 469–470
Photosynthesis
    algae and, 13, S25
    blue-green bacteria and, S25
    in chloroplasts, S135
    defined, 16, S134
    as an energy resource, 16
    in leaves, 466, S136
    by lichens, 13
    process of, 16
Physician, 126–127
Pioneer 10 spacecraft, S69, S71
Pistil, 473, 475, **475**
Planarian, S31

Planets
  elliptical orbit of, 402–403
  Kepler's theory of, 402
  model of motion of, 393–395
  orbits, 402
  in plane of the ecliptic, 395
  revolution of, 395
  rotation of, 395
  of Solar System, 411–416, **411–416**, 418–419, 421, S115–S116
Plant Kingdom
  classification of, 110, S27–S29
Plants
  adaptations in, 80, S134
  in a biosphere, 448–449, **448–449**
  classification of, S27–S29
  environmental needs of, 486
  flowers on, 473–478, **474**
  germination of seeds in, 450, **450**
  hydroponic growth of, 471–472
  light levels for growth, 487
  nutrients of, 456
  pesticide effect on, 484–485
  photosynthesis in, 466
  pollination of, 473–478, **477–478**
  reproduction of, 476, **476**
  root growth in, 481–482, **482**, S139–S141
  soil and, 456
  structure of, 467–468, S134–S141
  transpiration in, 466
  transplanting of, 461, **461**
  uses of, 487–489, **487–489**, S141–S144
  vegetative reproduction of, 479–480, **479**
  water and, 463–468
Plant reproductive cycle, 476, **476**
Plate tectonics
  plate boundaries and, 321, **321**
  rock recycling and, 358
  theories of, 321
  volcanoes and, 339
Pluto, 416, 421, S116–S117
Polar bear, 77, **77**, **86**
Polaris, 391
Pollen, 473, 476
Pollination
  cross-, 477–478
  defined, 473
  in flowers, 473–478

  as mutualism, 12
  in plant reproductive cycle, **476**
  self-, 477–478
Pollutants, S18
Pollution
  acid rain, 51–56
  consequences of, S19
  defined, S18
  effect on grebes, 145, **145**
  of the environment, S18–S19
Populations, 41–43, S3–S5
Porpoise, 86
Potassium, 469–470
Praying mantis, 77, **77**
Precambrian Era, 366, **368**, S36
Precipitation, S96
Pre-Columbian architecture, S84–S85
Predator, 18
Prey, 18
Primary consumer, 21
Principle of Uniformitarianism, S105
Producer, in a community, 16
Protists, classification of, S25
Protozoa, 14, **14**
Protozoan, description of, S25
Ptolemy, 389–390, **389**
Puffball, game of, 242–245
Pulsar, S125
Pumice, 339, **348**
Pyramid of the Sun, S84

Q

Quartz, S99–S100
Quartzite, **356**
Quasar, 436, 444, **444**, S125
Quinine, S144

R

Rabbits, 43
Radioactive decay, S108
Radio telescope, 444–445, **444**, S127–S128
Red shift, S124
Regeneration, S32
Relative age, S108
Relative time scale, 365, **365**
Renaissance, 299, **299**, 314–315
Reproductive group, S3
Reptiles, S33–S34
Resistance, S41
Resources
  abiotic, S6–S8
  natural, S17–S18
  nonrenewable, S18

  renewable, S18
Retrograde motion, 389–390
Revolution, 395
Rhyolite, **348**
Richter Scale of Magnitude, 329–330
Rock cycle, 357–359, S91–S93
Rockets, S129
Rocks
  classification of, 343, S93–S98
  formation of S90–S91
  fossil correlation and, S109–S110
  groups of, 344
  identification key for, 345
  igneous, 344, **344**, 346–350, **348**, **349**, S94–S95
  metamorphic, 344, **344**, 355–358, **356**, S97–S98
  numeric dating of, S108–S109
  plate tectonics and, 358
  recycled, 357
  relative age of, S107–S108
  rock cycle and, 357–359, S91–S93
  sedimentary, 344, **344**, 351–354, **352**, S95–S97
  telling time with, 365–368, **368**
  uses of, 341–342, **341–342**
Root
  cap, 482, S140
  growth, 481–482, **481–482**, S140–S141
  hairs, 463–464, **464**, S141
  systems, S139–S140
  tip, 482, S140
  in vegetative reproduction, 481
Rotation, 395
Roundworms, S31

S

Salt industry, 152–155
Salts
  characteristics of, S59
  defined, S59
  uses of, S59–S60
Sand, 456
Sandstone, **352**, 356, **356**, S95
Satellites, 413, S116–S117
Saturation, 176
Saturn, **413**, 416, S116–S117
Scavenger, 18
Secondary consumer, 21

Sedimentary rock
  classification of, S95–S97
  identification of, 345, **345**
  introduction to, 344
  formation of, 351–354, S95–S96
  lithification of, 354
  simulated formation of, 354, **354**, S91
Seed
  adaptations of, 82
  analyst, 451
  germination of, 450–452
  in plant reproductive cycle, 476, **476**
Segmented worms, S31
Seismic waves, 328
Seismograph, 329, **329**
Seismologist, 329
Selective breeding, S40
Self-pollination, 477–478
Sepal, 475, **475**, 476
Shale, **352**, 356, **356**, S95
Shear force, 273–274, **274**
Shock waves, 328
Silicates, S99–S100
Sill, 355
Silt, 456
Sirius, 431
Slate, 356, **356**, S98
Sliding boundary, 321, **321**
Snowshoe hare, 77, **77**
Soil
  contents of, 454, 456
  formation during succession, S10
  mixtures, 456, 458
  particles of, 455
  percolation rate of, 458
  plants and, 456
  sorting of, 456
  texture of, 455
  water and, 457
Solar System, 389, 395, S115–S118
Solar wind, S114
Solubility
  described, 148, S48
  graph of, 178
  table of, 177
Solutes
  effect on boiling point, 164–166
  crystals form in, 183, **184**
  concentrated, 170
  defined, 141, S46
  dilute, 170
  effect on freezing point, 169
  in hard water, 147
  osmosis and, 464
  salt as a, 152–155
  saturation with, 176
  solubility of, 177

Solution
  atmosphere as a, 179
  changes in, 152, 156
  concentrated, 170–171, S47
  defined, 135, S46
  desalination of, 156
  dilute, 170, S47
  in the environment, 179–188
  non-solution and, 134–139
  ocean as a, 180
  parts of, 140–141, **141**
  saturated, 176, S47
Solvents, 141, 144, S46
Space Plane, S130
Space Shuttle, S130
Species
  endangered, S19
  pioneer, S10
Speed
  defined, S62
  terminal, S66
Sponges, S30
Sputnik 1 spacecraft, S129
Stamen, 473, **475**
Stars
  absolute magnitude of, S119–S120
  binary, S122
  chart of, 387, **387**
  composition of, 423
  distance to, S118–S119
  formation of, 437
  forming black holes, 427
  history of, 425
  life of, 426–429, **428**
  nuclear reactions in, 423
  paralax and, S119
  Sun, 422–423, 426–427
  supernova, 426, **426**, 436
  temperature of, 424–425, S120–S121
  tracks of, 387
  types of, 426–427, S120–S123
  variable, S120, S122
Starling, 43
Stems
  growth of, S138–S139
  structure of, S137–S138
Stick insect, 76, **76**
Stigma, 475, **475**, 476
Stomata, 468, **468**, S135
Stonehenge, 341, **341**
Streak, of a mineral, S102
Strength
  of building materials, 278
  in engineering structures, 282–283
  measuring, 220–222
  in shapes, 279–281, **279–281**
  tensile, 278

Structure
  of an arch, 284–285, 286–287
  of a cantilever, 284
  of a dome, 284–285
Structures
  engineering, 282–291
  failure of, 267–268, **267–268**
  response to forces, 272–277
  strong shapes in, 279–281
  successful, 269, **269–270**
Stupa, S80
Style, 475
Succession
  defined, 36, S9
  environmental changes and, 36, **36**, 37
  at Mt. St. Helens, 36, **36**
  primary, S9–S11
  secondary, S11–S12
Suckerfish, 14, **14**
Sugar
  honey and, 166
  maple tree, 164
  as a solution, 164
Summer solstice, 401
Sun
  calendar wheel, 386, **386**
  characteristics of, 422–423, **422**
  composition of, S113–S114
  dial, 384
  energy production of, S114–S115
  formation of, 437
  life of, 426–427
Supernova, 426, 436, S123
Surtsey volcano, 337, **337**
Suspension, S49–S50
Suspension bridge, 288, **288**

T

Tacoma Narrows Bridge, 267, **267**, 291, **291**
Taj Mahal, S80–S81
Taxol, S144
Tectonic plates, S92
Telescope, S127–S128
Temperature
  of freezing water, 168
  range of tolerance for, 26–28
  saturation and, 177
  in solutions, 168–169, **168**
Tenochtitlán, S84
Tensile force, 273–274
Tensile strength, 278

Teotihuacán, S84
Tepee, S87
Tereshkova, Valentina, S129
Termite, 14, **14**
Terrestrial biomes, S12–S15
Texas coral snake, 79, **79**
Tihuanaco, S85
Time scale
  geologic, 366, **368**
  relative, 365, **365**
Tolerance
  light, 32–34
  moisture, 29–31
  temperature, 26–29
Top carnivore, 21
Torii, S83
Trajectory, S67
Transpiration, 466
Tropic of Cancer, 401
Tropic of Capricorn, 400
Tropical rain forest, S14
Truss, 282
  compared to beam, 283, **283**
Tsiolkovsky, Konstantin, S129
Tubular column, 281
Tundra, S12–S13
Turbellarians, S31
Tyndall Effect, 136–137, **136–137**, S51

U

Ulysses spacecraft, S130–S131
Unconformity, S107
Uniformitarianism, Principle of, S105
Uniformity, S105–S106
Universal indicator, S58
Uranus, 416, 421, S116–S117
Urban environment, 304–306, **304–305**

V

Variable stars
  absolute magnitude and, S120
  description of, S122
Variation, S38–S39
Vascular bundles, S137–S138
Vascular tissue, S28, S137
Vegetative reproduction, 479–480
Velocity, S63
Venus, **415**, 416, 421, S115–S116
Vernal equinox, 400
Vertebrates
  classification of, 100
  defined, 100, S32
  examples of, **104–105**

  subgroups of, 104, S32–S34
Veterinary science, 60–61
Volcano
  famous eruptions of, 336–337, **336–337**
  kinds of, 336–337
  location of, 338
  Mount Pinatubo, 335
  Mount Saint Helens, 332–334, **332–334**
  plate tectonics causing, 339
Voyager spacecraft, 411, **412**, 414, S117

W

Wallace, Alfred Russel, S38
Water
  conservation of, S22
  cycle, 158–159, **158–159**, S6–S7
  freezing point of, 168
  hard, 146
  percolation rate, 458
  plant interaction with, 463–468
  pollution, S18–S19
  salt, 174
  soft, 146
  soil and, 457
Water cycle, 158–159, **158–159**, S6–S7
Wegener, Alfred, 321
Weight
  compared to mass, 205, S76
  defined, 205
  measuring, 223, **223**
Wigwam, S87
Wilson, J. Tuzo, 321, **321**
Winter solstice, 400
Wisakedjak, 113
Woodland community, 44
Woodpecker, 70, **70**
Wright, Frank Lloyd, 302–303

Z

Zodiac, 393–394, **394**
Zoning laws, 305

PHOTO CREDITS

Abbreviated as follows: (t) top; (b) bottom; (l) left; (r) right; (c) center.

Table of Contents: iii(tr), Jeff Lepore/Photo Researchers, iii(b), HRW Photo by John Langford; iv(t), Robert Frerck/Woodfin Camp & Associates, iv(b), Nelson Max/Peter Arnold, Inc.; v(tr), Kip Peticolas/Photo Researchers, v(b), HRW Photo by John Langford; vi(t), L. Kemper/SuperStock International; vi(b), R. King/SuperStock International; vii(t), Robert Cassell/The Stock Market; vii(b), HRW Photo by John Langford; viii(t), FPG International; viii(b), Runk/Schoenberger/Grant Heilman Photography; ix, HRW Photo by John Langford; x(l), Telegraph Colour Library/FPG International; x(c), FPG International; x(r), HRW Photo by John Langford.

Unit 1: 2, John Livzey/Allstock; 9(tr), Stephen J. Krasemann/Valan Photos; 9(cr), Ontario Ministry of Natural Resources; 9(br), Philip Jon Bailey/SuperStock International; 9(bc), M. Keller//SuperStock International; 10, Jim Simmen/Allstock; 11(l), Ontario Ministry of Natural Resources, Toronto; 11(c), Thomas Kitchin/Valan Photos; 11(r), Bill Ivy; 12(l), Bill Ivy; 12(r), Bill Ivy; 13(l), John Fowler/Valan Photos; 13(tl), Gilles Delisel/Valan Photos; 13(tr), Charlie Ott/Photo Researchers; 13(bl), Harold V. Green/Valan Photos; 13(br), Charlie Ott/Photo Researchers; 16tr, Ontario Ministry of Natural Resources, Toronto; 16(inset), J. A. Wilkinson/Valan Photos; 16(tc), R. C. Simpson/Valan Photos; 16(bc), Bill Ivy; 16(br), Ontario Ministry of Natural Resources, Toronto; 17(tl), Stephen J. Krasemann/Valan Photos; 17(tcl), Bill Ivy; 17(tcr) Kennon Cooke/Valan Photos; 17(tr), Pat Louis/Valan Photos; 17(c), Bill Ivy; 17(cr), R .K. LaVal/Animals, Animals; 17(bl), HRW Photo by Richard Haynes; 17(cbl) Val Whelan/Valan Photos; 17(cbr), Bill Ivy; 17(br), Bill Ivy; 18(tl), Alan Carey/Photo Researchers; 18(tc), Peter Arnold, Inc.; 18(c), William & Marcia Levy/Photo Researchers; 18(bc), Jack Swedberg/Canapress Photo Service; 18(b), Bill Ivy; 19(t), Runk & Schoenberger/Grant Heilman Photography; 19(b), C. C. Lockwood/Animals Animals; 20(tl), Ontario Ministry of Natural Resources, Toronto; 20(c), Harold V. Green/Valan Photos; 20(tr), Ontario Ministry of Natural Resources, Toronto; 20(bl), Joseph Collins/Photo Researchers; 20(br), Bill Ivy; 24(l), Pat Louis/Valan Photos; 24(tr), John R. MacGregor/Peter Arnold, Inc.; 24(tcr), Bill Ivy; 24(bcr), Ontario Ministry of Natural Resources, Toronto; 24(br), Gregory G. Dimijian, M.D./Photo Researchers; 25(tl), Egon Bock/Department of Regional Industrial Expansion; 25(c), John Fowler/Valan Photos; 25(cr), Gary Crandall/Envision; 25(bc), Sid Leszczynski/Animals, Animals; 25(br), Ed Reschke/Peter Arnold, Inc; 25(br), Tom & Pat Leeson/Photo Researchers; 29(tr), Tom Murphy/SuperStock International; 29(cl), Harold V. Green/Valan Photos; 29(c), Stephen J. Krasemann/Valan Photos; 29(cr), Thomas Kitchin/Valan Photos; 29(bl), Stock Imagery/Canapress Photo Service; 29(bc), Alan D. Briere/SuperStock International; 29(br), Don McPhee/Valan Photos; 32(tl), Ontario Ministry of Agriculture and Food; 32(tr), Kenbrate/Photo Researchers; 32(cl), Emily Johnson/Envision; 32(cr), Emil Muench/Photo Researchers; 32(bl), Bill Ivy; 32(br), Bill Ivy; 34(tr), Richard Steedman/The Stock Market; 34(cr), A, Kaiser/SuperStock International; 34(bl), John Heseltine/Photo Researchers; 34(br), John M. Roberts/The Stock Market; 36(tr), Kirkendall/Spring; 36(bl), Curtis Willocks/Brooklyn Image Group; 36(bc), Robert Noonan/Photo Researchers; 36(br), Patti Murray/Earth Scenes; 37(tl), Jeff Henry/Peter Arnold, Inc.; 37(tr), Jeff Henry/Peter Arnold, Inc.; 37(bl), William E. Ferguson; 37(br), Ferrell Grehan/Photo Researchers; 40, Stephen J. Krasemann/Allstock; 43(t), B. Lyon/Valan Photos; 43(b), Victor Englebert/Canapress Photo Service; 44(l), Stephen J. Krasemann/Valan Photos; 44(r), Stephen J. Krasemann/Valan Photos; 44(r), Tom & Pat Leeson; 45(tl), Department of Regional Industrial Expansion; 45(tr), Ontario Science Centre; 45(rc), Bill Ivy; 45(c), Ontario Ministry of Natural Resources, Toronto; 45(bl), Michel Bourque/Valan Photos; 45(rbc), Ontario Science Centre; 45(bc), Ontario Ministry of Natural Resources, Toronto; 46(lc), Bill Ivy; 46(tr), Francois Morneau/Valan Photos; 46(bl), Michael P. Gadomski/Photo Researchers; 46(br), Bill Ivy; 47(tl), Bill Ivy; 47(bl), Bill Ivy; 47(br), Bill Ivy; 47(c), Bill Ivy; 47(cr), J. A. Wilkinson/Valan Photos; 51(t), Ontario Ministry of Natural Resources, Toronto; 51(c), Margaret McCarthy/Peter Arnold, Inc.; 51(b), Charlie Archambault/U.S. News & World Reports; 54(t), Ontario Ministry of Natural Resources, Toronto; 54(b), Canapress Photo Service; 55, Eric Beldowski/Canapress Photo Service; 56, AP/Wide World Photos, Inc.; 58(t), William & Marcia Levy/Photo Researchers; 58(bl), Alan D. Briece/SuperStock International; 58(cr), Emily Johnson/Envision; 58(br), Jim Simmen/Allstock, 59, John Livzey/Allstock; 60, Frank Siteman/Stock Boston; 61(t), David R. Frazier; 61(bl), Bohdan Hrynewych/Stock Boston; 61(br), Elena Rooraid/PhotoEdit; 62(t), Al Giddings/Images Unlimited; 62(b), Doug Menuez/Reportage; 63(t), Norbert Wu/Peter Arnold, Inc.; 63(b), Norbert Wu/Peter Arnold, Inc.

Unit 2: 64, D. Northcott/SuperStock International; 70(c), Bill Ivy; 70(cr), Michael Mitchell; 70bl, Breck P. Kent/Animals, Animals; 70(bcl), Michael Mitchell; 70(bcr), Gordon B. Smith/Canapress Photo Service; 70(br), Pat Morrow/Department of Regional Industrial Expansion; 71(bl), Michael Mitchell; 71(bcl), Steven Dalton/Canapress Photo Service; 71(bcr), Wayne Lankien/Valan Photos; 71(br), Canapress Photo Service; 72(tr), Bill Ivy; 72(c), Bill Ivy; 72(cr), Herman H. Giethoorn/Valan Photos; 72(bl), Kennon Cooke//Valan Photos; 72(bc), Stephen Dalton/AllStock; 72(br), Ted Grand/Department of Regional Industrial Expansion; 73(tl), Tom McHugh/AllStock; 73(tc), Milton Love/Peter Arnold, Inc.; 73(tr), Heather Angel Biofotos; 73(l), Bill Curtsinger/Photo Researchers; 73(cr), R. C. Simpson/Valan Photos; 73(br), James Watt/Animals Animals; 74(t), Peter Arnold, Inc.; 74(c), Luis Castaneda/The Image Bank; 74(b), Nuridsany et Perennou/Photo Researchers; 75(tl), Clayton Fogle/Allstock; 75(tr), Daniel Weiss/SuperStock International; 75(rc), Jeffrey L. Rotman/Peter Arnold, Inc.; 75(lc), S. Barrow/SuperStock International; 75(bl), SuperStock International; 75(br), Oxford Scientific Films/Canapress Photo Service; 76(tl), Anthony Bannister/Animals Animals; 76(tr), Bill Ivy; 76(bl), Bill Ivy; 76(br), Stouffer Production/Animals Animals; 77(tl), Mike Beedell/Department of Regional Industrial Expansion; 77(tc), K. G. Preston-Mafham/Animals Animals; 77(tr), Bill Ivy; 77(bl), Stephen J. Krasemann/Photo Researchers; 78(tl), Rob Talbot/TSW; 78(tr), Bill Ivy; 78(cr), Canapress Photo Service; 78(cl), Bill Ivy; 78(br), Hans Pfetschinger/Canapress Photo Service; 78(bl), Ontario Ministry of Natural Resources, Toronto; 79, Karl Maslowski/Canapress Photo Service; 86(t), Ontario Ministry of Natural Resources, Toronto; 86(b), Ontario Ministry of Natural Resources, Toronto; 87(t), Breck P. Kent/Animals Animals; 87(b), Peter Parks/Animals Animals; 92, Jack Dermid; 93, Peter Menzel/Stock Boston; 97(tl), Ontario Ministry of Natural Resources, Toronto; 97(tc), Bill Ivy; 97(tr), M. I. Walker/Photo Researchers; 97(cl), Bill Ivy; 97(c), Bill Ivy; 97(cr), James Kent/Canapress Photo Service; 97(bl), Philip A. Harrington/Peter Arnold Inc.; 97(bc), Martin Kuhnigk/Valan Photos; 97(br), Manley Features/SuperStock International; 98(moray-eel), Norbert Wu/Peter Arnold, Inc.; 98(penguins), George Holton/Photo Researchers; 98(snail), Y. Momautiuk/Photo Researchers; 98(snake), Suzanne L. Collins & Joseph T. Collins/Photo Researchers; 98(mussel), Andrew J. Martinez/Photo Researchers; 99(giraffe), Don Mason/The Stock Market; 99(frog), John R. MacGregor/Peter Arnold, Inc.; 99(dog), HRW Photo by Russell Dian; 99(grasshopper), HRW Photo by Russell Dian; 99(fish), S. Machado/SuperStock International; 99(chick), Schuster/SuperStock International; 99(lobster), Bob Abraham/The Stock Market; 99(salamander), D. Northcott/SuperStock International; 99(octopus), Stephen Frink/The Stock Market; 99(tiger), Reagan Bradshaw/The Image Bank; 99(crocodile), J. Deselliers/SuperStock International; 99(tarantula), Tom McHugh/Allstock; 100(millipede), Ricard Kolar/Animals Animals; 100(jellyfish), National Audubon Society/Photo Researchers; 100(starfish), Dale Calder/Royal Ontario Museum, Toronto; 100(clams), James Carmichael, Jr./Nature Photographer; 100(spider), Bill Ivy; 100(earthworm), Bill Ivy; 100(sand dollar), Biophoto Associate/Photo Researchers; 100(sea anemone), Stephen J. Krasemann/Valan Photos; 101(mollusk), Fred Braverman/Peter Arnold, Inc; 101(butterfly), Bill Ivy; 101(leech), J. Howard/Photo Researchers; 101(crayfish), John R. MacGregor/Peter Arnold, Inc.; 101(centipede), J. A. Wilkinson/Valan Photos; 101(spider), Bill Ivy; 101(clam), Zig Leszczynski/Animals Animals; 101(sea urchin & star fish), Doug Wechsler/Animals Animals; 101(sea creature), Sea Studios/Peter Arnold Inc.; 101(daddy long legs), Stephen Dalton/Animals Animals; 101(bettle), Pam Hickman/Valan Photos; 101(octopus), Paul J. Janosi/Valan Photos; 101(grasshopper), E. R. Degginger/Animals Animals; 101(crab), Dale Calder/Royal Ontario Museum, Toronto; 101(amoeba), E. R. Degginger/Animals Animals; 104(bat), Stephen Dalton/AllStock; 104(dragon fish), Bill Ivy; 104(loon), Stephen J. Krasemann/Peter Arnold, Inc.; 104(python), K. L. Switak/Photo Researchers; 104(tree frog), R. Andrew Odum/Peter Arnold, Inc.; 104(girl), R. Chen/SuperStock International; 104(frog), Bill Ivy; 105(seal), C. Allan Morgan/Peter Arnold, Inc.; 105(angel fish), C. Newbert/SuperStock International; 105(alligator), Don Brison/Envision; 105(owl), John Cancalosi/Peter Arnold, Inc.; 105(eft), Jeff Lepore/Photo Researchers; 105(duck), Department of Regional Industrial Expansion, Ottawa; 105(turtle), Ontario Ministry of Natural Resources, Toronto; 105(shark), Zig Leszczynski/Animals Animals; 107(mantee), Fred Bavendam/Peter Arnold, Inc.; 107(locust), Scott Camazine/Photo Researchers; 107(crab), Kevin Schafer/Martha Hill/Tom Stack & Associates; 107(snail), Y. Momautiuk/Photo Researchers; 107(horse), Gerard Lacz/Peter Arnold, Inc.; 108(millipede), G. I. Bernard/Animals Animals; 108(squirrel), Wayne Lankien/Valan Photos; 108(mollusk), Zig Leszczynski/Animals Animals; 108(starfish), J. A. Wilkinson/Valan Photos; 108(toad), Bill Ivy; 108(salamander), Bill Ivy; 109(sea turtle), Luiz C. Marigo/Peter Arnold, Inc.; 109(caterpillar), Zig Leszczynski/Animals Animals; 109(amoeba), Biophoto Associates/Photo Researchers; 109(sea creature), Harold V. Green/Valan Photos; 109(starfish), Zig Leszczynski/Animals Animals; 109(porpoise), Kennon Cooke/Valan Photos; 109(snake), Bill Ivy; 109(duck), Ontario Ministry of Natural Resources, Toronto; 109(penguins), Luiz C. Marigo/Peter Arnold, Inc.; 109(ostrich), Stephen J. Krasemann/Valan Photos; 110(t), Ontario Ministry of Natural Resources, Toronto; 110(b), Bill Ivy; 111(all), Bill Ivy; 114(l), Canapress Photo Service; 114(r), UNICEF; 115(l), UNICEF; 115(r), Harvey Lloyd/The Stock Market; 124(tl), James Carmichael, Jr./Nature Photographer; 124(tcl), Gerard Lacz/Peter Arnold, Inc.; 124(tcr), Steinhart Aquarium/Tom McHugh/Photo Researchers; 124(tr), Suzanne L. Collins & Joseph T. Collins/Photo Researchers; 124(bl), E. S. Ross; 124(bcl), Tom McHugh/Photo Researchers; 124(bcr), Tim Gibson/Envision; 124(br), Alisa Schulman/Marine Mammal Images; 125(tl), Stephen J. Krasemann/Peter Arnold, Inc.; 125(tcl), A. Briere/SuperStock International; 125(tcr), John Mitchell/Photo Researchers; 125(tr), Gary Crandall/Envision; 125(bl), R. King/SuperStock International; 125(bcl), Ken Graham/Allstock; 125(bcr), D. Cavagnaro/Peter Arnold, Inc.; 125(br), Dr. E. R. Degginger, FPSA; 125(b), D. Northcott/SuperStock International; 126(t), R. Heinzen/SuperStock International; 126(c), Ken Straiton/The Stock Market; 126(b), Curtis Willocks/Brooklyn Image Group; 127(t), David R. Frazier; 127(b), Hal Stucker/Brooklyn Image Group; 128(t), Don & Pat Valenti/DRK Photo; 128(b), Grant Heilman/Grant Heilman Photography; 129(t), Photo by Peter B. Kaplan (c) 1974; all rights reserved; Peter B. Kaplan Photography, Inc.; 129(b), Stephen J. Krasemann/DRK Photo.

Unit 3: 130, Guy Motil/West Light; 145(l), Tom Myers/Hot Shot Stock Photos; 145(r), Robert

S162

Landau/West Light; 174, A. B. Allen/Bruce Coleman, Inc.; 183(tl), Royal Ontario Museum, Toronto M24849; 183(c), Royal Ontario Museum, Toronto M3335; 183(cr), Royal Ontario Museum, Toronto M34014; 183(bl), Phillip A. Harrington/The Image Bank; 183(br), Royal Ontario Museum, Toronto M32303; 187, Guy Motil/West Light; 188(l), Hank Morgan/Photo Researchers; 188(r), Jay Freis/The Image Bank; 189, HRW Photo by Russell Dian; 190(t), Klaus D. Francke/Bilderberg/The Stock Market; 190(c), George H. Harrison/Grant Heilman Photography; 190(r), Anthony Bannister/Animals Animals; 191, Lefever/Grushow/Grant Heilman Photography.

Unit 4: 192, Tom Skrivan/The Stock Market; 194, Brent Bear/West Light; 196(all), C. McNeill/Miller Comstock; 198(t), Paul Kennedy/Leo de Wys, Inc.; 198(c), HRW photo by Bruce Buck; 198(cr), J. Imber/H. Armstrong Roberts, Inc.; 198(b), Focus on Sports; 199(tr), HRW photo by Bruce Buck; 199(b), Steve Smith/Wheeler Pictures; 207, Telegraph Colour Library/FPG International; 214, Michael Mitchell/Equipment Courtesy Sargent Welch Scientific Company; 215(t), HRW Photo by Peter Gonzales; 215(b), Michael Mitchell/Equipment Courtesy Sargent Welch Scientific Company; 216(l), Michael Mitchell/Equipment Courtesy Sargent Welch Scientific Company; 216(r), Michael Mitchell/Equipment Courtesy Sargent Welch Scientific Company; 218(l), Michael Mitchell/Equipment Courtesy Sargent Welch Scientific Company; 218(r), Michael Mitchell/Equipment Courtesy Sargent Welch Scientific Company; 219(t), Michael Newman/ PhotoEdit; 219(b), Michael Mitchell/Equipment Courtesy Sargent Welch Scientific Company; 220, Michael Mitchell/Equipment Courtesy Sargent Welch Scientific Company; 222(t), Michael Mitchell/Equipment Courtesy Sargent Welch Scientific Company; 222(b), Michael Mitchell/Equipment Courtesy Sargent Welch Scientific Company; 231(l), Michael Mitchell; 231(r), Michael Mitchell; 231(b), Pulp and Paper Research Institute of Canada.; 232(t), Michael Mitchell/Equipment Courtesy Pedlar Cycles; 232(b), Michael Mitchell/Equipment Courtesy Pedlar Cycles; 234(tr), Larry Sergeant/Comstock, Inc.; 234(tl), Nathan Bilow Photography; 234(cr), Thomas Zimmermann/Miller Comstock; 234(bl), Marva Production/The Image Bank; 234(bc), Rothwell/FPG International; 234(br), Ira Block/The Image Bank; 235(t), Andre Sima /Miller Comstock; 235(b), NASA; 236, NASA; 248(t), HRW Photo by Peter Gonzales; 248(both), Michael Mitchell; 249(both), Michael Mitchell; 256, Brent Bear/West Light; 257, Tom Skrivan/The Stock Market; 258, Stephen Derr/The Image Bank; 259(t), The Hamilton Spectator; 259(c), Jeff Smith/The Image Bank; 259(br), Phil Degginger; 260(t), Peter Gridley/FPG International; 260(b), NASA; 261(t), Bruce Frisch/Photo Researchers; 261(b), NASA.

Unit 5: 262, SuperStock International; 267, UPI/Bettmann Newsphotos; 268(l), Canapress Photo Service; 268(cr), K. Leland; 268(br), Martin Roetker/Taurus Photos; 269(tr), Mitchel Osborne/The Image Bank; 269(cr), John Butterhill; 269(l), Ontario Ministry of Natural Resources; 269(br), Miller Comstock; 270(tl), Stephen Dalton/ Animals Animals; 270(tr), Masterfile; 270(cl), Miller Comstock; 270(cr), AP/Wide World Photos; 270(bl),Nick Pavloff/ The Image Bank; 270(br),Bard Martin/ The Image Bank; 282, K. Leland; 285(t), Miller Comstock; 285(bl), Dallas & Tom Heaton/Miller Comstock; 285(br), H. Armstrong Roberts; 290(tl), Shostal Associates/SuperStock International; 290(tr), George Hunter/Miller Comstock; 290(cl), R. Waldkirch/Miller Comstock; 290(cr), Jeff Gnass/The Stock Market; 290(br), M. Schneiders/H. Armstrong Roberts; 291, Forest and Whitmire Photography/Washington State Highway Department Photos; 292(tl), Masterfile; 292(tr), William Marvin; 292(cl), Higuchi/Miller Comstock; 292(cr), D. Person; 294(t), Dilip Mehta; 294(b), Canapress Photo Service; 295(t), Mike Yamashita/Woodfin Camp & Associate; 295(b), H. Higuchi/Miller Comstock; 298, Hans Wolfe/The Image Bank; 299(tl), D. Nuccio/SuperStock International; 299(tr), A. Friedlander/SuperStock International; 299(b), R. Manley /SuperStock International; 300(tl), Ronald R. Johnson/The Image Bank; 300(tr), Marvin E. Newman/The Image Bank; 300(b), R. Kord/H. Armstrong Roberts; 301, Sandak/A Division of G.K. Hall & Co.; 302(l), Todd White/The Hayward Gallery, London; 302(t), "Translucent Variation on Spheric Theme," 1951. Collection Solomon R. Guggenheim Museum, NY, Robert E. Mates; 303(tl), The Bridgeman Art Library; 303(tr), Canapress Photo Service; 303(b), Sandak/A Division of G.K. Hall & Co.; 304(t), R. Kord/H. Armstrong Roberts; 304(b), Ralph Krubner/H. Armstrong Roberts; 305, SuperStock International; 306(l), Canapress Photo Service; 306(r), Alan Becker/The Image Bank; 307, Geri Engberg/The Stock Market; 309, Comstock, Inc.; 310, Royal Ontario Museum, Toronto; 311, SuperStock International; 312(t), SuperStock International; 312(b), Henley & Savage/The Stock Market; 313(t), Burton McNeely/The Image Bank; 313(b), William E. Ferguson; 314(t), SuperStock International; 314(b), Filippo Brunelleschi/Scala/Firenze/Art Resource, NY; 315(t), SuperStock International; 315(b), Jan Lukas/Art Resource, NY.

Unit 6: 316, Tom Algire/H. Armstrong Roberts; 318, Sharon Chester/Comstock, Inc.; 320, John N.A. Lott/Biological Photo Service; 321, Jack Marshall and Co. Ltd/Canapress Photo Service; 323(tr), Bruce F. Molnia/Terraphotographics; 323(cr), Ernest S. Booth/Earth Images; 323(bl), H. Gritscher/Peter Arnold, Inc.; 323(br), Galen Rowell/Peter Arnold, Inc.; 324(t), John K. Nakata/Terraphotographics; 324(b), Michael Mitchell; 325, G. K. Gilbert/U.S. Geological Survey; 326, Michael Mitchell; 327(t), R. E. Wallace/U.S. Geological Survey; 327(b), CBC Picture Service/ Canapress Photo Service; 329, Tom McHugh/Photo Researchers; 330(t), Larry Smith/H. Armstrong Roberts; 330(b), Terry Domico/Earth Images; 332(all), Gary Rosenquist/Earth Images; 333(r), Gary Rosenquist/Earth Images; 334(t), Canapress Photo Service; 334(b), Jim Shelton/Canapress Photo Service; 335(l), George Wedding/Canapress Photo Service; 335(c), Canapress Photo Service; 335(r), Canapress Photo Service; 336(t), The Bettmann Archive; 336(b), Franz Lazi/FPG International; 337(t), Canapress Photo Service; 337(b), Canapress Photo Service; 338, Geological Survey of Canada; 339(t), Geological Survey of Canada; 339(b), Geological Survey of Canada; 340(l), Geological Survey of Canada; 340(r), Canapress Photo Service; 341, Bernard Silberstein/Canapress Photo Service; 342(t), Alain Le Garsmeur/TSW; 342(bl), The Telegraph Colour Library/FPG International; 342(c), E. Faure/SuperStock International; 342(br), Barbara J. Miller/Biological Photo Service; 344(tr), Breck Kent/Earth Science; 344(cr), Breck Kent; 344(l), William E. Ferguson; 345(all), Michael Mitchell; 348(all), Michael Mitchell; 349(bl), Geological Survey of Canada 85965; 349(inset), Geological Survey of Canada 202486-H; 349(tr), Joseph Holmes; 349(br), Mike Chuang/FPG International; 352(all), Michael Mitchell; 355(both), Geological Survey of Canada; 356(b), Dr. E. R. Degginger, FPSA; 361(r), Royal Ontario Museum, Toronto; 361(bl), Ward's Natural Science Establishment; 361(bc), Ward's Natural Science Establishment; 361(br), Ward's Natural Science Establishment; 362(all), Bill Ivy; 367, Metropolitan Toronto Public Reference Library; 369, Grant Heilman; 370, G.R. Roberts; 371, Tom Algire/H. Armstrong Roberts; 372(t), Carl Bigras/Canapress Photo Service; 372(b), M. Thonig/H. Armstrong Roberts; 373, David Austen/AUSTRALIA/Stock Boston; 374, Murray Alcosser/The Image Bank; 375, SuperStock International.

Unit 7: 376, Henry Groskinsky/LIFE Magazine © Time Warner, Inc.; 378, Astrostock/Sanford; 380(t), David A. Hardy/Science Photo Library/Photo Researchers; 380(cl), NASA; 380(cr), Astronomical Society of the Pacific, The Planetary System; 380(c), Astronomical Society of the Pacific; 380(bl), Photri/The Stock Market; 380(br), Astronomical Society of the Pacific; 381(t), Dr. Seth Shostak/Science Photo Library/Photo Researchers; 381(c), Dr. Fred Espenak/Science Photo Library/Photo Researchers; 383, Lick Observatory; 386(t), Comstock; 386(br), Lick Observatory; 387(t), Lick Observatory; 387(b), Erich Lessing/Magnum Photos, Inc.; 388, Omikron Collection/Photo Researchers; 389(t), Dr. Jeremy Burgess/Science Photo Library/Photo Researchers; 389(b), George Lovi; 392, Bridgeman/Art Resource, NY; 395(l), Eunice Harris/Photo Researchers; 395(r), HRW Photo by Eric Beggs; 403, Ligature, Inc.; 406, Paolo Koch/Photo Researchers; 407, Astrostock/Sanford; 408(t), Astronomical Society of the Pacific; 408(c), Giraudon/Art Resource, NY; 409, Astronomical Society of the Pacific; 410(b), Otto Rogge/The Stock Market; 411(b), Astronomical Society of the Pacific; 412(t), Jet Propulsion Labs/NASA; 412(c), Astronomical Society of the Pacific; 413(l), Astronomical Society of the Pacific; 413(tr), Jet Propulsion Labs/NASA; 413(c), Astronomical Society of the Pacific; 413(rc), Jet Propulsion Labs/NASA; 413(br), Astronomical Society of the Pacific; 413(bl), Astronomical Society of the Pacific; 414(both), Astronomical Society of the Pacific; 415(l), NASA/Peter Arnold, Inc.; 415(r), Astronomical Society of the Pacific; 415(b), Astronomical Society of the Pacific; 416, Pay Pfortner/Peter Arnold, Inc.; 417, NASA; 419(both), Astronomical Society of the Pacific; 420, Art by Michael Whelan/National Geographic; 421(b), Astronomical Society of the Pacific; 422(c), NASA; 422(b), Hale Observatories/Science Source/Photo Researchers; 423(both), Astrostock/Sanford; 425, Astronomical Society of the Pacific; 426, Lick Observatory; 427(t), Frank P. Rossotto/The Stock Market; 427(b), Robin Scagell/Science Photo Library/Photo Researchers; 429, NASA; 432, Lick Observatory; 434(b), Science Source/Photo Researchers; 435(l), Jean Miele/The Stock Market; 436(b), Dana Berry/NASA; 440, Dennis Di Cicco/Peter Arnold, Inc.; 441(b), Henry Groskinsky/Life Magazine © Time Warner, Inc.; 442, HRW Photo by Michael Lyon; 443(tr), HRW Photo by John Langford; 443(bl), HRW Photo by John Langford; 444(tr), Bob Burch/Bruce Coleman, Inc.; 444(lc), Photri; 444(rc), Caltech Astronomy Library/Palomar Observatory.

Unit 8: 446, Photri, Inc.; 453(b), Kennon Cooke/Valan Photos; 454, Geological Survey of Canada 95660; 463, Clara Parsons/Valan Photos; 467(t), Barry L. Runk/Grant Heilman Photography; 467(b), Joyce Photographics/Valan Photos; 468(r), Val and Alan Wilkinson; 468(r), Runk and Schoenberger/Grant Heilman Photography; 468(bl), Arthur Strange; 472, G. I. Bernard/Earth Scenes; 473(t), HRW Photo by Eric Beggs; 473(b), Lefever/Grushow/Grant Heilman Photography; 474(tl), Irwin Barrett/Valan photos; 474(tr), Professor R.C. Simpson/Valan Photos; 474(bl), Gerhard Karmann/Valan Photos; 474(br), Gerhard Karmann/Valan Photos; 478, University of Alberta Department of Plant Science; 481, Doug Wechsler/Earth Scenes; 486, E. Lewin/The Image Bank; 493(tl), Donald Specker/Earth Scenes; 493(tr), Donald Specker/Earth Scenes; 495, Photri, Inc.; 496(tl), Photri, Inc.; 496(tr), HRW Photo by Eric Beggs; 497, Grant Heilman; 498(tl), R. L. Brinster and R. E. Hammer, School of Veterinary Medicine, University of Pennsylvania; 498(bl), Nelson Max/LLNL/Peter Arnold, Inc.; 498(br), Keith V. Wood, Ph.D.

SOURCEBOOK

Unit 1: S1, John Livzey/Allstock; S3, Art Wolfe/The Image Bank; S4, Steven Kaufman/Peter Arnold, Inc.; S5, Johnny Johnson/DRK Photos; S7, Brent Kent/Earth Scenes; S8, Stephen Krasemann/DRK Photos; S9(l), Larry Ulrich/DRK Photos; S9(r), Jim Brompton/Valan Photos; S10, Doug Allan/Earth Scenes; S11, Tom Bean; S13(l), Nancy Simmermann/TSW; S13(c), Leonard Rue/Animals Animals; S13(r), Breck Kent; S13(b), C. C. Lockwood; S14, Katherine Thomas/Taurus Photos; S15(t), Dudley Foster/Woods Hole Oceanographic Institute; S15(c), D. P. Wilson/Science Source/Photo Researchers; S15(b), D. P. Wilson/Science Source/Photo Researchers; S16, O S F/Animals Animals; S17(t), David Stoecklein/After-Image; S17(b), John James Audubon, National Audubon Society Photo Researchers; S18, George Hall/Woodfin Camp & Associates; S19(t), Mark Sherman/Bruce Coleman, Inc.; S19(bc), Yoram Lehmann/Peter Arnold, Inc.; S20(tl), Australian News & Information Bureau; S20(tr), Tom McHugh/Photo Researchers; S20(b), HRW Photo/Yoav Levy/Phototake; S21(c), Bruce Kurosaki/After-Image; S21(br), Richard Choy/Peter Arnold, Inc.; S22(t), Jim Whitmer/Nawrocki Stock Photo; S22(b), Peregrine Fund.

Unit 2: S23, SuperStock International; S23(tl), S23(tr), David M. Phillips/Visual Unlimited; S23(bl), T. E. Adams; S23(br), T. E. Adams/Click/Chicago; S25(t), Manfred Kage/Peter Arnold, Inc.; S25(c), Eric Graves/Photo Researchers; S25(b), Arthur M. Siegelman; S26(l), Manfred Kage/Peter Arnold, Inc.;

S26(c), Walt Anderson/Tom Stack & Associates; S26(r), Walter Dawn; S27(cl), D. Smiley/Peter Arnold, Inc.; S27(cr), M. P. Kahl/Photo Researchers; S28(t), W. H. Hodge/Peter Arnold, Inc.; S28(b), Michael P. Gadomski/Bruce Coleman, Inc.; S28(bl), Rod Plank/Tom Stack & Associates; S28(bc), Stephen J. Krasemann; S28(br), E. S. Ross; S29(cl), Zig Leszczynski/Earth Scenes; S28(c), E. R. Degginger/Bruce Coleman, Inc.; S28(cb), E. R. Degginger; S30(t), Jeff Simon; S30(b), Kim Taylor/Bruce Coleman, Inc.; S31(t), Carl Roessler; S31(bl), Runk/Schoenberger/Grant Heilman Photography; S31(br), Oxford Scientific Films/Animals Animals; S32(l), Bob & Clara Calhoun/Bruce Coleman, Inc.; S32(c), Rod Borland/Bruce Coleman, Inc.; S32(r), Runk/Schoenberger/Grant Heilman Photography; S33(l), Runk/Schoenberger/Grant Heilman Photography; S33(c), Tom McHugh/Photo Researchers; S33(r), Dave Woodward/Tom Stack & Associates; S34(tl), Zig Leszczynski/Animals Animals; S34(tr), S34(bl), John Markham/Bruce Coleman, Inc.; S34(br), M. Timothy O'Keefe/Bruce Coleman, Inc.; S36, Breck Kent; S37, Summerhays/Photo Researchers; S38(t), Andrew S. Dalsimer/Bruce Coleman, Inc.; S38(b), M. Austerman/Animals Animals; S39, Heather Angel Biofotos; S40(t), Ray Hillstrom/Root Resources; S40(b), HBJ Photo; S40(c), Edith Hughes; S41(r), HBJ Photo; S44, Precision Graphics.

Unit 3: S45, Guy Motil/West Light; S46, HRW Photo by Dennis Fagan; S46(t), HRW Photo by Dennis Fagan; S49(b), Paul Silverman/Fundamental Photography; S50(tl), (tr), Yoav Levy/Phototake; S50(bl), HRW Photo by Richard Haynes; S51(t), HRW Photo by Richard Haynes; S51(b), HBJ Photo by Richard Haynes; S53, Gary Milburn/Tom Stack & Associates; S54(tl), Ted Horowitz/The Stock Market; S54(tc), Erik Anderson/Stock Boston; S54(tr), Brownie Harris/The Stock Market; S54(cl), E. R. Degginger; S56, Gabe Palmer/After Image; S57(tr), Grapes/Michaud/Photo Researchers; S57(b), HRW Photo by Dennis Fagan; S58(tl), HRW Photo by Dennis Fagan; S58(b), Bill Weedmark/Panographics; S59, E. R. Degginger; S60, H. Wendler/The Image Bank.

Unit 4: S61, Tom Skrivan/The Stock Market; S62, Bill Ross/Woodfin Camp & Associates; S63(cr), Joe McNally/Wheeler Pictures; S63(br), Daniel Zisinsky/Photo Researches; S64(t), Ken Biggs/The Stock Market; S64(b), The Stock Market; S65, HRW Photo by Eric Beggs; S66, Pat Rogers/The Image Bank; S67(t), Eunice Harris/Photo Researchers; S67(b), Park Street Photography; S69(t), (c), Loomis Dean, Life Magazine © 1954, Time, Inc.; S69(b), NASA; S70(t), General Motors Corp.; S71(t), NASA; S71(b), HBJ Photo by Rodney Jones; S72(both), HRW Photo by Elizabeth Hathon; S73(tr), Sportech Productions; S73(b), HBJ Photo; S75(t), The Bettmann Archive; S75(b), Richard Megna/Fundamental Photographs; S76, NASA.

Unit 5: S77, SuperStock International; S78, Roger Wood/TSW; S79(tr), Robert Frerck/The Stock Market; S79(bl), (br), SuperStock International; S80(tc), Paolo Koch/Photo Researchers; S80(bl), Dan DeWilde; S81(tc), Cameramann International, Ltd.; S81(cr), Richard Gorbun/Leo deWys, Inc.; S81(br), Steve Vidler; S82(t), Tony Stone Worldwide; S82(b), Susan McCartney/Photo Researchers; S83(tr), Cameramann International, Ltd.; S83(tc), SuperStock International; S83(br), Tony Stone Worldwide; S83(bc), SuperStock International; S84(tl), James Hackett/Leo deWys, Inc.; S84(bl), James Hackett/Leo deWys, Inc.; S84(br), SuperStock International; S85(tr), Van Phillips/Leo deWys, Inc.; S85(cr), Mark Sherman/Bruce Coleman, Inc.; S85(bl), Olaf Soot/TSW; S86(cl), L. K. Townsend; S86(c), Art Grossman; S86(b), David Muench Photography; S87, Marc & Evelyn Bernheim/Woodfin Camp & Associates; S88, Rhodesian National Tourist.

Unit 6: S89, Tom Algire/H. Armstrong Roberts; S90, J. H. Robinson; S91(l), E. R. Degginger; S91(r), Michael & Barbara Reed/Earth Scenes; S92, Bud Titlow/Naturegraphics/Tom Stack & Associates; S93(l), E. R. Degginger; S93(r), Breck P. Kent/Earth Scenes; S94(tl), Gianni Tortoli/Photo Researchers; S94(cl), Pat Lanza Field/Bruce Coleman, Inc.; S94(cr), E. R. Degginger; S94(bl), E. R. Degginger; S95(tr), Breck P. Kent/Earth Scenes; S95(c), E. R. Degginger; S95(cr), E. R. Degginger; S95(bl), E. R. Degginger; S95(br), Breck P. Kent/Earth Scenes; S96(t), E. R. Degginger; S96(c), E. R. Degginger; S96(cl), E. R. Degginger; S96(cr), Breck P. Kent/Earth Scenes; S96(bl), S. D. Halperin/Animals Animals; S97(tl), Horst Schafer/Peter Arnold, Inc.; S97(cl), Paolo Koch/Photo Researchers; S97(cr), Robert Whitney; S97(tr), Robert Whitney; S97(b), (all), E. R. Degginger; S98(tl), Art Kane/The Image Bank; S98(cl), E. R. Degginger/Earth Scenes; S98(cr), Breck Kent; S99, Mary Root/Root Resources; S100(cl), E. R. Degginger; S100(c), Mary Root/Root Resources; S100(cr), E. R. Degginger; S100(bl), Breck P. Kent/Earth Scenes; S101(tr), Paul Silverman/Fundamental Photographs; S101(c), J. Beckett/American Museum of Natural History; S102(tl), Paul Silverman/Fundamental Photographs; S102(c), Paul Silverman/Fundamental Photographs; S102(bl), Fundamental Photographs; S103(t-b), E. R. Degginger; S103, Breck P. Kent/Earth Scenes; S103, E. R. Degginger; S103, E. R. Degginger; S103, E. R. Degginger; S103, E. R. Degginger; S104(tl), Breck P. Kent/Earth Scenes; S104(tc), Breck P. Kent/Earth Scenes; S104(cl), (cr), E. R. Degginger; S104(bl), E. R. Degginger; S105(tl), HRW Photo by Eric Beggs; S105(b), Park Street Photography; S106, Craig Hamill/The Stock Market; S107(t), Vivian M. Peevers/Peter Arnold, Inc.; S107(r), James Tallon; S109, Robert P. Carr/Bruce Coleman, Inc.

Unit 7: Henry Groskinsky/LIFE Magazine © Time Warner, Inc.; S112(t), NASA; S112(b), NASA; S114, Jack Finch/Science Photo Library/Photo Researchers; S117(t), Jet Propulsion Laboratory; S117(c), Science Source/Photo Researchers; S117(br), Julian Baum/Science Photo Library/Photo Researchers; S118(t), NASA; S118(b), Anglo-Australian Telescope Board, 1977; S122(t), Yerkes Observatory; S122(b), John Sanford/Science Photo Library/Photo Researchers; S123(t), National Optical Astronomy Observatories; S123(c), Dennis Diccicco/Peter Arnold, Inc.; S123(b), Bill Iburg/Science Photo Library/Photo Researchers; S124(t), NASA; S124(cl), Hale Observatory; S124(c), National Optical Astronomy Observatories; S124(cr), Kitt Peak National Observatory; S124(tr), NASA; S125(t), Chris Butler; S125(c), National Optical Astronomy Observatories; S126, Lawrence Migdale; S127(t), Art Resource, NY; S127(b), The Granger Collection, New York; S128(tl), National Radio Astronomy Observatory/AU; S128(tr), Peter Menzel/Wheeler Pictures; S128(cl), (cr), NOAO; S129(t), Culver Pictures; S129(b), Photri, Inc.; S130(t), NASA; S130(b), (c), NASA; S131(t), NASA/ARC; S131(bl), (br), Precision Graphics.

Unit 8: S133, Photri, Inc.; S134(t), Biomedia Associates; S134(c), E. S. Ross; S134(b), Zig Leszczynski/Animals Animals; S136, P. Dayanandan/Photo Researchers; S138(tl), E. R. Degginger; S138(tr), P. Dayanandan/Photo Researchers; S140, Runk/Rannels/Grant Heilman Photography; S141, Runk/Schoenberger/Grant Heilman Photography; S142, HRW Photo by Russell Dian; S143(t), Billy E. Barnes/Stock Boston; S143(c), Michael Philips Manheim/Photo Researchers; S143(b), William Hopkins; S144, Zig Leszczynski/Animals Animals.

ART CREDITS

Victoria Bruck 6, 40, 50, 52, 142, 148, 168, 343, 354, 369, 381, 382, 402, 424, 435, 462, 485
Les Case 22, 50, 92, 138, 139, 152, 153, 268, 328, 379, 394, 397, 400, 401, 411, 423, 431, 432, 434, 439, 441
Brenda Clark 206
Alan Clarke 264, 265, 296, 297
Heather Collins 28, 90, 91, 147, 150, 157, 159, 161, 169, 171, 175, 176, 181, 185, 200, 201, 202, 219, 220, 228, 229, 230, 233, 234, 238, 239, 242, 243, 244, 245, 250, 251, 252, 253, 254, 319, 320, 353, 354, 357, 365
Holly Cooper 133, 152, 153, 167, 224, 255, 355, 359, 369, 384
Sam Daniel 346, 347, 350, 358, 364
Nancy Easun 38, 39, 94
Helen Fox 450, 461, 470, 485, 489
David Griffin 14, 15, 53, 79, 119, 122, 141, 151, 166, 167, 186, 275, 351, 375, 384, 390, 391, 399, 412-413, 421, 433, 456, 459, 471, 472, 477, 479, 484, 487, 488
Linda Hendry 21, 57

Andrew Hickson 452, 464
Francis Kagege 113
Michael Krone 84, 85, 93, 132, 134, 337, 351, 363, 365, 385, 391, 398, 404, 418, 433, 438, 492, 493
Vesna Krstanovich 26, 27, 146, 149, 181, 204, 205, 246, 247
Susan Krykorka 154, 155, 164, 165
Ligature, Inc. 387
Emmanuel Lopez 272, 278
Jock MacRae 271, 272, 274, 276, 277, 279, 280, 281, 283, 284, 286, 287, 288, 289, 448, 450, 469, 491
Laurie Marks 150, 151, 162, 163
Michael Martchenko 207, 208, 226, 344
Bernard Martin 463, 464, 465, 477, 483
Jon McKee 266, 293
Rebecca Merrilees 494
Julian Mulock 4, 5, 14, 15, 48, 49, 66, 67, 68, 69, 80, 81, 82, 89, 96, 115, 117, 118, 212, 213, 223, 254, 360, 368, 449, 457, 458, 460, 466, 469, 480, 482
Debi Perna 209
Larry Raymond 52
Ric Riordon 142, 158, 159, 201, 203, 210, 211
Joan Rivers S113
David Simpson 225, 241
TSI 385, 410, 428, 430
Gary Undercuffler 197, 203, 206, 227, 254
Angela Vaculik 453, 475, 476, 479
Henry Van Der Linde 120, 121
Peter Van Gulik 13, 33, 35, 136, 137, 138, 140, 141, 169, 170, 171, 172, 173, 179, 180, 184, 217, 233, 234, 237, 240, 321, 322, 330, 331, 341, 355
Edythe Wright 138, 139

ACKNOWLEDGMENTS

For permission to reprint copyrighted material, grateful acknowledgment is made to the following sources:

Dr. Ivan H. Crowell: "A Legend: Wisakedjak Names All Creatures" by Dr. Ivan H. Crowell.

Harcourt Brace Jovanovich, Inc.: From *HBJ Earth Science* by Cesare Emiliani, et.al. Copyright © 1989 by Harcourt Brace Jovanovich, Inc. From *HBJ Physical Science* by William G. Lamb, et al. Copyright © 1989 by Harcourt Brace Jovanovich, Inc.

Heritage Canada: From an article on citizens' group participation in conservation of old buildings and communities by Phyllis Lambert from *Canadian Heritage*, August 1980. Copyright © 1980 by Heritage Canada.

Holt, Rinehart and Winston, Inc.: From *Holt Earth Science* by William L. Ramsey, et al. Copyright © 1986, 1982, 1978 by Holt, Rinehart and Winston, Inc. From *Holt General Science* by William L. Ramsey, et al. Copyright © 1988, 1983, 1979 by Holt, Rinehart and Winston, Inc. From *Holt Life Science* by William L. Ramsey, et al. Copyright © 1986, 1982, 1978 by Holt, Rinehart and Winston, Inc. From *Holt Physical Science* by William L. Ramsey, et al. Copyright © 1988, 1986, 1982, 1978 by Holt, Rinehart and Winston, Inc.

Life Magazine Company: From "The Red Planet May be the Next Giant Step for Mankind" by Brad Darrach and Steve Petranek from *Life*, May 1991. Copyright © 1991 by Life Magazine Company.

Ray Lincoln Literary Agency, Elkins Park House, 107-B, Elkins Park, PA, as agent for Marian T. Place: From "Search and Rescue" from *Mount St. Helens: A Sleeping Volcano Awakes* by Marian T. Place. Copyright © 1981 by Marian T. Place.

Melody Lane Publications, Inc.: Two lines from "Deep in the Heart of Texas" by Don Swander and June Hershey. Copyright © 1941 by Melody Lane Publications, Inc. Copyright renewed. International copyright secured. All rights reserved.

Macmillan Publishing Company: "Bats" from *The Bat-Poet* by Randall Jarrell. Copyright © 1963, 1964 by Macmillan Publishing Company.

Scott Meredith Literary Agency on behalf of Carl Sagan: From *Cosmos* by Carl Sagan. Copyright © 1980 by Carl Sagan. All rights reserved.

National Geographic Society: "Mars" from *National Geographic Picture Atlas of Our Universe* by Roy A. Gallant. Copyright © 1980 by National Geographic Society.

Texas A&M University, Texas Agricultural Extension Service: Adapted from "Texscape for Conservation" by Richard L. Duble and William C. Welch.